Seismology of the Sun
and the Distant Stars

NATO ASI Series

Advanced Science Institutes Series

A series presenting the results of activities sponsored by the NATO Science Committee,
which aims at the dissemination of advanced scientific and technological knowledge,
with a view to strengthening links between scientific communities.

The series is published by an international board of publishers in conjunction with the
NATO Scientific Affairs Division

A	Life Sciences	Plenum Publishing Corporation
B	Physics	London and New York
C	Mathematical	D. Reidel Publishing Company
	and Physical Sciences	Dordrecht, Boston, Lancaster and Tokyo
D	Behavioural and Social Sciences	Martinus Nijhoff Publishers
E	Engineering and	The Hague, Boston and Lancaster
	Materials Sciences	
F	Computer and Systems Sciences	Springer-Verlag
G	Ecological Sciences	Berlin, Heidelberg, New York and Tokyo

Seismology of the Sun and the Distant Stars

edited by

D. O. Gough

Institute of Astronomy and Department of Applied Mathematics
and Theoretical Physics, University of Cambridge,
Cambridge, UK

D. Reidel Publishing Company

Dordrecht / Boston / Lancaster / Tokyo

Published in cooperation with NATO Scientific Affairs Division

Proceedings of the NATO Advanced Research Workshop on
Seismology of the Sun and the Distant Stars
Cambridge, U.K.
June 17-21, 1985

Library of Congress Cataloging in Publication Data

NATO Advanced Research Workshop on Seismology of the Sun and the Distant Stars (1985 :
    Cambridge, Cambridgeshire)
    Seismology of the sun and the distant stars.

    (NATO ASI series. Series C, Mathematical and physical sciences; vol. 169)
    "Proceedings of the NATO Advanced Research Workshop on Seismology of the Sun
and the Distant Stars, Cambridge, U.K., June 17-21, 1985"-T.p. verso.
    "Published in cooperation with NATO Scientific Affairs Division."
    Includes index.
    1. Sun–Congresses. 2. Stars–Congresses. 3. Seismology–Congresses. I. Gough,
D. O. II. North Atlantic Treaty Organization. Scientific Affairs Division. III. Title.
IV. Series: NATO ASI series.
Series C, Mathematical and physical sciences; vol. 169.
QB520.N36 1985   551:2'2'09994   85-31186
ISBN 90-277-2196-3

Published by D. Reidel Publishing Company
P.O. Box 17, 3300 AA Dordrecht, Holland

Sold and distributed in the U.S.A. and Canada
by Kluwer Academic Publishers,
190 Old Derby Street, Hingham, MA 02043, U.S.A.

In all other countries, sold and distributed
by Kluwer Academic Publishers Group,
P.O. Box 322, 3300 AH Dordrecht, Holland

D. Reidel Publishing Company is a member of the Kluwer Academic Publishers Group

TABLE OF CONTENTS

PREFACE

The history of modern helioseismology is only ten years old. In 1975
F-L Deubner separated for the first time the spatial and temporal properties of the solar five-minute oscillations, and was thus able to
measure the dispersion relation for high-degree acoustic modes (p modes).
The diagnostic value of these observations was appreciated immediately.
Indeed, by comparing the observed relation with computations that had
already been carried out by R.K. Ulrich, and subsequently by H. Ando and
Y. Osaki, it was recognised that contemporary solar models that had been
constructed with the low observed neutrino flux in mind were too hot in
their outer layers. Moreover, their convection zones were too shallow.
Since that time the observations have improved. There is now good reason
to suppose that a sufficiently careful analysis will lead to a direct
determination of the helium abundance in the solar convection zone,
especially when foreseeable further improvements in the observations
have been achieved. The data will also provide useful diagnostics of
the uncertain equation of state of partially ionized plasmas, and they
might also enable us to measure the large-scale structure of the convective flow.
    Low-degree solar p modes were first recognised by G.R. Isaak and
his colleagues in 1979 from whole-disk observations. Many of them penetrate to the centre of the sun, and therefore sense conditions in the
energy-generating core. A long continuous interval of observation permitted G. Grec, E. Fossat and M.A. Pomerantz to isolate and identify the
degrees of the modes, and somewhat different observations by P.H. Scherrer,
the late J.M. Wilcox and their colleagues extended the data to higher
degrees. From a comparison of the observed frequencies with the eigenfrequencies of solar models constructed from stellar evolution theory, it
appeared that the theory of the chemical evolution of the core was roughly
correct: at least, a substantial dilution of the reaction products had not
been affected by mixing with the hydrogen-rich environment. Moreover, it
was possible to estimate the primordial solar helium abundance, which is
of considerable cosmological interest.
    More recently, spatially resolved measurements by T. Duvall and
J.W. Harvey have yielded frequencies of p modes of both low and intermediate degree. These have permitted the computation of the sound speed
throughout much of the solar interior. By bridging the gap between the
low and the high degree, the observations indisputably established the
orders of the low-degree modes whose frequencies were known already from
the observations with little or no spatial resolution. Rotational splitting of sectoral modes yielded the angular velocity in the vicinity of
the equatorial plane, suggesting a value for the sun's gravitational
quadrupole moment J_2 that is consistent with the analyses of planetary
orbits and General Relativity. T.M. Brown, at the workshop, presented
data that permits the determination of the sun's angular velocity away

from the equatorial plane, which essentially confirms the previous
estimate of J_2.

Low-degree modes have been observed using detectors with no spatial
resolution. Consequently, in principle the methods can be used for
other stars. Of special interest are the stars that pulsate in many
modes simultaneously, for these are the most likely to provide fruitful
diagnostic information. Several attempts have already been made, and in
the last years data have been collected that are enabling us to confront
the predictions of stellar structure theory. For example, D. Kurtz and
J. Seeman have reported six modes of oscillation of the Ap star HD 24712
which have raised some interesting questions whose answers are not
immediately obvious in the light of current thinking. Also E. Fossat
and his collaborators and R. Noyes and his collaborators have found
evidence for asymptotically spaced high-order p-mode frequencies in the
more solar-like stars α Centauri A and ε Eridani, whose spacing provides
estimates of the sound travel time from the centre to the surface of the
star. The value for α Centarui A appears to be inconsistent with what
one would have deduced from previous astronomical analyses. In addition,
very interesting data have been obtained from B stars and ZZ Ceti stars.
These pioneering investigations have assured us that our hopes for
studying the seismology of the distant stars might be realised.

The workshop gathered theorists and observers who are actively
working on helioseismology together with some observers who have a knowledge of techniques that can be used for measuring low-amplitude stellar
oscillations. Its purpose was twofold. First, it was to provide an
opportunity for helioseismologists to discuss their latest researches on
the sun, and to plan future coherent programmes of research. Hopefully
the discussion would heighten the stellar astronomers' awareness of the
considerable diagnostic power of seismic observations. The second purpose
was to stimulate programmes to search for stars that support a spectrum
of oscillations, with a view to carrying out subsequent systematic
observations on those that promise to yield useful results.

The timeliness of the meeting is evident. The experience of the
helioseismologists is now sufficiently mature that an effective fertili-
zation of astronomers could lead to a new powerful method for studying
the structures of stars in the solar neighbourhood. Moreover, the few
recent seismological results that have been obtained from stars other
than the sun indicate that available instrumental sensitivity is probably
adequate for the task.

The workshop appeared to achieve its immediate objectives, though
these proceedings record only the lesser part of the activities: the
formal presentations which concentrate on what has already been attained.
I am extremely grateful for the help I received from my scientific
colleagues on the Organising Committee: Harvey Butcher, Jørgen
Christensen-Dalsgaard, Jack Harvey and Jüri Toomre, and for the practical
support from Jean Burris, Margaret Harding, Michael Ingham, Sally Roberts
and Norah Tate.

On behalf of all the participants, I thank the Director of the
Scientific Affairs Division of NATO, and the Institute of Astronomy in
the University of Cambridge, for granting financial support for a
fruitful Advanced Research Workshop.

Douglas Gough
October, 1985

LIST OF PARTICIPANTS

Anderson, B.	University of Oslo, Norway
Baade, D.	European Southern Observatory, F.D.R.
Barker, P.	University of Western Ontario, Canada
Belvedere, G.	Osservatorio Astrofisico di Catania, Italy
Brown, T.M.	High Altitude Observatory, USA
Butcher, H.	Kapteyn Sterrewacht, The Netherlands
Cacciani, A.	Osservatorio Astronomico di Roma, Italy
Campos, L.M.B.C.	Instituto Superior Tecnico, Lisbon, Portugal
Christensen-Dalsgaard, J.	Nordita, København, Denmark
Connes, P.	Service d'Aeronomie du CNRS, France
Däppen, W.	High Altitude Observatory, USA
Delache, P.	Observatoire de Nice, France
Demarque, P.	Yale University, USA
Deubner, F-L.	Universität Würzburg, F.D.R.
Domingo, V.	ESTEC, The Netherlands
Duncan, D.K.	Mt. Wilson & Las Campanas Observatories, USA
Duvall, T.	NASA, Goddard Space Flight Center, Greenbelt MA, USA
Ellis, N.J.	DAMTP, Cambridge, UK
Elsworth, Y.	University of Birmingham, UK
Fossat, E.	Université de Nice, France
Frandsen, S.	Universitet Aarhus, Denmark
Gabriel, M.	Institut d'Astrophysique, Liège, Belgium
Goode, P.	New Jersey Institute of Technology, USA
Gough, D.O.	University of Cambridge, UK
Hill, F.	National Solar Observatory, USA
Isaak, G.R.	University of Birmingham, UK
Kuhn, D.	Princeton University, USA
Kurtz, D.	University of Cape Town, South Africa
Lynden-Bell, D.	Institute of Astronomy, Cambridge, UK
Noyes, R.W.	Center for Astrophysics, Harvard, USA
O'Donoghue, D.	University of Cape Town, South Africa
Osaki, Y.	University of Tokyo, Japan
Paternò, L.	Osservatorio Astrofisico di Catania, Italy
Perdang, J.	Institut d'Astrophysique, Liège, Belgium
Pietraszewski, K.A.R.	Imperial College, London, UK
Praderie, F.	Observatoire de Paris, France
van der Raay, H.B.	University of Birmingham, UK
Reay, K.	Imperial College, London, UK
Rhodes, E.J.	University of Southern California, USA
Ribes, E.	Observatoire de Meudon, France
Ring, J.	Imperial College, London, UK
Roxburgh, I.W.	Queen Mary College, London, UK
Roberts, B.	University of St. Andrews, UK
Scherrer, P.H.	Stanford University, USA

Spruit, H.   Max-Planck Institut, F.D.R.
Toomre, J.   Joint Institute for Laboratory Astrophysics, USA
Ulrich, R.K.   University of California, Los Angeles, USA
Warner, B.   University of Cape Town, South Africa
Weiss, N.O.   DAMTP, Cambridge, UK
Zahn, J-P   Observatoires du Pic-du-Midi et de Toulouse, France

PROPERTIES OF SOLAR OSCILLATIONS

Juri Toomre

Joint Institute for Laboratory Astrophysics
Department of Astrophysical, Planetary and Atmospheric Sciences
University of Colorado
Boulder, Colorado 80309 USA

ABSTRACT. Many of the oscillations that can be observed in the atmosphere of the Sun are resonant acoustic or gravity modes of the interior. Accurate measurement of the frequencies of these p and g modes permits deductions about the internal structure and dynamics of this star. Some of the methods of interpretation, involving a close interplay between observation and theory, can be carried over to the study of more distant stars.

1. INTRODUCTION

Our nearest star, the Sun, is oscillating in an intricate manner. Observations carried out in the solar atmosphere reveal disturbances that are the superposition of a broad range of acoustic and gravity waves. Many of these waves represent global modes of oscillations of the deep solar interior which are able to extend upward into the atmosphere where they can be seen. These oscillations are observed there as Doppler shifts of spectral lines or as associated intensity fluctuations. Careful analysis of the frequencies of these seismic disturbances of the Sun, much like those of the Earth, affords the possibility of studying the internal structure and dynamics of a star for the first time.

Helioseismology promises to let us probe the inside of this star in sufficient detail to test the premises of stellar structure theory, and this has implications for much of astrophysics. The rapid advances in helioseismology have been achieved by a direct interaction between observations and theory. We will here introduce some of the basic elements of helioseismology so as to provide a background for the more detailed considerations of this workshop. We are concerned with bridging the subjects of helioseismology and asteroseismology, and most of the issues to be discussed briefly here are common to studies of both the Sun and the more distant stars. Comprehensive reviews of helioseismology are available and should be consulted (cf. Gough 1983, 1984c, d, e, 1985; Toomre 1984; Deubner & Gough 1984; Christensen-Dalsgaard 1984b, c; Christensen-Dalsgaard, Gough & Toomre 1985; Leibacher, Noyes, Toomre & Ulrich 1985; Brown, Mihalas & Rhodes 1986),

1

D. O. Gough (ed.), Seismology of the Sun and the Distant Stars, 1–22.
© *1986 by D. Reidel Publishing Company.*

as well as assessments of asteroseismology (Christensen-Dalsgaard 1984d). Detailed theoretical consideration of how the stratification of the Sun affects its frequencies (the forward problem) is provided in these proceedings by Christensen-Dalsgaard (1986). The intricate subject of inversion techniques to deduce the interior structure and dynamics given oscillation frequencies and their splittings (the inverse problem), and making provisions for noise in the data, is discussed by Gough (1984a, 1985). The sensitivity of inversion procedures for dealing with rotational splitting data is considered by Christensen-Dalsgaard & Gough (1984a).

We offer here only a brief tour of helioseismology in order to introduce some notation and some basic concepts so that the discussions during the workshop will have a common thread. The strength of this workshop is the diversity of professional experiences brought by the participants working on a range of advanced instrumentation, observation and theory in astrophysics. We wish to exploit these strengths by encouraging discussions of how the interior structure of more distant stars may be probed by studying their oscillations, motivating these by turning to recent discoveries about the insides of our nearest star as gleaned from the study of its vibrations. Of course, related work on more distant stars is hampered by the tyranny of photons due to the faintness of the source, by the lack of information about spatial structures on the unresolved stellar disk, and by the challenges of obtaining the necessary long intervals of observing time on a large telescope. Yet relatives of the solar 5-minute oscillations may already have been detected in the stars HR 1217, α Centauri A and ϵ Eridani (e.g. Kurtz & Seeman 1983; Fossat et al. 1984; Noyes et al. 1984). The white dwarfs of the ZZ Ceti variety (cf. McGraw 1979, Robinson 1979, Liebert 1980, O'Donaghue & Warner 1986) also display oscillations with a variety of frequencies that can be used to test the interior structure of these remnant stars (e.g. Dziembowski 1977, 1979). Thus our considerations here are more than academic. Nevertheless, advances in asteroseismology may well require nearly dedicated large telescopes or more specialized "light buckets" of considerable collecting area, and new breeds of stable and sensitive Doppler analyzers and efficient detectors. None of these are simple matters to implement, yet the scientific rewards are likely to be great, if we are to judge by what has already been learned about the Sun.

The Sun, of spectral class G2, is a common main-sequence star. It is in the midst of the longest phase of its evolution, powered by thermonuclear reactions which convert hydrogen to helium in its high-temperature core. That reactive core occupies the inner 25% by radius of the Sun. The energy so released is diffused radiatively outwards. Conditions in the outer 30% of this star are such as to favor turbulent convection. Vigorous fluid motions there suffice to carry the energy upward to the relative cool surface, or photosphere, where it is radiated into space. The thermal diffusion time, which is about 10^7 years for this star, is far less than times typical of nuclear evolution. Thus the Sun has usually been thought to be in thermal balance, with the rate of energy production in the core just balanced by the rate at which energy is radiated away from the atmosphere. However, it may be that such a placid balance is only true over long averages in time. Helioseismology should enable us to determine whether the Sun is indeed simply in thermal equilibrium, or whether there is evidence of recent mixing of the core that can lead to a temporary imbalance in energy production and losses. From such tests may emerge solutions to the solar neutrino problem: current theories predict a threefold greater flux of neutrinos from nuclear reactions in the Sun than is actually observed.

If the Sun is any guide, such ordinary stars on the main sequence are quite dynamic objects. The Sun displays strong differential rotation with latitude at its surface that probably results from the redistribution of angular momentum by the largest scales of convection called the giant cells. The Sun appears to build and organize magnetic fields in a cyclic fashion with a period of about 11 years, which suggests that a magnetic dynamo is at work. At the most active phases of the Sun, such magnetic fields erupt into the surface layers where they appear as active regions that may lead to a prodigious release of energy as a flare. Yet even in periods of relative inactivity, the magnetic field is seemingly ordered and reordered by convection sweeping the field into networks formed at the periphery of supergranulation cells. Although considerable theoretical effort has gone into studying how turbulent convection, rotation and magnetic fields interact with each other, we are still far from really understanding how this dynamic star functions, except possibly in the broadest brush strokes. Helioseismology offers ways to examine some of the interior dynamics of this ordinary star, and the results are likely to be of great help in guiding the theories.

The Sun possesses both acoustic waves, where pressure is the restoring force, and internal gravity waves, where buoyancy is the dominant force. The acoustic or pressure waves can yield resonant modes of oscillation which are designated as p modes, whereas the gravity waves can yield g modes. The p modes are distinguished by short periods (between 3 minutes and 1 hour) and the g modes by longer periods (exceeding 40 minutes). A given mode of oscillation is trapped within a specific region of the Sun, with its energy confined mainly to that region. For p modes the trapping is between an upper turning point just beneath the atmosphere where an acoustic wave is reflected back downwards by the rapid decrease in the density, and a lower turning point where the wave has been progressively turned around by the increase of sound speed with depth. Acoustic waves with larger horizontal wavelengths are refracted more gradually and thus travel to greater depths. As indicated in Figure 1, some waves are confined to shallow regions just below the surface, whereas others are able to penetrate almost down to the center of the Sun. The characteristic period of a resonant acoustic wave within each such effective cavity depends upon its travel time between the turning points, and this is controlled by the varying sound speed c at which that wave propagates. Thus the temporal frequency ν of the acoustic mode is an integral measure of c within the depth range of the cavity.

For gravity modes, wave propagation is possible only in the atmosphere and in regions below the convection zone where buoyancy can be a restoring force rather than a destabilizing one. A gravity mode is trapped in regions where its frequency ν is less than the local buoyancy frequency N. The characteristic period of a g mode depends upon its travel time through the region of trapping, and this is controlled by the variation of N with depth. By determining the frequencies of p and g modes with precision, we can probe either the variations of sound speed c or of buoyancy frequency N over different intervals within the Sun. Since c and N measure differing aspects of the stratification and chemical composition, helioseismology can serve to

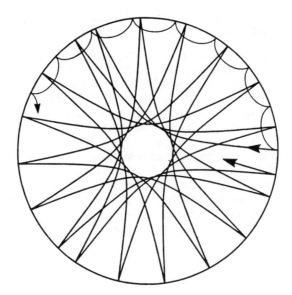

Figure 1. Ray paths of typical acoustic waves within the interior of the Sun. The circle represents the surface of the star in this full cross-section view. The more deeply penetrating rays are for waves with $n/\ell = 5$; the waves confined to the shallower cavity have $n/\ell = 1/20$. [Adapted from Gough 1983]

unravel much about the internal structure of this star, provided a sufficient variety of observable modes is available. Indeed, observations of the p modes have already made it possible to estimate that the thickness of the convection zone is 30% of the solar radius (Gough 1977, Ulrich & Rhodes 1977, Berthomieu *et al.* 1980).

Stellar structure theory is able to rationalize the observed relations between luminosity, mass, radius and age of the Sun. However, there are fundamental aspects of the interior structure which are far from certain. For instance, the abundance of elements and their variations with depth within the Sun needs to be tested. Stars like the Sun are composed essentially of hydrogen and helium, with their proportional abundances by mass denoted as X and Y. These two elements largely determine the equation of state of the stellar material. Most other stable elements are present also, though only in trace amounts; the overall mass fraction of these metals is denoted by Z. Although Z may be only a few percent, the metals make major contributions to the opacity that impedes the transport of radiation from the center to the surface. Regretably, spectroscopic observations of the Sun or the sampling of its wind by spacecraft only impose weak limits on what are appropriate values for these elemental abundances. Nor do theoretical solar models place clear constraints on the abundances: models can be built for a range of X and Y in the outer envelope with little basis for preferring one over another. Of all observational techniques, the accurate measurement of the oscillation frequencies of the Sun is beginning to provide the most accurate values for X and Y in the outer envelope. It would appear that solar models with $Y \simeq 0.25$ and $X \simeq 0.73$ are needed to account for the observed frequencies of 5-minute oscillations confined to the outer envelope (e.g. Christensen-Dalsgaard & Gough 1980b, 1981). Such helium abundances however make the neutrino problem more severe, for a lower value of Y ($\simeq 0.19$) had offered one way of appreciably reducing the predicted neutrino flux so as to bring it into accord with the observed flux.

Knowing the chemical composition of the Sun also has bearing on cosmological theories. According to usual stellar evolution theory, material in the outer envelope of main-sequence stars has not been mixed with products of the nuclear processing in the core. By determining X and Y in the solar envelope, the oscillation frequencies are thereby providing estimates of the abundance of helium in the proto-Sun (cf. Gough 1983b). The values of Y determined so far by helioseismology are consistent with predictions from cosmology about the proportionate amount of helium produced by the Big Bang.

The solar oscillations also permit us to study dynamical processes deep within this star. Advection of the waves by rotation or large-scale flows should result in a splitting of the frequencies of the oscillations (cf. Ledoux 1949; Cowling & Newing 1949; Gough 1978, 1981, 1982, 1984b; Gough & Toomre 1983). Such frequency fine structure has been detected, and has led to the probing of the differential rotation of the Sun both with depth and latitude. Judging from much younger stars, it is believed that the Sun was once rotating much more rapidly than it is today. At its surface, the Sun now rotates once in about 25 days at its equator, while requiring 33 days at the poles. Because the Sun is losing angular momentum through its wind of escaping plasma, it must be in the process of being spun down. It has seemed reasonable to presume that the braking of rotation would be pronounced in the convection zone, and that the angular velocity probably increases with depth below that outer zone. However, preliminary results from helioseismology suggest that the angular velocity actually appears to decrease slowly with depth over much of its interior, though the core may be rotating considerably more rapidly than the surface (e.g. Duvall & Harvey 1984, Leibacher 1984, Brown 1985). Knowing the internal rotation profile allows us to deduce the shape of the Sun's gravitational profile, with departures from spherical symmetry measured by the quadrupole component J_2. The values of J_2 most recently estimated from helioseismology (Duvall et al. 1984) offer no difficulties for the theory of general relativity (as tested by the advance of the perihelion of Mercury), though earlier studies would dispute this (e.g. Dicke 1970, 1982; Hill et al. 1982).

It is crucial to understand just how such a rotation profile is established within the Sun, for the redistribution of angular momentum within stars as they age is a key element in theories of stellar evolution. Moreover, all attempts to model the solar magnetic dynamo require knowing the rotation profile with as much detail as possible. Helioseismology holds out the promise that the probing of differential rotation with depth and latitude can be sufficiently refined to give real guidance in the formulation of convection and magnetic dynamo models. In a related manner, the frequency fine structure has also permitted the detection of flows and thermal structures below the surface that appear to be associated with the giant cells (cf. Hill, Toomre & November 1983; Hill, Gough & Toomre 1984a, b). Such cells are likely to have a dominant role in controlling the dynamics of that convection zone and therefore much of what is observed as solar activity at the surface.

2. THE 5-MINUTE OSCILLATIONS

The Doppler shifts of oscillations detected in the solar atmosphere are the result of interference between about 10^7 resonant modes of vibration. These oscillations were discovered by Leighton (1960) and Leighton, Noyes & Simon (1962), who noted that typically half the solar disk shows patches that are oscillating intermittently with periods close to 5 minutes and amplitudes of about $1000\,\mathrm{m\,s}^{-1}$. These patches have the appearance of wave packets which persist for about 6 or 7 periods and have spatial coherence scales of about 30 Mm (with the solar radius being about 700 Mm).

For some time these patches were thought to be local resonances excited by convective elements overshooting into the atmosphere. But then the real explanation emerged: Ulrich (1970), and independently Leibacher & Stein (1971), proposed that the 5-minute oscillations are the superposition of coherent acoustic modes within wave cavities below the solar surface. That prediction was confirmed observationally by Deubner (1975) who made Doppler velocity measurements spanning many hours over an equatorial strip on the solar disk. Fourier transforms in longitude and time resulted in a power spectrum showing a sequence of ridges in power, much as would be expected for resonant cavities. This formed the beginning of helioseismology, for the patches result from interference between many resonant acoustic modes, each with modest amplitudes of about $10\,\mathrm{cm\,s}^{-1}$.

Yet the excitation of these modes is still an unresolved issue. The modes may be driven stochastically by turbulence in the convection zone, with the excitation and damping proceeding almost continuously in a spatially distributed manner. The modes may also be able to extract energy from the radiation field by something like the κ-mechanism. Although we have little observational guidance about variations in phases, amplitudes or lifetimes of the many modes, their modest amplitudes do make it possible to describe individual modes quite simply using linear theory.

3. CLASSIFICATION AS NORMAL MODES

The individual solar oscillations can be described as small perturbations about a spherical equilibrium state. This allows the oscillations to be separated into normal modes, with the radial component of the velocity of a mode being expressed in spherical polar coordinates (r, θ, ϕ) as

$$V_{n\ell}(r)\; P_\ell^m(\cos\theta)\; \cos(m\phi - 2\pi\nu t)\;, \tag{1}$$

where P_ℓ^m is the associated Legendre function of degree ℓ and azimuthal order m, and ν is the temporal cyclic frequency. The coordinate r is the radial distance from the center of the Sun, θ is colatitude and ϕ is longitude.

Figure 2 shows how individual modes might appear at a given instant in time to an observer. The spatial pattern of Doppler velocities for a mode consists of

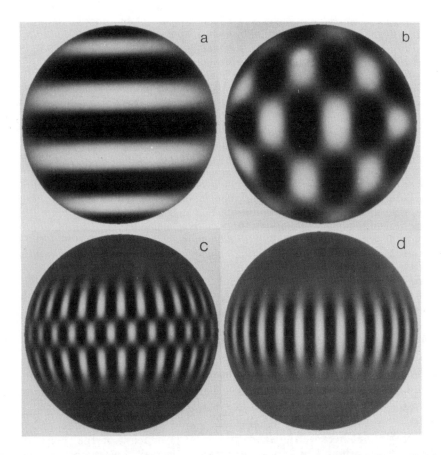

Figure 2. Doppler velocities as they may appear on the solar disk for individual acoustic modes of resonance. The shading displays line-of-sight velocities, with approaching flows shown as dark regions and receding ones as light regions. Panel (a) is an axisymmetric zonal mode ($m = 0$) of degree $\ell = 10$; (b) and (c) are tesseral modes with $(\ell, m) = (20,10)$ and $(32,30)$; (d) is a sectoral mode ($m = \ell$) of degree 30 which is confined to an equatorial strip. [Adapted from Gough 1984b]

regions of approaching flow (shown as light tones) alternating with receding ones (dark tones). Increasing the degree ℓ or order m tends to yield increasingly complex spatial patterns for the velocities, as is suggested by the tesseral modes ($m \neq \ell$). For zonal modes ($m = 0$) the nodal lines of the spherical harmonics are lines of latitude; for sectoral modes ($m = \ell$) they are lines of longitude. With increasing ℓ the sectoral modes are confined to an ever narrower equatorial strip. A convention has also developed in helioseismology that roughly groups the modes into three classes according to their degree: *low-degree* modes for $\ell \lesssim 3$, *intermediate-degree* modes for $4 \lesssim \ell \lesssim 100$, and *high-degree* modes for $100 \lesssim \ell \lesssim 1000$. Of course motions on the real Sun are the superposition of many such modes, and the resulting Doppler

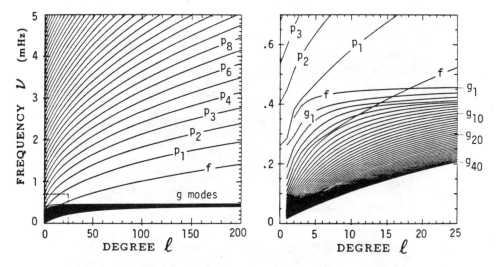

Figure 3. Eigenfrequencies $\nu_{n\ell}$ obtained from a theoretical solar model for p and g modes as the degree ℓ is varied. The discrete frequencies of modes of like order n are connected by straight lines. The right panel is an expanded view for low frequencies and degrees of the left panel. The plotting of g modes was stopped arbitrarily at $n = 40$. [From Model 1 of Christensen-Dalsgaaard 1982]

velocity patterns are much more complicated.

The functions $V_{n\ell}$ are the eigenfunctions of a linear system of ordinary differential equations, with the eigenvalues $\nu_{n\ell}$ being the frequencies of the normal modes. Were the Sun to be perfectly spherical, $V_{n\ell}$ amd $\nu_{n\ell}$ would be independent of m. This comes about because the azimuthal order m depends on the choice of axes of the coordinate system, and with spherical symmetry all choices must be equivalent. Such a degeneracy in m can be broken by any departure from spherical symmetry, such as is achieved by rotation or magnetic fields, thereby leading to splitting or fine structure in the frequency spectrum. There are two discrete spectra of modes for given values of ℓ and m, corresponding in turn to the acoustic modes and the gravity modes. The modes can be arranged in ascending order of frequency (for the p modes) and period (for the g modes), and labelled with the integer n, the radial order of the mode. The eigenfunctions $V_{n\ell}(r)$ typically possess n zeros or nodes. (A shorthand notation has evolved, with acoustic modes of order $n = 2$ designated as p_2 and gravity modes similarly as g_2). In addition there is an f mode whose frequency is between those of p modes and g modes. The f mode is essentially a surface gravity wave whose frequency at large ℓ is independent of the stratification of the Sun; except when ℓ is small, it has no nodes in its radial structure.

The computation of theoretical eigenfrequencies $\nu_{n\ell}$ and eigenfunctions $V_{n\ell}$ is performed numerically using an equilibrium model of the Sun. The frequencies can be calculated to a precision as high as the observational accuracy. Figure 3 shows frequencies obtained from a typical solar mode. For clarity, frequencies with like values of n are joined by straight lines as the degree ℓ is varied, yielding a large fanlike structure. For the p modes, the frequencies ν increase steadily with ℓ at fixed n. In contrast, the frequencies for the g modes approach a finite limit at high ℓ which corresponds to the maximum buoyancy frequency within the solar interior; in this model that frequency is 0.47 mHz. Inspection of Figure 3a emphasizes that the p modes are nearly uniformly spaced in *frequency* as n is varied at fixed low ℓ; that spacing becomes wider and more variable with n at larger values of ℓ. Figure 3b indicates that the g modes become increasingly crowded in frequency both with increasing ℓ and n. A distinguishing characteristic of g modes is that they are nearly uniformly spaced in *period* as n is varied, and this contributes to the severe crowding in ν.

Figure 3 also reveals that similar temporal frequencies ν can be achieved with a variety of modes, each mode possessing different spherical harmonic structure as measured by ℓ and sampling quite different regions within the Sun. Figure 4 demonstrates this by showing the variation with radius of scaled vertical velocity amplitudes for several p and g modes. The three examples of acoustic modes on the right have frequencies of 3.3 mHz (or periods of about 5 minutes). Increasing the degree ℓ leads to the modes becoming increasingly confined to a region close to the surface; the similar frequencies are achieved by suitably decreasing the order n as ℓ in increased. The gravity modes on the left have frequencies of 0.10 mHz (periods of 165 minutes), and increasing ℓ and n yields more complicated structure near the core of the Sun. All the g modes possess an upper reflection point near the base of the convection zone (shown shaded), and their amplitudes become increasingly attenuated with increasing ℓ within that zone. If the g modes were to have comparable amplitudes in the deep interior, it is likely that only g modes of low ℓ can be detected in the solar atmosphere.

4. ACOUSTIC MODES OF INTERMEDIATE DEGREE

It would be appropriate to inquire whether the actual oscillations of the Sun concur with the pattern of frequencies suggested by the normal mode analysis. A remarkable confirmation was obtained by the observation of modes of low and intermediate degree by Duvall & Harvey (1983). They were able to study the zonal modes of the Sun by averaging light in the east-west direction and appropriately projecting Doppler velocities onto Legendre functions. The resulting power spectrum is shown in Figure 5, and is similar to the theoretical eigenfrequencies presented in Figure 3a. The order of each ridge can be determined unambiguously by comparing this power spectrum with ones obtained for high degree modes where the f and p_1 ridges are present. The ridges can be followed to modes with degree as low as unity, thereby identifying the radial order n in low ℓ observations. Duvall & Harvey (1983) also provide a detailed list of frequencies measured from their power spectrum, and additional ones are given in Harvey & Duvall (1984a, b).

Another aspect of such data sets was pointed out by Duvall (1982), showing that the frequencies from the many ridges can all be plotted as the single curve displayed in Figure 6. That realization paved the way for the inversion of frequency data to deduce the speed of sound throughout much of the solar interior (cf. Christensen-Dalsgaard et al. 1985).

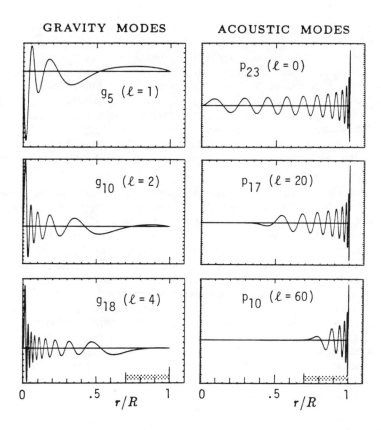

Figure 4. Vertical component of velocity as obtained from a theoretical solar model for several acoustic and gravity modes of oscillation. Shown is variation with proportional radius r/R within the Sun, scaled by $r\rho^{1/2}$, where $\rho(r)$ is density. The three g modes on the left all have frequencies of about 0.10 mHz (periods of 165 minutes); increasing ℓ and n yields modes with more complicated structure near the solar core. The three p modes on the right have frequencies of about 3.3 mHz (periods of 5 minutes); the p modes are increasingly confined to a region close to the surface as ℓ is increased. The ordinate scales are arbitrary; the horizontal line is at zero. The shaded bands indicate the extent of the convection zone. All the g modes possess an upper reflection point near the base of the convection zone, and the modes are evanescent and thus decay within that zone. [From Model 1 of Christensen-Dalsgaard 1982]

5. ACOUSTIC MODES OF LOW DEGREE

The large fan-like structure of power with frequency in Figure 5 supports the notion that the Sun is indeed resonating with a large number of normal modes. Yet sorting out one mode from another in such observations has required spatially resolved measurements on the solar disk, combined with suitable spatial transforms or pro-

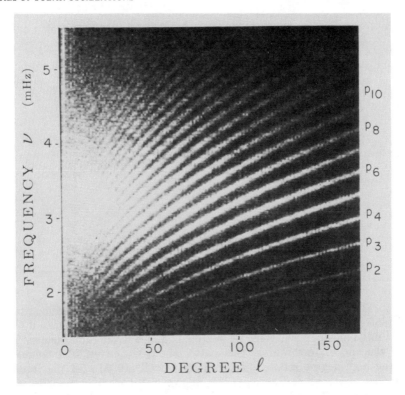

Figure 5. Power spectrum with frequency ν and degree ℓ of zonal p modes of oscillation observed by Duvall & Harvey (1983), with lighter tones representing greater power. The radial orders n of the modes of intermediate degree contributing to the power in the ridges were identified by comparison with observations of high-degree oscillations in which the f modes, whose frequencies are essentially independent of solar structure, were readily apparent.

jections onto spherical harmonics. This is not possible in observation of oscillations on other stars where we cannot image the disk; some inspiration for what might be achieved there is provided by results obtained in helioseismology using integrated sunlight.

The first detection of 5-minute oscillations in light integrated from the entire disk of the Sun was obtained by Isaak and his colleagues making Doppler measurements using resonance scattering from a gaseous sodium or potassium cell (Claverie *et al.* 1979); they detected a uniformly spaced sequence of frequencies. Shortly thereafter, Fossat and his colleagues used a similar resonance detector, with its high spectral stability and low noise level, to observe the Sun from the South Pole during austral summer, in order to avoid or minimize effects of day-night gaps in the data which otherwise produce confusing sidelobes in the frequency spectra. They obtained an uninterrupted record of 120-hours duration (Grec, Fossat & Pomerantz 1980, 1983), and their power spectrum is shown in Figure 7.

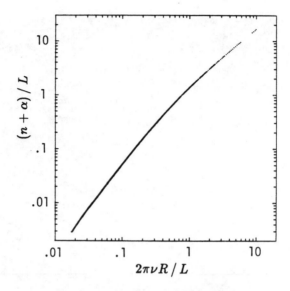

Figure 6. The observed frequencies ν of solar p modes can be reduced to a single curve if plotted as $(n + \alpha)/L$ vs. the surface phase velocity $2\pi\nu R/L$ of the modes (shown in units of Mm s^{-1}). Here $L^2 = \ell(\ell+1)$ and n is the radial order. The value $\alpha = 1.57$ was chosen to minimize the scatter about the curve. The frequency data include the high-degree modes discussed by Duvall (1982) and the intermediate-degree modes of Duvall & Harvey (1983) shown in Figure 5.

The forest of regularly spaced peaks in Figure 7 can be understood by realizing that only modes of low degree ℓ will contribute to the Doppler signal averaged over the disk; in the South Pole observations, only modes with $\ell \lesssim 3$ are visible (cf. Christensen-Dalsgaard & Gough 1980a, 1982). Turning to Figure 3a, we might expect power at discrete frequencies ν corresponding to $\ell = 0$, 1, 2 and 3 for a succession of different orders n. Since the power in Figure 7 is greatest in the frequency range of about 2.5 to 4.5 mHz, Figure 3a suggests that only modes with $n \gtrsim 12$ are likely to contribute. Just why modes at lower ν are not in evidence in the power spectrum must be a consequence of preferential excitation, but this is an issue we do not yet understand.

It was convenient that approximate asymptotic analysis was available to help interpret the full-disk observations. Low-degree modes (with $\ell \ll n$) should satisfy dispersion relations (Vandakurov 1967; see also Gough 1983a) of the form

$$\nu_{n\ell} = \left[n + \frac{1}{2}\left(\ell + \frac{1}{2}\right) + \alpha \right]\nu_0 + \varepsilon_{n\ell}\,, \tag{2}$$

where

$$\nu_0 = \left(2 \int_0^R \frac{dr}{c} \right)^{-1} \tag{3}$$

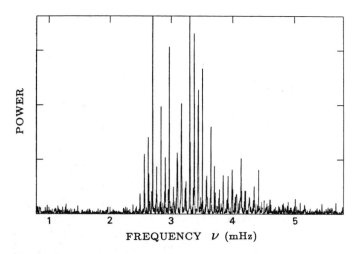

Figure 7. Power spectrum in frequency ν obtained by Grec, Fossat & Pomerantz (1980) from the South Pole of low-degree acoustic modes as observed in Doppler velocities averaged over the solar disk.

is the reciprocal of the sound travel time from the surface at radius R to the core and back, and α is a parameter of order unity which is related to the effective polytropic index of the outer layers of the solar envelope. The next term in the expansion was determined by Tassoul (1980) to be

$$\varepsilon_{n\ell} \simeq -\nu_0 \frac{a\,\ell(\ell+1) - b}{n + \frac{1}{2}\ell + \alpha} , \qquad (4)$$

involving the positive constants a and b which are of order unity. The functional form of equation (2), given that ε is a small correction to the leading term for $n \gg \ell$, indicates that increasing ℓ by 2 and decreasing n by 1 (with $n+\frac{1}{2}\ell$ unchanged) leaves the frequency $\nu_{n\ell}$ almost the same. Thus modes of degrees 0 and 2 and modes of degrees 1 and 3 should contribute alternately to the groupings of peaks in the South Pole power spectrum, with the spacing between these groups being $\frac{1}{2}\nu_0$. Tassoul (1980) also implied that ε is a measure of conditions close to the center of the Sun, and the results predicted that the separation Δ in frequencies between modes of degree 0 and 2 for the same value of $n + \frac{1}{2}\ell$ is just 3/5 of the separation of the corresponding modes of degree 1 and 3. That ratio is independent of the structure of the Sun.

Detailed comparison of the frequencies in the South Pole data with those obtained from solar models was at first facilitated by plotting the frequencies of peak power in the manner of Scherrer *et al.* (1983) as an echelle diagram, with that form suggested by the dispersion relation (2). Theoretical models of the Sun suggest $\nu_0 \simeq 136 \ \mu$Hz, and therefore the South Pole spectrum is divided into segments of that length starting at an arbitrary frequency, and the segments are presented

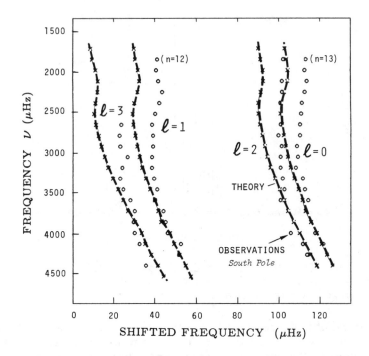

Figure 8. Echelle diagram used to compare the frequencies of low-degree acoustic modes of oscillation as obtained from a theoretical solar model (Ulrich & Rhodes 1983) and from full-disk Doppler observations carried out from the South Pole (shown in Figure 7). The observed power spectrum has been divided into segments of $\nu_0 = 136\mu$Hz starting at an arbitrary frequency, the peaks in each segment identified and plotted here with an open circle at a frequency shifted from the beginning of each segment, and the segments displayed one beneath another. The corresponding theoretical frequencies at the indicated low degrees ℓ and successive radial orders n are joined by the dashed curves.

directly beneath one another to yield Figure 8. This manner of display emphasizes that modes of odd and even degree fall into separate groups of frequency, and that successive segments at higher ν corresponds to modes of higher radial order n. The separation Δ of the peaks within the odd and even groupings in ℓ indeed satisfied the 3:5 ratio. Figure 8 also presents frequencies for these low degree modes as determined from a solar model by Ulrich & Rhodes (1983).

The differences in frequency between observations and theoretical predictions in Figure 8 are of order 5 μHz, or about 0.2%, and thus greater than the observational uncertainties estimated to be 2 μHz. Although various other solar models yield slightly different sets of predicted frequencies, none matches the observations consistently within their errors (cf. Christensen-Dalsgaard 1982; Shibahashi, Noels & Gabriel 1983, 1984; Ulrich & Rhodes 1984; Christensen-Dalsgaard & Gough 1984b; Noels, Scuflaire & Gabriel 1984; Rhodes, Ulrich & Brunish 1984). In many

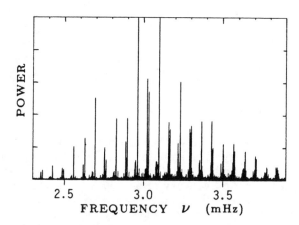

Figure 9. Power spectrum with frequency ν of low-degree full-disk Doppler observations carried out in 1981 over a period of 3 months at Tenerife and Hawaii, providing up to 22 hours of coverage per day. The greatest amplitude is about 15 cm s^{-1}. The separations in frequency, Δ, between modes with like $n + \frac{1}{2}\ell$ are just visible. [Adapted from Claverie *et al.* 1984]

ways such small differences in frequencies between observation and theory serve to affirm that the models are essentially correct. Still, there are aspects in stellar structure theory that probably need modification, whether they be with equations of state, or with opacities, or with elemental abundances and their distributions, or with possible intermittent mixing that modifies the stratification. Despite the unresolved differences, it would appear that models with an initial solar helium abundance of $Y = 0.25 \pm 0.02$ are reasonable. Further, these low-degree data also have placed constraints on the structure of the core. Some have argued that the neutrino problem could be alleviated by mixing core material with its surroundings (e.g. Schatzman *et al.* 1981). However, that would also increase the sound speed in the core because of the increased abundance of hydrogen. A standard solar model without such mixing of core and envelope material is able to reproduce the observed properties of the frequency separation Δ, whereas a chemically homogeneous model predicts a Δ that is 50% greater than the observed value.

The most precisely determined frequencies of solar oscillations may be those obtained by Isaak and his colleagues from full-disk Doppler observations made from a two-station network sited at Hawaii and Tenerife (Claverie *et al.* 1984). A power spectrum of their observations spanning 3 months in time is shown in Figure 9. Such data suggest that the spectral features may be narrower at the lower frequencies, arising possibly from longer lifetimes of those modes. Another route for making observations with high precision of oscillations in integrated solar intensity, rather than in Doppler velocity, has been provided by the Active Cavity Radiometer (ACRIM) on the Solar Maximum Mission satellite (e.g. Willson & Hudson 1981, Woodard & Hudson 1983).

The types of close interplay between observational method and theory evident in all these solar studies of low-degree modes certainly suggest techniques to be applied to other stars. In particular given the tyranny of photons, the power spectra in asteroseismology are likely to be quite noisy; at first one may be able only to pull out the frequencies by using theoretical predictions about their patterns with degree ℓ and order n.

6. GRAVITY MODES

The possible detection and identification of gravity modes has presented major challenges in helioseismology. The amplitudes of the modes appear to be modest, and the inherently long periods of the g modes require observations spanning many months or even years if good frequency resolution is to be achieved to sort out the closely spaced frequencies of these modes. However, if the core of the Sun is stably stratified, the g modes offer a powerful diagnostic of the innermost regions of the Sun, for they approach much closer to the center than the p modes.

There is a long-standing g mode candidate. Observations of full-disk Doppler velocities from Crimea and Stanford have revealed an oscillation with a period of 160.01 minutes (e.g. Severny et al. 1976, Kotov et al. 1978, Scherrer et al. 1979, Scherrer & Wilcox 1983). The nature of that oscillation mode is still uncertain, but if indeed only g modes of low degree are able to survive attentuation by the convection zone, then it may correspond to a g_{10} mode with $\ell = 2$.

The frequencies of the g modes are expected to have a distinctive pattern. Low-degree g modes of high order n satisfy dispersion relations (cf. Vandakurov 1967; Zahn 1970; Ledoux & Perdang 1980; Gough 1984e) of the form

$$\left(n + \frac{1}{2}\ell + \beta\right)\nu_{n\ell} \simeq L/P_0 \,, \tag{5}$$

where

$$P_0 = 2\pi^2 \left(\int\limits_0^{r_c} N\frac{dr}{r}\right)^{-1} \,, \tag{6}$$

$L = [\ell(\ell+1)]^{1/2}$, and β is a constant of order unity related to the nature of variation of the buoyancy frequency N close to the base of the convection zone at $r = r_c$. Such asymptotic analysis predicts that the periods of gravity modes are uniformly spaced as the order n varies. The spacing is P_0/L and hence decreases with increasing degree ℓ or L. For typical solar models, the value of P_0 is 33 to 36 minutes.

The signature of g modes as suggested by the asymptotic relation (5) has not been confirmed unambiguously by observation. There have been tentative identifications of sequences of peaks in power that appear uniformly spaced in period (e.g. Delache & Scherrer 1983, Severny et al. 1984, Fröhlich & Delache 1984; Isaak et al. 1984, Scherrer 1984). The values of P_0 inferred from these observations are rather greater than obtained from standard solar models. Such differences in P_0 might suggest that the gradients in chemical composition, and thus N, in the theoretical models should be somewhat reduced, and indeed a modest degree of diffusive mixing could accomplish this. However, such mixing is probably not consistent with the observed frequency separation Δ of the low-degree p modes.

It is recognized that some of the sequences of peaks may be questionably identified as g modes: the splitting in frequencies induced by rotation is comparable with the separation between g modes of different order and degree; similarly, confusion arises from the sidelobes introduced into the power spectra by day-night interruptions. Moreover, since the ℓ values of the modes cannot be determined from the observations, the frequencies of some of the observed peaks may be identified with g modes obtainable from standard solar models (Gabriel 1984, Gough 1984e). We probably have to await further observations before any firm conclusions about the structure of the Sun can be drawn from g modes.

7. CONCLUSIONS

Helioseismology has made rapid advances in the past few years, though so far most of the deductions about the internal structure and dynamics of the Sun have been based on observations of p modes. Such progress has been helped by several factors. There is no paucity of photons when observing this nearby star, and thus a variety of instruments measuring only a fraction of the photons from a spectral line can function to provide a good signal-to-noise ratio in Doppler observations. It is also possible to spatially resolve the oscillation modes by imaging the solar disk, and thus many of the modes can be identified with reasonable confidence, especially if optimized spatial filters are employed (cf. Christensen-Dalsgaard 1984a). Further, the frequency resolution has been steadily improved by extended observations carried out from a network of stations or by taking care in dealing with data gaps when combining observations spanning many days or even months from a single station. Moreover, theoretical interpretations based on the forward problem have been aided by the maturity of stellar pulsation and structure theory, while those based on the inverse problem have been able to employ inversion techniques largely developed in geophysics.

The close coupling between observation and theory in helioseismology has provided measures of the depth of the convection zone and of its helium content, of the sound speed and the rate of rotation throughout much of the solar interior, and of the possible presence of giant convection cells. Yet we must not be too complacent about the progress. The neutrino problem remains, and we are still far from understanding how the magnetic dynamo operates within the Sun. We are just learning how compressible convection in a rotating shell can drive various forms of differential rotation. The solar core also provides real challenges to theory, especially if it were to be rotating much more rapidly than the surface and to possess strong magnetic fields. Furthermore, we are uncertain about the excitation and damping of the oscillation modes themselves, or just what effect these processes have on their frequencies. This may have important bearing on trying to reconcile the differences in frequencies between observation and theory. And finally, we have much to learn about inversion techniques to be used to deduce the chemical composition and detailed stratification of the solar interior, for unlike in dealing with frequency splitting to study rotation, the inverse problem for composition and stratification is intrinsically nonlinear. For all these reasons we know that we have just begun our work in helioseismology. We have made enough progress in the subject to be confident that it is a rich endeavor, but we have hardly begun to sort out the intricacies of this supposedly simple and nearby star.

Fundamental advances in helioseismology will require networks of ground-based observatories to ameliorate the severe effects of day-night gaps in the data. Several modest networks using non-imaging Doppler analyzers are already operational, and plans to expand them or implement others are well under way. Judging from spectra such as the one displayed in Figure 9, these are promising ventures. It is also essential to observe a wide range of modes of higher degree, and this requires imaging capability, such as is being incorporated into the Doppler instruments to be used by the Global Oscillations Network Group (GONG) from about six sites around the world. The GONG effort, to be largely funded by the U.S. National Science Foundation and coordinated and implemented by the National Solar Observatory of NOAO, will require active participation and attention from a wide team of scientists from many countries, and the data will be available to all. The GONG data sets extending over several years of observation should resolve oscillation modes with degrees up to $\ell \simeq 250$. Even though identical instruments are to be employed at the different sites, there are likely to be major challenges in learning how to best combine imaged data sets which are affected by complicated atmospheric seeing distortions. And given the size of such data sets, the tasks of data reduction and efficient analysis by forward and inverse techniques are also formidable.

The ground-based networks must be complemented by observations carried out from space (cf. Noyes & Rhodes 1984, Toomre 1984). The atmospheric seeing distortions have a most serious impact on measurements of the high-degree modes needed to study dynamics within the convection zone. We must observe these modes with a high signal-to-noise ratio, and attain this within a sufficiently short time interval so that the underlying convection cells do not change or move out of the field of view. This is not likely to be achieved from the ground, and thus observations from space are required. To this end several instruments to study solar oscillations are now being considered by the European Space Agency and by the National Aeronautics and Space Administration for flight on the satellite SOHO. That satellite is to be positioned in a fully sunlit orbit at the Lagrangian point between the Earth and the Sun. That L_1 point has a very small radial velocity with respect to the Sun, and thus would permit Doppler measurements of the oscillations with unprecedented sensitivity. The SOHO satellite is likely to provide uninterrupted observations whose quality is to be limited only by the intrinsic "solar noise" from random surface velocities and intensity fluctuations due mainly to solar convection and magnetic fields. Such observations from space also hold out the best promise for measuring the frequencies of the elusive g modes; as observed from the ground the g modes may be overpowered by atmospheric noise, yet they have the greatest potential for studying conditions in the core of the Sun. An imaging solar oscillations instrument in space is an essential complement to the network of ground observatories, and such combined efforts should let us approach the limits of what can in principle be achieved by helioseismology.

Scientific discussions with Drs Jørgen Christensen-Dalsgaard, Tom Duvall, Douglas Gough and Jack Harvey have been very helpful, as has been their generosity in providing material for some of the illustrations used here. This work was supported in part by the National Aeronautics and Space Administration through grants NSG-7511 and NAGW-91 and by the Air Force Geophysics Laboratory through contract F19628-82-K-0008.

8. REFERENCES

Berthomieu, G., Cooper, A.J., Gough, D.O., Osaki, Y., Provost, J., and Rocca, A. 1980, in *Nonradial and Nonlinear Stellar Pulsation: Lecture Notes in Physics* **72** (eds. H.A. Hill and W.A. Dziembowski; Springer, New York), p. 307.

Brown, T.M. 1985, *Nature* **317**, 591.

Brown, T.M., Mihalas, B.W., and Rhodes, E.J. Jr. 1986, in *Physics of the Sun* (eds. P.A. Sturrock *et al.*; Reidel, Dordrecht) (*Part* 1).

Christensen-Dalsgaard, J. 1982, *Mon. Not. Roy. Astron. Soc.* **199**, 735.

Christensen-Dalsgaard, J. 1984a, in *Solar Seismology from Space* (eds. R.K. Ulrich *et al.* ; NASA Jet Propulsion Laboratory Publ. *84-84*, Pasadena), p. 219.

Christensen-Dalsgaard, J. 1984b, in *The Hydromagnetics of the Sun* (European Space Agency *SP-220*, Noordwijkerhout), p. 3.

Christensen-Dalsgaard, J. 1984c, in *Theoretical Problems in Stellar Stability and Oscillations* (eds. M. Gabriel and A. Noels; Institut d'Astrophysique, Liège), p. 155.

Christensen-Dalsgaard, J. 1984d, in *Space Research Prospects in Stellar Activity and Variability* (eds. A. Mangeney and F. Praderie; Paris Observ. Press, Meudon).

Christensen-Dalsgaard, J. 1986, these proceedings.

Christensen-Dalsgaard, J., Duvall, T.L., Jr., Gough, D.O. Harvey, J.W., and Rhodes, E.J., Jr. 1985, *Nature* **315**, 378.

Christensen-Dalsgaard, J., and Gough, D.O. 1980a, in *Nonradial and Nonlinear Stellar Pulsation: Lecture Notes in Physics* **72** (eds. H.A. Hill and W.A. Dziembowski; Springer, New York), p. 184.

Christensen-Dalsgaard, J., and Gough, D.O. 1980b, *Nature*, **288**, 544.

Christensen-Dalsgaard, J., and Gough, D.O. 1981, *Astron. Astrophys.* **104**, 173.

Christensen-Dalsgaard, J., and Gough, D.O. 1982, *Mon. Not. R. Astron. Soc.* **198**, 141.

Christensen-Dalsgaard, J., and Gough, D.O. 1984a, in *Solar Seismology from Space* (eds. R.K. Ulrich *et al.* ; NASA Jet Propulsion Laboratory Publ. *84-84*, Pasadena), p. 79.

Christensen-Dalsgaard, J., and Gough, D.O. 1984b, in *Solar Seismology from Space* (eds. R.K. Ulrich *et al.* ; NASA Jet Propulsion Laboratory Publ. *84-84*, Pasadena), p. 199.

Christensen-Dalsgaard, J., Gough, D.O., and Toomre, J. 1985 *Science* **229**, 923.

Claverie, A., Isaak, G.R., McLeod, C.P., van der Raay, H.B., and Roca Cortes, T. 1979, *Nature*, **282**, 591.

Claverie, A., Isaak, G.R., McLeod, C.P., van der Raay, H.B., Palle, P.L., and Roca Cortes, T. 1984, *Mem. Soc. Astron. Ital.* **55**, 63.

Cowling, T.G., and Newing, R.A. 1949, *Astrophys. J.* **109**, 149.

Delache, P., and Scherrer, P.H. 1983, *Nature* **306**, 651.

Dicke, R.H. 1970, *Ann. Rev. Astron. Astrophys.* **8**, 297.

Dicke, R.H. 1982, *Solar Phys.* **78**, 3.

Deubner, F.-L. 1975, *Astron. Astrophys.* **44**, 371.

Deubner, F.-L., and Gough, D.O. 1984, *Ann. Rev. Astron. Astrophys.* **22**, 593.

Duvall, T.L. Jr. 1982, *Nature* **300**, 242.

Duvall, T.L. Jr., Dziembowski, W., Goode, P.R., Gough, D.O., Harvey, J.W., and Leibacher, J.W. 1984, *Nature* **310**, 22.

Duvall, T.L., Jr., and Harvey, J.W. 1983, *Nature* **302**, 24.

Duvall, T.L., Jr., and Harvey, J.W. 1984, *Nature* **310**, 19.

Dziembowski, W. 1977, *Acta Astron.* **27**, 1.

Dziembowski, W. 1979, in *White Dwarfs and Variable Degenerate Stars, IAU Colloq.* **53** (eds. H.M. Van Horn and V. Weidemann; Univ. Rochester, New York), p. 359.

Fossat, E., Grec, G., Gelly, B., and Decanini, Y. 1984, *C. R. Acad. Sci. Paris Ser. B* **299**, 17.

Fröhlich, C., and Delache, P. 1984, *Mem. Soc. Astron. Ital.* **55**, 99.

Gabriel, M. 1984, *Astron. Astrophys.* **134**, 387.

Gough, D.O. 1977, in *The Energy Balance and Hydrodynamics of the Solar Chromosphere and Corona* (eds. R.M. Bonnet and P. Delache; de Bussac, Clermont-Ferrand), p. 3.

Gough, D.O. 1978, in *Proc. Workshop on Solar Rotation* (eds. G. Belvedere and L. Paterno; Univ. Catania Press), p. 255.

Gough, D.O. 1981, *Mon. Not. Roy. Astron. Soc.* **196**, 731.

Gough, D.O. 1982, *Nature* **298**, 334.

Gough, D.O. 1983a, *Phys. Bull.* **34**, 502.

Gough, D.O. 1983b, in *Proc. ESO Workshop on Primordial Helium* (eds. P.A. Shaver, D. Knuth and K. Kjär; European Southern Observ., Munich), p. 117.

Gough, D.O. 1984a, in *Solar Seismology from Space* (eds. R.K. Ulrich *et al.* ; NASA Jet Propulsion Laboratory Publ. *84-84*, Pasadena), p. 49.

Gough, D.O. 1984b, *Phil. Trans. R. Soc. London Ser.* A **313**, 27.

Gough, D.O. 1984c, *Mem. Soc. Astron. Ital.* **55**, 13.

Gough, D.O. 1984d, *Adv. Space Res.* **4**, 85.

Gough, D.O. 1984e, in *Solar Physics and Interplanetary Traveling Phenomena* (eds. B. Chen and C. de Jager; Yunnan Observatory, Kunming).

Gough, D.O. 1985, *Solar Phys.* **100**, p. 65-99.

Gough, D.O., and Toomre, J. 1983, *Solar Phys.* **82**, 401.

Grec, G., Fossat, E., and Pomerantz, M. 1980, *Nature*, **288**, 541.

Grec, G., Fossat, E., and Pomerantz, M. 1983, *Solar Phys.* **82**, 55.

Harvey, J.W., and Duvall, T.L. Jr. 1984a in *Solar Seismology from Space* (eds. R.K. Ulrich *et al.* ; NASA Jet Propulsion Laboratory Publ. *84-84*, Pasadena), p.165.

Harvey, J.W., and Duvall, T.L. Jr. 1984b, in *Theoretical Problems in Stellar Stability and Oscillations* (eds. M. Gabriel and A. Noels; Institut d'Astrophysique, Liège), p. 209.

Hill, F., Gough, D.O., and Toomre, J. 1984a, *Mem. Soc. Astron. Ital.* **55**, 153.

Hill, F., Gough, D.O., and Toomre, J. 1984b, in *Solar Seismology from Space* (eds. R.K. Ulrich *et al.* ; NASA Jet Propulsion Laboratory Publ. *84-84*, Pasadena), p. 95.

Hill, F., Toomre, J., and November, L.J. 1983, *Solar Phys.* **82**, 411.

Hill, H.A., Bos, R.J., and Goode, P.R. 1982, *Phys. Rev. Lett.* **49**, 1794.

Isaak, G.R., van der Raay, H.B., Palle, G., Roca Cortes, T., and Delache, P. 1984, *Mem. Soc. Astron. Ital.* **55**, 91.

Kotov, V.A., Severny, A.B., and Tsap, T.T. 1978, *Mon. Not. R. Astron. Soc.* **183**, 61.

Kurtz, D.W., and Seeman, J. 1983, *Mon. Not. R. Astron. Soc.* **205**, 11.

Ledoux, P. 1949, *Mém. Soc. Roy. Sci. Liège Collect.* 4^0 **9**, 263.

Ledoux, P., and Perdang, J. 1980, *Bull. Soc. Math. Belg.* **32**, 133.

Leibacher, J.W. 1984, in *Theoretical Problems in Stellar Stability and Oscillations* (eds. M. Gabriel and A. Noels; Institut d'Astrophysique, Liège), p. 298.

Leibacher, J.W., Noyes, R.W., Toomre, J., and Ulrich, R.K. 1985, *Scientific American*, **253**, No. 3, p. 48.

Leibacher, J.W., and Stein, R.F. 1981, *Astrophys. Lett.* **7**, 191.

Leighton, R.B. 1960, *Proc. IAU Symp.* **12**, 321.

Leighton, R.B., Noyes, R.W., and Simon, G.W. 1962, *Astrophys. J.* **135**, 474.

Liebert, J. 1980, *Ann. Rev. Astron. Astrophys.* **18**, 363.

McGraw, J.T. 1979, *Astrophys. J.* **229**, 203.

Noels, A., Scuflaire, R., and Gabriel, M. 1984, *Astron. Astrophys.* **130**, 389.

Noyes, R.W., Baliunas, S.L., Belserene, E., Duncan, D.K., Horne, J., and Widrow, L. 1984, *Astrophys. J. Lett.* **285**, L23.

Noyes, R.W., and Rhodes, E.J. Jr. (eds.) 1984, *Probing the Depths of a Star: The Study of Solar Oscillations from Space* (NASA Jet Propulsion Laboratory Publ. *400-237*, Pasadena).

O'Donaghue, D., and Warner, B. 1986, these proceedings.

Rhodes, E.J. Jr., Ulrich, R.K., and Brunish, W.M. 1984, *Mem. Soc. Astron. Ital.* **55**, 37.

Robinson, E.L. 1979, in *White Dwarfs and Variable Degenerate Stars, IAU Colloq.* **53** (eds. H.M. Van Horn and V. Weidemann; Univ. Rochester, New York), p. 343.

Schatzman, E., Maeder, A., Angrand, R., and Glowinski, R. 1981, *Astron. Astrophys.* **96**, 1.

Scherrer, P.H. 1984, *Mem. Soc. Astron. Ital.* **55**, 83.

Scherrer, P.H., and Wilcox, J.M. 1983, *Solar Phys.* **82**, 37.

Scherrer, P.H., Wilcox, J.M., Christensen-Dalsgaard, J., and Gough, D.O. 1983, *Solar Phys.* **82**, 75.

Scherrer, P.H., Wilcox, J.M., Kotov, V.A., Severny, A.B., and Tsap, T.T. 1979, *Nature* **277**, 635.

Severny, A.B., Kotov, V.A., and Tsap, T.T. 1976, *Nature* **259**, 87.

Severny, A.B., Kotov, V.A., and Tsap, T.T. 1984, *Nature* **307**, 247.

Shibahashi, H., Noels, A., and Gabriel, M. 1983, *Astron. Astrophys.* **123**, 283.

Shibahashi, H., Noels, A., and Gabriel, M. 1984, *Mem. Soc. Astron. Ital.* **55**, 163.

Tassoul, M. 1980, *Astrophys. J. Suppl.* **43**, 469.

Toomre, J. 1984, in *Solar Seismology from Space* (eds. R.K. Ulrich *et al.* ; NASA Jet Propulsion Laboratory Publ. *84-84*, Pasadena), p. 7.

Ulrich, R.K. 1970, *Astrophys. J.* **162**, 993.

Ulrich, R.K., and Rhodes, E.J. Jr. 1977, *Astrophys. J.* **218**, 521.

Ulrich, R.K., and Rhodes, E.J. Jr. 1983, *Astrophys. J.* **265**, 551.

Ulrich, R.K., and Rhodes, E.J. Jr. 1984, in *Solar Seismology from Space* (eds. R.K. Ulrich *et al.* ; NASA Jet Propulsion Laboratory Publ. *84-84*, Pasadena), p. 371.

Vandakurov, Yu. V. 1967, *Astron. Zh.* **44**, 786.

Willson, R.C., and Hudson, H.S. 1981, *Astrophys. J. Lett.* **244**, L185.

Woodard, M., and Hudson, H.S. 1983, *Nature* **305**, 589.

Zahn, J.-P. 1970, *Astron. Astrophys.* **4**, 452.

THEORETICAL ASPECTS OF HELIO- AND ASTEROSEISMOLOGY

J. Christensen-Dalsgaard
NORDITA, København, Denmark
HAO, Boulder, Colorado, and
Astronomisk Institut,
Aarhus Universitet
DK-8000 Aarhus C, Denmark[1]

ABSTRACT. The dependence of adiabatic oscillations, particularly their frequencies, on the structure of the sun or other stars is analyzed, on the basis of asymptotic analysis and numerical computations. As a particular example I consider the effects of mixing in the solar core; mixing sufficiently strong to bring the computed neutrino flux into agreement with the observations appears to be inconsistent with the oscillation observations. I also discuss the effects of modifications to the upper, superadiabatic part of the convection zone.

1. INTRODUCTION

The object of helio- and asteroseismology is to get information about the interiors of the sun and other stars from analysis of oscillations observed on the surfaces of these objects. This requires an understanding of the relations between the properties of the stellar interior and those of the observed oscillations, in particular their frequencies. The understanding has several different levels. One must be able to compute oscillation frequencies for a given model of the star, with an accuracy that matches the accuracy of the observations. In this way the model can be tested against the observations. If it should fail the test, as it almost certainly will, one must study the way in which the frequencies are modified by changes to the model, and carry out the appropriate changes. This can be done by means of carefully controlled numerical experiments, where a single property of the model is changed, keeping all other properties precisely fixed. However more systematic procedures can be devised, whereby the frequency changes, for sufficiently small changes in the properties of the model, can be expressed as linear functionals of these changes. In addition important guidance may be obtained from approximate, asymptotic expressions for the frequencies. Finally these techniques can lead to proper inverse analyses, whereby the properties of the model, or the changes required to an initial trial model, are inferred directly from the observations.

1) Permanent address.

D. O. Gough (ed.), Seismology of the Sun and the Distant Stars, 23–53.
© 1986 by D. Reidel Publishing Company.

The present paper is mainly concerned with the forward problem, i.e. the calculation of frequencies for given models and of changes in the frequencies resulting from changes in the models. In particular the observed properties of solar oscillations, and their interpretation, are not discussed. Neither is any attempt made at giving a comprehensive review of the field. Inverse methods are treated by Gough (these proceedings; see also Gough 1985). A general introduction to the seismology of the sun and the distant stars is given by Toomre (these proceedings). In addition the large number of recent reviews on helioseismology may be consulted (e.g. Brown, Mihalas & Rhodes 1986; Deubner & Gough 1984; Gough 1984a, b, 1985a, b; Fossat 1985; Harvey 1985; Christensen-Dalsgaard 1984b, 1984c; Christensen-Dalsgaard, Gough & Toomre 1985), as well as Christensen-Dalsgaard (1984a) on asteroseismology.

2. ASYMPTOTIC DESCRIPTION OF THE OSCILLATIONS

Based on earlier work by Lamb (1932) Gough obtained a general asymptotic differential equation for adiabatic stellar oscillations (cf. Deubner & Gough 1984). It may be written as

$$\frac{d^2\Psi}{dr^2} + K(r)\Psi = 0 \; , \tag{2.1}$$

where r is distance from the centre, and the dependent variable is

$$\Psi = c^2\rho^{\frac{1}{2}}\text{div}\delta\underset{\sim}{r} \; , \tag{2.2}$$

c being the adiabatic sound speed, ρ density, and $\delta\underset{\sim}{r}$ the displacement in the oscillation. Furthermore K is given by

$$K = \frac{\omega^2}{c^2}\left[1 - \frac{\omega_c^2}{\omega^2} - \frac{S_\ell^2}{\omega^2}\left(1 - \frac{N^2}{\omega^2}\right)\right] \; ; \tag{2.3}$$

here ω is the oscillation frequency,

$$S_\ell = \frac{cL}{r} \; , \tag{2.4}$$

where $L = \sqrt{\ell(\ell+1)}$, ℓ being the degree of the oscillation; N is the Brunt-Väisälä frequency and ω_c is a generalized acoustical cut-off frequency,

$$\omega_c^2 = \frac{c^2}{4H^2}\left(1 - 2\frac{dH}{dr}\right) \; , \tag{2.5}$$

where H is the density scale height.

It is obvious that the behaviour of a mode of oscillation depends on the value of its frequency relative to the characteristic frequencies

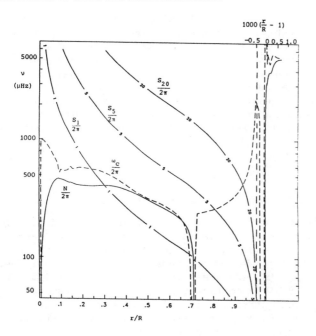

Figure 1.   Characteristic frequencies for a model of the present
sun. The dashed curve shows the generalized acoustical cut-off
frequency ω_c (cf. equation 2.5). The continuous curve shows the
buoyancy frequency N, and the remaining curves show the characteristic acoustical frequencies S_ℓ, for ℓ = 1, 5 and 20. The abscissa
is fractional radius r/R, where R is the photospheric radius. Notice
the change of scale near r/R = 1.

S_ℓ, N and ω_c. On Fig.1 these frequencies are plotted for a typical model
of the present sun. Note that K can also be written as

$$K = \frac{\omega^2}{c^2} \left(1 - \frac{\omega_{\ell,+}^2}{\omega^2} \right)\left(1 - \frac{\omega_{\ell,-}^2}{\omega^2} \right) , \tag{2.6}$$

which defines the characteristic frequencies $\omega_{\ell,+}$ and $\omega_{\ell,-}$. They have
been shown, e.g. by Deubner & Gough (1984). It is easy to verify that
in the interior of the model, and for ℓ greater than about 2, $\omega_{\ell,+} \approx S_\ell$
and $\omega_{\ell,-} \approx N$; near the surface $\omega_{\ell,+} \approx \omega_c$ whereas $\omega_{\ell,-}$ is very small.
   A mode of oscillation is said to be *trapped* in the region where
the eigenfunction oscillates as a function of r. From equations (2.1)
and (2.6) follows that this occurs where

$$\omega > \omega_{\ell,+}, \qquad \omega > \omega_{\ell,-} \tag{2.7a}$$

or

$$\omega < \omega_{\ell,+}, \qquad \omega < \omega_{\ell,-} . \tag{2.7b}$$

As $\omega_{\ell,+} \gg \omega_{\ell,-}$ in most of the model it follows from equation (2.3) that in the former case the behaviour of the mode is dominated by S_ℓ, with N making only a small contribution. Such modes are controlled by the pressure fluctuations, and are therefore labeled p modes. In the limit where N can be neglected they are simply standing acoustic waves. Similarly the modes satisfying equations (2.7b) are dominated by N, i.e. by buoyancy; they are labeled g modes. In the limit of very low frequency S_ℓ can be neglected, and the modes are simply standing gravity waves.

From JWKB theory follows that the oscillation frequencies approximately satisfy

$$\omega \int_{r_1}^{r_2} \left[1 - \frac{\omega_c^2}{\omega^2} - \frac{S_\ell^2}{\omega^2}\left(1 - \frac{N^2}{\omega^2} \right) \right]^{\frac{1}{2}} \frac{dr}{c} \approx \pi(n + \varepsilon); \qquad (2.8)$$

here r_1 and r_2 are consecutive *turning points* or zeros of K, such that K > 0 for $r_1 < r < r_2$; n is the radial order of the mode, [2] and ε is a phase constant which depends on the exact conditions at the turning points (see also Ellis, these proceedings; Gabriel, these proceedings). For simple turning points JWKB analysis shows that $\varepsilon = -1/2$; Christensen-Dalsgaard (1984b) found that this provides a reasonable approximation to actual computed frequencies of solar oscillation.

For g modes the turning points are defined, appoximately, by $\omega = N$; thus at very low frequencies the modes in the sun extend from near the centre to near the base of the convection zone, whereas at higher frequency the modes are confined close to the maximum in N. The maximum value N_{max} of N gives an upper limit to the frequencies of g modes. For p modes, on the other hand, the lower turning point is approximately at $r = r_t$, where r_t satisfies

$$\frac{c(r_t)}{r_t} = \frac{\omega}{L} . \qquad (2.9)$$

Thus with increasing degree or decreasing frequency the turning point moves closer to the surface. The upper turning point is defined by $\omega = \omega_{\ell,+} \approx \omega_c$. The behaviour of $\omega_{\ell,+}$ near the surface, displayed on Fig.2, is fairly complicated; much of this complication is associated with the superadiabatic region near the top of the convection zone, as illustrated by $|N|$ which is also shown on the figure. As a result modes with frequencies higher than about 2000 μHz penetrate almost to the surface, whereas modes with frequencies below this value are trapped somewhat deeper in the sun. If this were to have observable consequences for the frequencies it might offer a diagnostics of the superadiabatic part

---

[2] Strictly speaking n is defined such that the number of zeros in Ψ in the trapping region is n-1. It may be shown that for high-order p-modes and high-order g-modes n is then also the order of the mode, which in these cases reduces to the total number of zeros in the radial displacement. For modes of lower order the definition of a global order of the mode becomes more complicated (e.g. Unno et al. 1979; Christensen-Dalsgaard 1980; Deubner & Gough 1984).

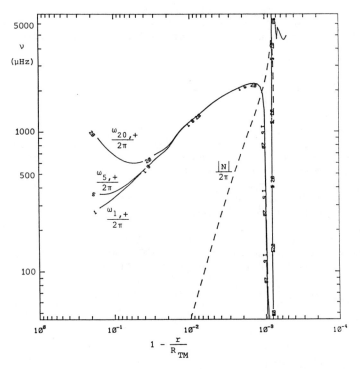

Figure 2. Characteristic frequencies near the surface of a model of
the present Sun. The solid curves show $\omega_{\ell,+}$, for $\ell = 1$, 5 and 20.
The dashed curve gives the modulus $|N|$ of the buoyancy frequency
(notice that N is imaginary in a convectively unstable region).
The abscissa is the distance beneath the temperature minimum, relative to the solar radius, on a logarithmic scale. The rapid variation in $\omega_{\ell,+}$ occurs in the strongly superadiabatic region just beneath the photosphere (see also Gough 1985a).

of the convection zone, which would be of obvious interest to studies of
convection. This question is addressed further in section 5.3 below. It
should be noted, however, that a simple application of JWKB theory is
highly questionable under circumstances like this where the coefficients
in the equations vary rapidly.

Outside the trapping region the eigenfunctions generally decrease
approximately exponentially; in such regions the modes are said to be
evanescent. Thus for p modes of high degree which are trapped close to
the surface, the amplitude in the deep interior of the star is very
small. The trapping is particularly important for g modes in stars which,
like the sun, have deep outer convection zones; for large ℓ the trapping
is so efficient that the modes are unlikely to be observed on the surface
(Dziembowski & Pamjatnykh 1978; Christensen-Dalsgaard, Dziembowski &
Gough 1980; see also Section 4 below).

For modes in the 5 min region, with frequencies typically exceeding
2000 µHz, ω_c is only significant very close to the surface, and the inte-

gral in equation (2.8) may be approximated in such a way as to remove it. Furthermore, N_{max} in the sun is only about 470 µHz; thus for these modes we can neglect N^2/ω^2 as a first approximation. Then the relation for the frequencies reduces to

$$\int_{r_t}^{R} \left[1 - \frac{c^2}{r^2} \frac{L^2}{\omega^2} \right]^{\frac{1}{2}} \frac{dr}{c} = \frac{\pi(n+\alpha)}{\omega} \qquad (2.10)$$

where R is the photospheric radius, and α is a new phase constant related to ε. This relation may also be derived directly from the dispersion relation for plane sound waves by using ray theory (Gough 1984c) or JWKB analysis (Christensen-Dalsgaard et al. 1985b). A relation of this form was first obtained by Duvall (1982) from observed data on high-degree 5 min modes. It was used by Christensen-Dalsgaard et al. to infer the variation of sound speed in much of the sun (see also Gough, these proceedings).

For low-degree, high-order p and g modes a higher order asymptotic approximation was obtained by Tassoul (1980), extending earlier work by, e.g. Vandakurov (1967). For p modes the result is

$$\nu_{n,\ell} \approx (n + \frac{\ell}{2} + \varepsilon^{(p)})\Delta\nu - (L^2 + \delta) \frac{A}{n + \frac{\ell}{2} + \varepsilon^{(p)}} \qquad (2.11)$$

where

$$\Delta\nu = \left[2 \int_0^R \frac{dr}{c} \right]^{-1} \qquad (2.12)$$

and $\varepsilon^{(p)}$ is yet another phase constant. The leading term in this expression, but with ℓ replaced by $L - 1/2$ can be obtained by expanding the integral in equation (2.10) at the centre of the star. The coefficient A in the second-order term is related to the variation of the sound speed near the centre of the sun. For a given n_0 equation (2.11) may be expanded near $n = n_0$ in terms of $x \equiv n + \frac{\ell}{2} - n_0$ as

$$\nu_{n,\ell} = \nu_0 - L^2 D_0 + (\Delta\nu_0 + L^2 d_0)x + O(x^2) \qquad (2.13)$$

(Scherrer et al. 1983; Christensen-Dalsgaard 1984a; Gough 1984a). Here D_0 is proportional to A and therefore measures conditions in the solar core.

To lowest order equation (2.11) predicts that $\nu_{n,\ell} \approx \nu_{n-1,\ell+2}$. This apparent degeneracy is lifted by the second term, resulting in a small separation

$$\delta\nu_{n,\ell} = \nu_{n,\ell} - \nu_{n-1,\ell+2} , \qquad (2.14)$$

related to D_0 by

$$\overline{\delta\nu_{n,\ell}} \approx (4\ell + 6)D_0 \qquad (2.15)$$

where the overbar denotes average in n around n_0.

For low degree, high-order g modes one similarly obtains, to lowest asymptotic order, that

$$T_{n,\ell} \equiv \frac{1}{\nu_{n,\ell}} \approx \frac{T_0}{L} \left(n + \varepsilon_\ell^{(g)} \right) ,$$ (2.16)

where

$$T_0 = \frac{2\pi^2}{\int_{r_1}^{r_c} N \frac{dr}{r}} ;$$ (2.17)

here the phase constant $\varepsilon_\ell^{(g)}$ may or may not contain the term $\frac{\ell}{2}$, depending on whether or not the core of the star is radiative, r_c is the radius at the base of the outer convection zone and r_1 is either 0 (for a radiative core) or the radius at the outer edge of a possible convective core.

For stars not too dissimilar to the sun, the variation of S_ℓ with position is superficially similar to that shown on Fig.1. N, however, depends quite sensitively on the mass and the evolutionary status of the star. Fig.3 (from Christensen-Dalsgaard 1984a) shows N in a sample of

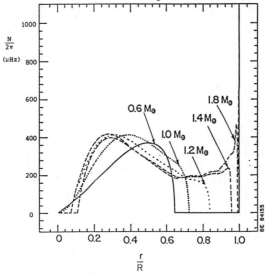

Figure 3. The buoyancy frequency N in ZAMS models, against fractional radius r/R. The masses of the models (in solar units) are indicated (from Christensen-Dalsgaard 1984a).

zero-age main sequence stars with varying mass. It is evident that with increasing mass the outer convection zone shrinks, eventually leading to a secondary maximum in N near the surface, whereas the stars develop convective cores for masses higher than about 1.2 $M_\odot$. With increasing age of the stars a gradient in the hydrogen abundance in the stellar core is built up, causing an increase in N. Such effects should be clearly visible in spectra of stellar g modes, should these be observed.

3. COMPUTATION OF SOLAR AND STELLAR OSCILLATIONS

The "standard problem" in stellar evolution and oscillation theory may
be defined as follows:

(1)   Compute spherically symmetrical stellar models, using a reasonably
      accurate representation of the physics of the matter in the star,
      and some version of local mixing length theory to describe convec-
      tion; do not take into account rotation, magnetic fields, mixing of
      material outside convection zones, or the effects of turbulent pres-
      sure.

(2)   Compute linear adiabatic oscillation frequencies for the models re-
      sulting from (1), neglecting again rotation and magnetic fields, as
      well as turbulent pressure.

Clearly the results of such calculations cannot be accurate descriptions
of the oscillations of real stars. In particular nonadiabaticity in the
outer layers of the stars has significant effects on the oscillation
frequencies (e.g. Christensen-Dalsgaard & Frandsen 1983a; Gough 1984a).
However no definite theoretical procedures exist to deal with such com-
plicating features. In contrast, the calculations described above provide
a fully determined route from the assumed physics of the stellar interiors to the oscillation frequencies. Starting from this basis one must
then attempt to determine whether the unavoidable discrepancies between
the observed and the computed frequencies are caused by deficiences in
the physics (e.g. the equation of state or the opacity), or by the simplifying assumptions, or both.

To these sources of error one must add numerical inaccuracy. The
numerical tasks defined by (1) and (2) are straightforward: in the former
case the solution of a system of coupled non-linear differential equations in space, combined with the time evolution of the composition; in
the latter the solution of a system of coupled ordinary differential
equations, with the oscillation frequency appearing as an eigenvalue.
Well-established numerical procedures exist for dealing with such problems. The main difficulty in seismological calculations is the need for
very high numerical accuracy. Present observations of solar p mode frequencies have relative precisions as high as 10^{-4} (van der Raay, Palle &
Roca Cortes, these proceedings), and one would obviously like the numerical accuracy of the computed frequencies to match this precision; in fact
significant changes in the assumed physics generally result in fairly
small changes in the frequencies, so that the high accuracy is needed to
draw meaningful conclusions on the basis of the observed frequencies. The
demands on precision must be met both by the evolution calculation and
by the calculation of oscillation frequencies; the difficulties may be
larger in the latter case, due to the fact that the observed modes are
typically of high radial order. Thus the computations should use a large
number of suitably distributed mesh points, and an integration scheme of
high numerical order; alternatively the variational principle for the
eigenfrequency may be used to improve the precision of the result (see
Christensen-Dalsgaard 1984b for a further discussion of these methods,
and references).

The variational principle requires some comments. As a consequence of the variational property the error in the computed frequency is quadratic in the error in the eigenfunction used to evaluate the variational integral. Thus the variational frequency ν_V should be less sensitive to numerical errors *in the calculation of the oscillation* than the eigenfrequency ν_E. However ν_V is obviously affected by errors in the equilibrium model. In particular the derivation of the variational expression assumes that the equilibrium model is in hydrostatic equilibrium. If this is not satisfied, due to numerical errors in the evolution calculation, ν_E and ν_V do not agree, even in the limit of infinitely high precision in the oscillation calculation. A further complication arises from the fact that different formulations must be used for the variational integral for radial (i.e. $\ell = 0$) and non-radial (i.e. $\ell > 0$) modes. In principle the general non-radial formulation (Chandrasekhar 1964) can be used for radial modes, but this apparently leads to severe cancellation and hence loss of numerical accuracy; thus for radial modes the formulation originally developed by Ledoux & Pekeris (1941) must be used. There is thus the risk that errors in the equilibrium model may affect ν_V differently for radial and non-radial modes.

That this can indeed be a problem was recently demonstrated by Fossat (1985). He compared observations of the separation $\delta\nu_{n,\ell}$ (cf. equation 2.14) with various theoretical calculations. For "standard" solar models the results were generally in good agreement with the observations; however the values obtained from Christensen-Dalsgaard (1982) were systematically higher than the rest by a few μHz. As $\delta\nu_{n,\ell}$ may be used to

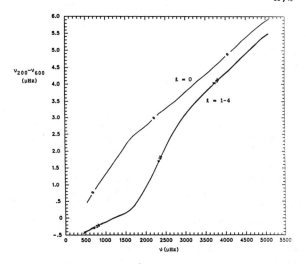

Figure 4. Differences between variational frequencies for models of the present sun computed with 200 and 600 mesh points, shown against frequency, for modes with $\ell = 0-4$. For $\ell > 0$ the differences are essentially independent of ℓ.

diagnose the core of the sun, it is important that this discrepancy be
resolved.

To investigate the effect of numerical errors in the evolution calculation I have computed a model with the same physics, and using the
same numerical techniques, as Model 1 of Christensen-Dalsgaard (1982)[3]
but with the number N_{ev} of points in the evolution calculation being
600; these were distributed as in Model 1, with roughly half the points
in the convection zone. This was compared with a model computed with
$N_{ev} = 200$. For the oscillation calculation the models were reset by cubic
interpolation to a mesh suitable for p modes, with 600 points. Fig.4
shows differences between the variational frequencies obtained for these
two models, for $\ell = 0$-4. There is a substantial difference between the
effect of the numerical error in the model for $\ell = 0$ and for $\ell > 0$. The
difference between the eigenfrequencies falls between the two curves
shown on the figure and is largely independent of ℓ.

The effects on $\delta\nu_{n,\ell}$, for $\ell=0$, of these errors are shown on Fig.5. The
values obtained from ν_E for $N_{ev}=200$ and 600 are indistinguishable, and agree
well with the other theoretical results presented by Fossat(1985); on the

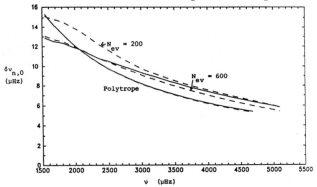

Figure 5. Frequency separation $\delta\nu_{n,0} = \nu_{n,0} - \nu_{n-1,2}$. The continuous
curves are based on eigenfrequencies, and the dashed curves on variational frequencies. Results are shown for models of the present sun,
computed with 200 and 600 mesh points, as well as for a polytrope
with n = 1.5, computed with very high numerical accuracy.

other hand the value obtained from ν_V for $N_{ev}=200$ is up to 2.2 μHz higher.
For $N_{ev}=600$ the values obtained from ν_E and ν_V agree to within about
0.3 μHz; thus numerical error in the equilibrium calculation is obviously responsible for the discrepancy at $N_{ev} = 200$. For comparison the figure also shows $\delta\nu_{n,\ell}$ calculated for a complete polytropic model of index
1.5, with the same mass and radius as the sun. This model, and its oscillation frequencies, were calculated with high numerical accuracy. As expected ν_E and ν_V give very similar results in this case. - It is interesting, if fortuitous, that these results for the polytropic model so

---

[3]Since the publication of that paper an error was discovered in the expansion of the hydrogen abundance at the centre of the model. This has
been corrected in all the calculations reported here. Its effect on the
frequencies of p-modes was substantially less than 1 μHz.

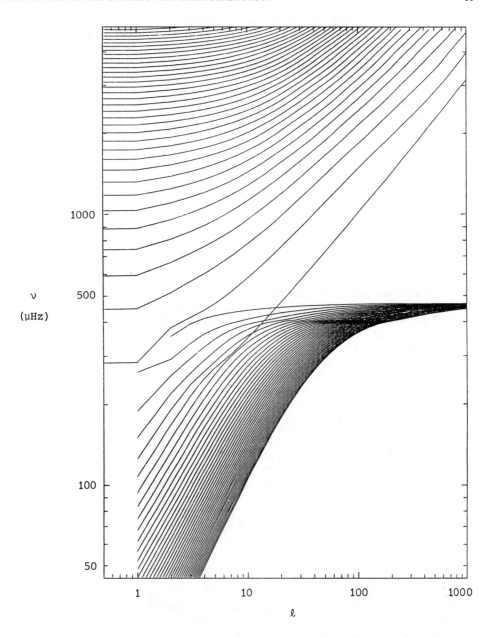

Figure 6. Adiabatic oscillation frequencies for normal model of the present sun, as functions of the degree ℓ. For clarity points corresponding to modes with a given radial order have been connected.

closely resemble those for a realistic model of the sun.

Given this difficulty with the variational method, its utility for computing accurate frequencies may be questioned. However it is still the case that the accuracy of the computed *frequencies*, as opposed to frequency separations, is considerably higher for ν_V than for ν_E. Furthermore the problem gives a very useful indication of the fact that the accuracy of the original equilibrium model is insufficient. Thus, although the use of the variational principle can be avoided by employing a larger number of mesh points in the oscillation calculation or a higher order numerical scheme, it would still provide a valuable test of the accuracy of the results.

4. COMPUTED PROPERTIES OF SOLAR AND STELLAR OSCILLATIONS

To illustrate the spectrum of possible modes in the sun, Fig.6 shows frequency as a function of degree for a selection of computed adiabatic oscillations. The p mode spectrum has been truncated at $\nu = 5000$ µHz; for frequencies higher than that, the effects of nonadiabaticity, and of the uncertain structure of the atmosphere, probably become important. The g mode spectrum has been stopped arbitrarily at a radial order of 40. The separation of the modes into two distinct groups is quite obvious from the figure, with the f and p mode frequencies increasing roughly as a power of ℓ for high ℓ, whereas the g mode frequencies tend to a finite limit, determined by the maximum N_{max} in the buoyancy frequency. However it is also clear that at moderate degrees and frequencies the two groups overlap. Thus the f modes, which at high degree are essentially surface gravity waves, extend into the g mode region. Although the curves appear to cross in this region, a closer inspection reveals that instead the modes undergo *avoided crossings* (e.g. Christensen-Dalsgaard 1980). It might also be noticed that the weak maximum in N near $r/R = 0.4$, $\nu = 400$ µHz, although too faint to trap any g modes, still causes an accumulation of modes at that frequency.

An important global property of a mode of oscillation, besides its frequency, is its integrated energy. Here we consider the normalized energy

$$E_{n,\ell} = \frac{\int_0^R [\xi_r^2(r) + \ell(\ell+1)\xi_t^2(r)]\rho r^2 dr}{4\pi M[\xi_r^2(R) + \ell(\ell+1)\xi_t^2(R)]} \tag{4.1}$$

where ξ_r and ξ_t are the radial and tangential components of the displacement, M is the total mass of the star and R is its photospheric radius. Thus $16\pi^2 E_{n,\ell}$ is the ratio between the actual kinetic energy of the oscillation, and the kinetic energy of the star if it were to be moving with the rms surface velocity. Fig.7 shows $E_{n,\ell}$ for p modes in the present sun. The dominant feature is the strong decrease with increasing frequency up to about 3000 µHz. This may be related to the mode trapping, and the behaviour of $\omega_{\ell,+}$, discussed in Section 2. At low frequencies there is a substantial region, whose width increases with decreasing frequency, where the mode is evanescent; this increases the ratio between

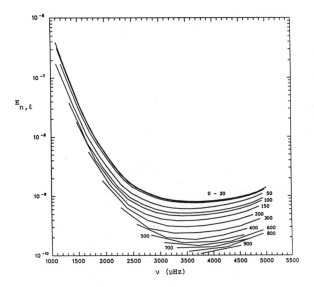

Figure 7. Normalized energy $E_{n,\ell}$ (cf. equation 4.1) for p modes in model of the present sun, as a function of the cyclic frequency ν. For clarity points corresponding to modes of given degree ℓ have been connected; the curves are labelled by ℓ.

the interior and the surface amplitude, and hence increases $E_{n,\ell}$. At high frequencies, on the other hand, the mode is oscillatory essentially to the surface, and the surface and interior amplitudes are comparable. It is also noticeable that the energy decreases substantially with increasing ℓ at fixed frequency; the increase in ℓ causes the mode to be confined closer to the surface, and hence to involve a smaller fraction of the mass of the star.

The normalized energy of g modes in the sun, illustrated on Fig.8, shows a much more dramatic variation, spanning almost 30 orders of magnitude! The principal reason for this is the trapping of the modes at high degrees and frequencies near the maximum in the buoyancy frequency N. This is particularly pronounced for frequencies exceeding the secondary maximum in N at 400 μHz. The very low values in $E_{n,\ell}$ found at selected frequencies for higher values of ℓ occur when the g modes take on the character of surface gravity waves; this happens when their frequencies are near the extension of the f mode frequencies shown on Fig.6. It is interesting that this phenomenon can be traced in the energy to a degree as low as 3.

The frequencies and energies of p modes in other stars resembling the sun are qualitatively similar to those shown here. However the properties of g modes show a stronger variation with stellar parameters, reflecting the difference in N shown on Fig.3. To illustrate this Fig.9 show the normalized energy for a 1.8 $M_\odot$ ZAMS, with a pronounced maximum

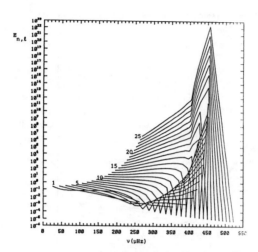

Figure 8. Normalized energy $E_{n,\ell}$ (cf. equation 4.1) for g modes in model of the present sun, as a function of the cyclic frequency ν. For clarity points corresponding to modes of given degree ℓ have been connected; selected curves are labelled by ℓ.

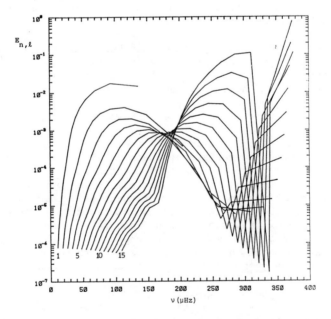

Figure 9. Normalized energy $E_{n,\ell}$ for g modes in a 1.8 $M_\odot$ ZAMS model. See caption to Fig.8.

in N very near the surface and a second maximum in the interior. This is probably responsible for the complicated variation of its energy. A detailed analysis of this behaviour in terms of the asymptotics of the eigenfunctions would be interesting, but has so far not been attempted.

The increase in the energy of p modes has a large part of the responsibility for the increase in their growth- or damping times with decreasing frequency which has been computed (e.g. Ando & Osaki 1975; Christensen-Dalsgaard & Frandsen 1983a; Kidman & Cox 1984) and observed (Grec, Fossat & Pomerantz 1983; a similar, but better defined trend, has been reported by Isaak). In addition, it is mainly responsible for the decrease at low frequency in the amplitudes predicted for stochastic excitation of these modes by convection (Goldreich & Keeley 1977; Christensen-Dalsgaard & Frandsen 1983b; Gough 1985a), although this decrease may be less pronounced than for the observed amplitudes. For the g modes it is hardly conceivable that the variation of the energy is without effect on their surface amplitudes. Certainly the energy must constrain the modes that are likely to be observed on the stellar surface (Dziembowski, Paternò & Ventura 1985). As a very simple application of this idea one might estimate the observed amplitude for fixed energy in the mode; this may give some indication of the relative probability that a given mode is observed. In such a calculation one can furthermore take into account the sensitivity of a given observing scheme, as determined by its spatial response function (e.g. Christensen-Dalsgaard & Gough 1982). For the sun such estimates have been made by Gough (1985b) and by Berthomieu & Provost (cf. Delache 1984). Fig.10 shows the predicted amplitudes in whole-disk observations of the sun, for an assumed energy of about 10^{34} erg. Due to the decrease in the sensitivity at higher ℓ, and the increase in the normalized energy, the expected amplitude is very small for degrees higher than about 5. Similar results for the 1.8 $M_\odot$ ZAMS model, using the same energy, are shown on Fig.11. Here the amplitudes are considerably higher, especially at low frequencies, reflecting the lower normalized energies. Thus in this sense g modes are easier to excite in a higher-mass star than in the sun; however, in the absence of a theory for the excitation mechanism, it would be rash to speculate that they might also be more likely to be observed.

It is convenient to characterize computed oscillation spectra by parameters derived from the asymptotic analysis discussed in Section 2. Such parameters may also be determined from observations that are insufficient to obtain accurate frequencies or precise identifications of the modes; they are therefore likely to provide the initial data for asteroseismology. For low-degree p modes the expansion in equation (2.13) provides a parametrization of the frequencies. Christensen-Dalsgaard (1984a) applied this to a selection of stellar models; Fig.12 shows the results for the parameters $\Delta\nu_0$ and D_0. For this limited set of models there appears to be a separation of the effects of evolution, which modifies the core of the model and hence D_0, from the effects of changing the mass. More extensive calculations are required to test whether this can provide a useful two-dimensional stellar classification, given the additional uncertainty in other stellar parameters such as the composition.

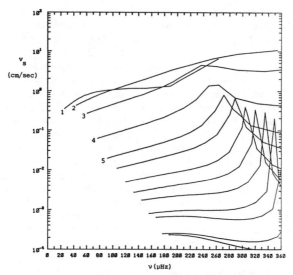

Figure 10. Estimated velocity amplitudes of g modes for the present sun, as observed in integrated light, if a fixed modal energy of about 10^{34} erg is assumed. The amplitudes are shown as functions of the frequency ν, and for clarity points corresponding to a given degree ℓ have been connected; selected curves are labelled by ℓ.

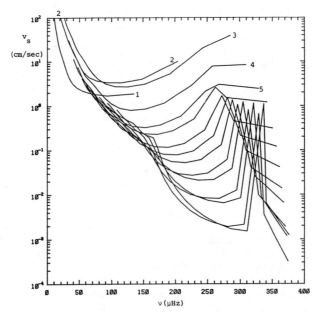

Figure 11. Estimated velocity amplitudes of g modes for a 1.8 $M_\odot$ ZAMS star, as observed in integrated light, if a fixed modal energy of about 10^{34} erg is assumed. See caption to Fig.10.

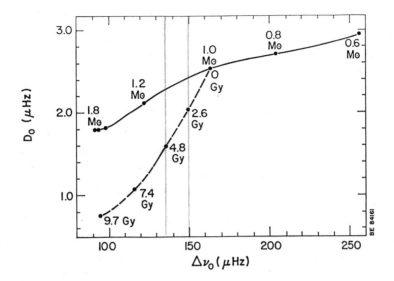

Figure 12. The location of the ZAMS (——————) and of a 1 M$_\odot$ evolution sequence (-------) in a ($\Delta\nu_0$, D_0) diagram. Here $\Delta\nu_0$ and D_0 are parameters characterizing the asymptotic behaviour of high-order, low-degree p mode frequencies; they were obtained by fitting the asymptotic relation (2.13), with n_0 = 22, to computed frequencies, including modes of degree 0 to 3 and with n + ℓ/2 between 14 and 30. The position of selected models have been indicated along the curves, labelled with their masses (in solar units) or ages (in 10^9 years). From Christensen-Dalsgaard (1984a).

5. MODIFICATIONS OF SOLAR MODELS AND THEIR FREQUENCIES

5.1. Introductory remarks

A considerable amount of insight in the relation between a solar (or stellar) model and its oscillation frequencies can be obtained from asymptotic relations such as equation (2.6). Nevertheless, given the uncertain precision of such relations for any given model, their use must be combined with calculations of oscillation frequencies under different assumptions. The results of such calculations can often be interpreted with the help of the asymptotic behaviour of the modes.

It is obvious that an important goal of helioseismology is to obtain a model that is an accurate representation of the real sun. It is tempting to try to attain this goal by computing different models, varying their parameters in such a way as to match the observed frequencies as closely as possible, while at the same time including the best representation of the physics that we believe we understand. However it is not obvious that this is the optimal procedure, given the present state of our knowledge about the solar interior. In fact, even if we were to pro-

duce a model that fitted all the available data to within their stated
accuracy, this would not teach us very much, unless we knew the level of
confidence to assign to the result. To address these questions we need
to know how the solar oscillation frequencies respond to given changes
in the model. These changes can be made without demanding that they produce a model that is "better", in the sense of fitting the data more
closely. However it is very important that the changes be made under
well-defined circumstances, so that the results can be interpreted cleanly. This requires that only one aspect of the model be investigated at a
time.

The computed models should obviously have the correct radius and
luminosity, as well as be of the right mass. In normal stellar evolution
calculations this is achieved by varying a parameter α_C determining the
efficiency of convection, usually the ratio between the mixing length
and the pressure scale height, and the initial hydrogen abundance X_0,
until a good fit to the observed radius and luminosity is obtained. In
computations testing the effects of modifications to the model, the models
should have radii and luminosities that agree so closely that the remaining differences have a negligible effect on the frequencies compared
with the effect being investigated. Ideally a full evolution calculation
should be carried out for each new set of assumptions, with a complete
iteration for α_C and X_0; in practice this might be regarded as too costly
in terms of computing resources.

A more economical approach is to compute static models (i.e. neglecting the release of gravitational energy in the luminosity equation) with
a prescribed hydrogen profile. Specifically it is assumed that the hydrogen abundance, as a function of the mass fraction q, is

$$X(q) = \chi X_i(q), \qquad (5.1)$$

where X_i is an initial, given hydrogen profile; the scale factor χ, and
α_C, are then adjusted to obtain the proper luminosity and radius. I have
tested this procedure by applying it to a model of the present sun, without any changes in the assumed physics; the scale factor χ required devi-
ated from 1 by less than 10^{-5}, and the p mode frequencies of the resulting model were within about 0.1 μHz of those of the original evolution
model. More generally, it is probably sufficiently accurate for any
changes in the model which do not affect the distribution of hydrogen
burning significantly. In addition it allows the computation of models
with a specified distribution of chemical composition.

I have used this technique to investigate the effects of two types
of modifications to the solar model: i) Mixing in the solar core, and
ii) Changes in the superadiabatic boundary layer at the top of the convection zone. The parameters χ and α_C were adjusted until the photospheric
radius and luminosity agreed to within a relative accuracy of 10^{-5} with
the observed values, which were taken to be 6.9599×10^{10} cm and
3.826×10^{33} erg s^{-1} respectively. The computed frequencies were compared
with the static model computed directly from a normal evolution model,
as indicated above.

To interpret the results of these calculations, it is useful first
to discuss a simple property of solar models, and a general relation for

the frequency change associated with changes in the model. It is easy to show that for an adiabatically stratified layer, with constant adiabatic exponent $\gamma = (\partial \ln p / \partial \ln \rho)_{ad}$, with constant interior mass and assuming vanishing surface pressure, the squared sound speed is given by

$$c_{an}^2(r) = (\gamma - 1) R g_s \left(\frac{R}{r} - 1\right), \tag{5.2}$$

where g_s is the surface gravity of the model. Thus the dependence of sound speed on depth is given entirely by γ and the surface gravity, and is independent of other details of the structure of the model. Except in a relatively thin layer near its top the solar convection zone approximately satisfies these conditions; thus it is perhaps not surprising that the sound speed in the convection zone is close to that obtained from equation (5.2).[4] Fig.13 shows the departure of the actual squared sound speed from c_{an}^2.

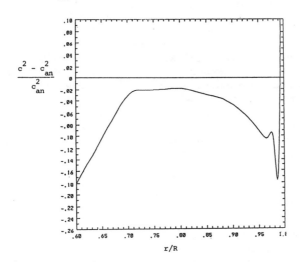

Figure 13. Relative deviation of squared sound speed from the simple approximation in equation (5.2), in the upper part of a model of the present sun, against fractional radius r/R.

From this property one would expect the sound speed in the convection zone of solar models to be relatively unaffected by modifications to the model, except perhaps near the surface, where the departure from c_{an} is greater. This is confirmed by the models discussed here, as well as by other model calculations (e.g. Ulrich & Rhodes 1977; Christensen-Dalsgaard 1984c). It has the unfortunate consequence that p modes, whose frequencies are mainly determined by the sound speed, are less sensitive

---

[4] A more refined version of this relation is discussed by Gough (1984b) and Däppen and Gough (these proceedings).

to changes in the model than might otherwise have been expected.

The eigenfrequencies of adiabatic oscillation can be treated as eigenvalues of a Hermitian operator (e.g. Eisenfeld 1969; Christensen-Dalsgaard 1981). It is really this property that provides the variational principle for the frequencies. More generally it may be shown from first-order perturbation theory applied to the oscillation equations that the change $\delta\omega$ in the eigenfrequency of a given mode of oscillation, associated with a small change in the equilibrium model or in the physics of the oscillation can be written as

$$\delta\omega = \frac{\int_0^R (\delta\vec{\chi}) \cdot (\delta\mathcal{L}\, \delta\vec{\chi})\, dr}{E}$$ (5.3)

where $\delta\mathcal{L}$ is a formal operator which is determined by the modifications in the model or the treatment of the oscillations, and E is the normalized energy introduced in equation (4.1). It might be noticed that this equation also describes the damping or excitation of the mode in the quasi-adiabatic approximation (e.g. Unno et al. 1979).

Clearly equation (5.3) depends for physical content on the precise form of $\delta\mathcal{L}$ which may be quite complex. However, generally speaking it implies that for modes with higher energy it is more difficult to modify the frequencies, which is intuitively very reasonable. This suggests that a measure of the change in the model (or the error implied by differences between computed and observed frequencies) is the energy scaled frequency difference $E\,\delta\omega$. A convenient alternative, which avoids the large variation with frequency in the energies for p modes (cf. Fig.8), is to scale the frequency difference by

$$Q_{n,\ell} = \frac{E_{n,\ell}}{\bar{E}_0(\nu_{n,\ell})} ,$$ (5.4)

where $\bar{E}_0(\nu_{n,\ell})$ is the energy of radial modes, interpolated to the actual frequency $\nu_{n,\ell}$ of the mode considered.

A highly simplified version of equation (5.3), for high-order p modes, can be obtained from equation (2.10) by noting that $c^2 L^2/(r^2\omega^2) \ll 1$ over most of the range of integration; it may then be shown that the frequency change $\delta\nu$ associated with a change δc in the sound speed in the model is approximately given by

$$\frac{\delta\nu}{\nu} = \frac{\int_{r_t}^R \frac{\delta c}{c} \frac{dr}{c}}{\int_{r_t}^R \frac{dr}{c}} .$$ (5.5)

Thus the relative frequency change is simply an average of the relative change in the sound speed, weighted by the time spent by the mode (when regarded as a superposition of traveling waves) in a given region of the sun.

5.2. Effects of mixing of the solar core

Solar models affected by mixing due to an imposed turbulent diffusion
have been studied by Schatzman et al. (1981), with a view towards understanding the solar neutrino problem as well as the observed surface abundances of ^7Li and ^3He. These models have also been invoked by Berthomieu,
Provost & Schatzman (1984) to explain a possible observational determination by Delache & Scherrer (1983) of the g mode period separation T_0
(cf. equation 2.17). However the models computed by Schatzman et al.,
although full evolution models, were not precisely calibrated to the
solar radius and luminosity. Furthermore little information was given
about the changes in the models relative to normal solar models. Thus
it may be interesting to study a similar model in more detail.

To do so I have computed a static model of the present sun, with an
initial estimate $X_i(q)$ of the hydrogen profile obtained from the curve
for Re* = 100 on Fig. 2 of Schatzman et al.[5] To get the correct luminosity X_i had to be rescaled by a factor $\chi = 0.998$; that this is so close
to unity indicates that the physics used in the calculation of the static
model is probably fairly consistent with that used by Schatzman et al.
Fig. 14 shows differences in various quantities between this model and
the normal static model. In addition Fig. 15 shows the relative difference in squared sound speed. It is obvious that the largest changes are
confined to the core where X is modified. Furthermore the temperature
change is obviously very small; because of the high temperature sensitivity of the nuclear reactions, only a very modest modification of the
temperature in the core is permitted in the readjustment of the model to
the modified composition. Thus the change of the squared sound speed is
given to a high degree of accuracy by the change in the mean molecular
weight, and hence tracks the change in X. It should also be noticed that
the changes in pressure p and density ρ are almost exactly equal in the
convection zone. Thus here the sound speed is nearly unchanged. This is
in accordance with the discussion following equation (5.2); the peak in
$\delta c^2/c^2$ near the surface coincides with the ionization zones of H and He,
where the adiabatic exponent varies, and where therefore the assumptions
behind equation (5.2) are violated.

Differences between selected p mode frequencies are shown on Fig. 16.
These closely reflect the difference in sound speed. Modes of degree 0
and 1 penetrate so close to the centre that they are affected by the increase in the sound speed where material has been mixed and the hydrogen
abundance consequently increased, relative to the normal model. This
positive $\delta c^2/c^2$ compensates for the negative $\delta c^2/c^2$ in the bulk of the
radiative interior, leading to near zero frequency differences. At higher
degrees the modes are confined outside the core, and the $\delta \nu$ is dominated
by a balance between the negative $\delta c^2/c^2$ in the radiative interior and
the positive $\delta c^2/c^2$ near the surface. Note that the latter, despite its
insignificant appearance on Fig. 15, has a substantial effect on the frequencies because it occurs where the sound speed is low (cf. equation

---

[5] Re* is a parameter characterizing the strength of the mixing; Re* = 100
gives a neutrino flux which is consistent with the observed value.

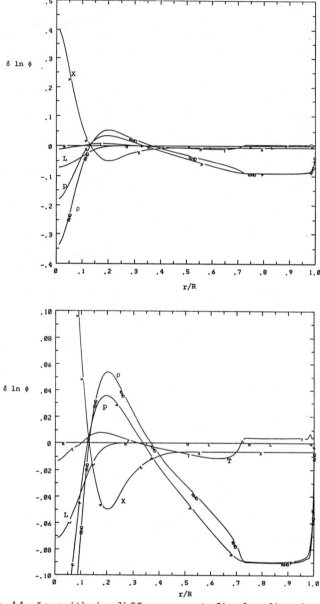

Figure 14. Logarithmic differences at fixed radius in pressure p, density ρ, temperature T, hydrogen abundance X, and luminosity L, between a static model of the present sun with core mixing and a normal model. The hydrogen profile of the mixed model was obtained from the Re* = 100 model of Schatzman et al. (1981). The abscissa is fractional radius r/R.

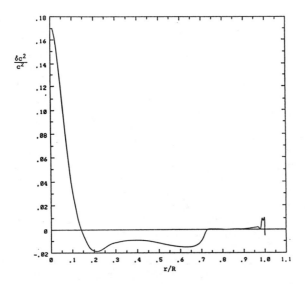

Figure 15. Relative difference in the squared sound speed between the mixed solar model considered on Fig. 14 and a normal solar model. The abscissa is fractional radius r/R.

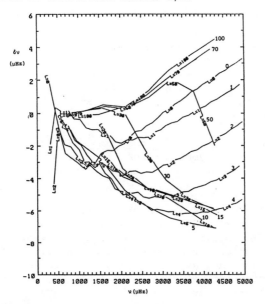

Figure 16. Adiabatic frequency differences between the mixed solar model considered on Fig. 14 and a normal solar model, against frequency ν. For clarity points corresponding to the same degree ℓ have been connected; the values of ℓ are indicated.

5.5). Finally, at degrees exceeding 50-70 (depending on frequency) the modes are confined to the convection zone, and only the positive $\delta c^2/c^2$ contributes to $\delta\nu$.

The strong variation with ℓ in $\delta\nu$ at low ℓ causes a substantial change in the separation $\delta\nu_{n,\ell}$; this is also predicted by the asymptotic equation (2.11), when the precise form of A is taken into account (e.g. Provost 1984). This may be expressed in terms of the fitting parameter D_0 which is listed in Table I, both for the normal model and for the mixed model, which was calculated as described in the caption to Fig. 12. It is obvious that mixing causes a substantial increase in D_0. For comparison a similar fit to the available observed data (Grec et al. 1983; Claverie et al. 1981; Duvall & Harvey 1983) gives $D_0 = 1.43$ µHz. This is close to, but slightly lower than, the value for the normal model, and is inconsistent with the value for the mixed model. Thus a degree of mixing as strong as considered here appears to be ruled out by the oscilation observations. Similar results were obtained by Cox & Kidman(1984). Also shown is the fit $\Delta\nu_0$ to the mean spacing between modes of adjacent orders; this is hardly changed by the mixing.

Mixing reduces the buoyancy frequency N and hence increases the period spacing T_0 for g modes. To investigate this effect I have computed g modes with degree 1 and 2, and fitted them to the first-order asymptotic expression (2.16), in the manner of Christensen-Dalsgaard (1984a). The resulting values of T_0 are also given in Table I. They are close to the values ob-

TABLE I

Model	$\Delta\nu_0$	D_0	T_0
Normal model	136.3 µHz	1.595 µHz	35.37 min
Schatzman et al.	136.5 µHz	2.045 µHz	56.85 min

Properties of a normal solar model and a model affected by mixing. $\Delta\nu_0$ and D_0 describe the asymptotic behaviour of p modes (cf. equation 2.13) and T_0 is related to the asymptotic spacing of g modes (cf. equation 2.16).

tained by Berthomieu et al. (1984). From observations of long-period oscillations Delache & Scherrer (1983) inferred that $T_0 = 38.6$ min, somewhat higher than the values for normal solar models. By interpolating linearly between the results in Table I it appears that mixing with Re* $\approx$ 15 would be sufficient to bring the theoretical value of T_0 into agreement with this result. The corresponding value of D_0 is 1.66 µHz; thus the discrepancy with the p mode observations is increased, although possibly not by a significant amount. Better observational data, or a more careful analysis of the data in hand, are required to decide whether any amount of mixing is consistent with the observations of p modes.

5.3. Modifications of the superadiabatic region

The use of mixing length theory is justifiably regarded as a major uncertainty in present solar and stellar evolution calculations. In the
sun, however, the uncertainty is largely confined to a region very near
the surface. Because of high efficiency of convection it is probable that
most of the convection zone is nearly adiabatically stratified. Then the
details of the treatment of convection only directly affects the structure of the convection zone near its top, where there is a transition
between convective and radiative transport, and the value of α_C serves
to determine the change in entropy from the atmosphere, where the conditions can be observed, to the adiabatic region where the entropy is
nearly constant. The structure of the remainder of the convection zone
and of the radiative interior, and hence the radius of the model, is almost entirely determined by the value of the entropy in the adiabatic
part of the convection zone (cf. Gough & Weiss 1976). Thus, if a different description of convection were to be chosen, it too would have to be
calibrated to yield the same adiabat in the interior of the convection
zone, and the changes to the structure of the model should be confined
very close to its surface. The same is true for other modifications in
the physics of the model confined to its outermost layers; these must be
compensated for by changes in α_C, to preserve the correct radius, and
one would expect little change in the interior of the model.
     It is obvious that for other stars, where radius calibration is not
available, the calculation of the entropy change from atmosphere to interior is highly uncertain; it is normally based on the determination of
α_C from solar models. Thus a better understanding of convection is of
major importance. It is thus of interest to investigate to what extent
the details of the superadiabatic region can be probed by solar oscillations. To do so requires a formulation for calculating the superadiabatic gradient with at least two free parameters, of which one is used
to calibrate the model. The ordinary local mixing length theory (e.g.
Böhm-Vitense 1958) contains several undetermined parameters beside the
mixing length; however once the model is calibrated its structure is almost completely insensitive to the values of these parameters (cf. Gough
& Weiss 1976). Thus here I have chosen to use an essentially arbitrary
formulation which however bears some superficial resemblance to the results of mixing length theory for an appropriate choice of parameters.
Specifically I have used

$$\nabla - \nabla_{ad} = f(\log p)\,(1 - \frac{\nabla_{ad}}{\nabla_R}), \tag{5.5}$$

$$f(x) = \alpha_C(x - 5.1)\exp[-\beta(x - 5.1)];$$

here $\nabla = d\log T/d\log p$, $\nabla_{ad} = (\partial\log T/\partial\log p)_{ad}$ and ∇_R is the value of ∇
that would be required to carry the energy purely by radiation. The para-
meter β determines the extent of the superadiabatic region, and for a
given value of β α_C is adjusted to obtain the correct radius. It is obvious that this description has little merit beyond its flexibility, and
certainly no physical content.

Here I consider a fairly extreme case, where β was chosen to be 3, which required $\alpha_C = 1.21$. Fig.17 illustrates the structure of the outermost layers of the model. As on Fig.2 the dashed curve shows $|N|$, which

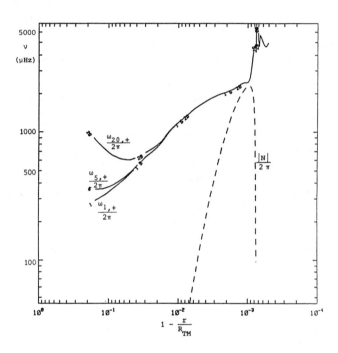

Figure 17. Characteristic frequencies $\omega_{\ell,+}$ and $|N|$ near the surface of a model of the present sun with artificially modified superadiabatic gradient (see caption to Fig.2).

is proportional to $\nabla - \nabla_{ad}$, and the continuous curves show the characteristic frequency $\omega_{\ell,+}$. It is evident that the superadiabatic region in this model is broader and less pronounced than in the normal model; this has led to a smoother variation of the density scale height H, and hence in the $\omega_{\ell,+}$ which in this region are essentially given by ω_c. However except in the outer 1 per cent of the model there is very little change in $\omega_{\ell,+}$. As shown on Fig.18 there is a substantial change in the sound speed, but this again is restricted to the outer layers of the convection zone. Finally Fig.19 shows that, as expected, the changes in the interior layers of the model are extremely small.

The effect of this modification on the p mode frequencies are shown on Fig.20a and b. The frequency differences, probably caused by the change in sound speed near the surface, are quite large, and show considerable variation with ℓ. However, as illustrated by Fig.20b, this variation is almost entirely caused by the variation in mode energy with ℓ . In fact the structure of the eigenfunctions near the surface is predominantly a function of frequency and depends little on ℓ (e.g. Chri-

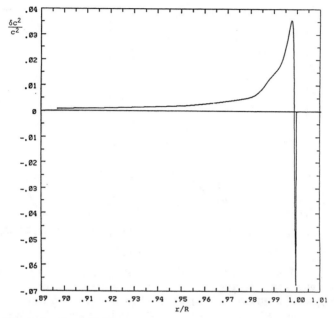

Figure 18. Relative difference in squared sound speed between a model of the present sun with artificially modified superadiabatic gradient and a normal model, against fractional radius r/R.

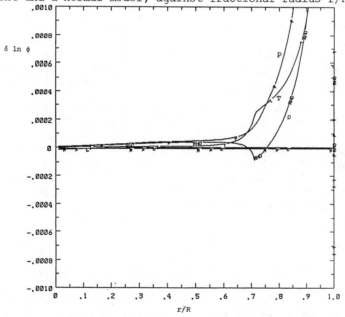

Figure 19. Relative differences between pressure p, density ρ and temperature T in the interior of the models considered on Fig. 18.

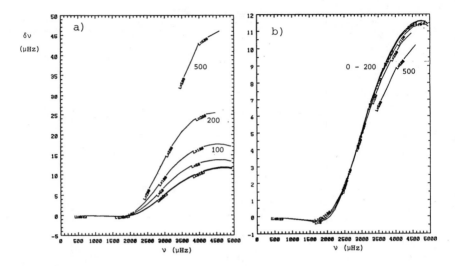

Figure 20. Adiabatic frequency differences, between a model of the
present sun with artificially modified superadiabatic gradient and
a normal model, against frequency. For clarity points corresponding
to the same degree ℓ have been connected; selected values of ℓ are
indicated. On a) the raw frequency differences $\delta\nu$ are plotted;
b) shows the energy scaled differences $Q_{n,\ell}\delta\nu$, where $Q_{n,\ell}$ is defined in equation (5.4).

stensen-Dalsgaard et al. 1985a); hence, since the perturbation in the
model is confined close to the surface, the numerator in equation (5.3)
is essentially independent of ℓ at a given frequency, and the variation
with ℓ in $\delta\nu$ is caused by the change in the energy. The only exceptions
are the modes with $\ell = 500$, which are confined so close to the surface
that they do not feel the entire contribution from δc.

This example thows that a sufficiently drastic modification of the
convection zone has a significant effect on the oscillation frequencies.
To determine the extent to which such modifications can be probed by the
details of the frequency changes requires an understanding of how the
changes shown on Fig.20b are related to the modifications in the model.
This I have so far not attempted to analyze. It is clear, however, that
observations of modes of very high degree, which may begin to resolve
the interesting region, would be of great value.

6. CONCLUSION

Procedures for calculating solar and stellar models and their oscillation
frequencies are now reasonably well established. However there are still
discrepancies between different calculations that must be removed. Efforts to do so are now under way. The results of Section 3 show that,
at least for one study, insufficient care had been taken in the evolution
calculation.

In the case of the sun we are beginning to get an indication of how the oscillation frequencies are affected by changes in the solar models. It has to some extent been possible to rationalize the results of numerical calculations be referring to the asymptotic properties of the oscillations; however we are still far from a genuine understanding of these relationships. Further numerical work, together with more careful asymptotic analysis, are needed. The procedures used here, viz the computation and comparison of models calculated with different parameters, is clearly a step in that direction; but an exhaustive study of all possible modifications would clearly be prohibitively time consuming. More systematic procedures can be set up, whereby the response of the model and the frequencies to a given perturbation can be expressed as weighted integrals of the perturbations (Däppen, 1983); such techniques should be employed.

The study of stellar oscillations involves probing a far greater range of parameters than for the sun. Studies of this nature are now beginning (e.g. Demarque, these proceedings), and must be pursued. It is important to establish the quality and quantity of oscillation observations required to begin probing the structure of stellar interiors, given that even the basic parameters of the stars, such as their masses and radii, are generally not accurately known (cf. Frandsen, these proceedings).

The present paper has only considered adiabatic oscillations, under the simplifying assumptions described in Section 3. It is essential that we try to get a better understanding of the departures from these assumptions. Thus there may be indications that the failure of present solar models to reproduce the observed frequencies may be due to errors in the calculation made very close to the surface (Christensen-Dalsgaard, in preparation), and these errors may well be caused by one or more of the effects neglected in standard calculations. To the extent that these effects are related to the excitation or damping of the oscillations, their study would profit greatly from observational data on the excitation of similar oscillations in other stars.

ACKNOWLEDGEMENTS: D.O. Gough is thanked for organizing such a fruitful and enjoyable workshop, as well as for many discussions on the subjects of this paper. I am grateful to W. Däppen for reading and commenting on an earlier manuscript, and to H. Nielsen for the competent typing. I thank R.M. MacQueen for hospitality at the High Altitude Observatory, National Center for Atmospheric Research (which is funded by the National Science Foundation) where much of the paper was written. The computations reported here were supported by the Danish Natural Science Research Council.

References

Ando, H. & Osaki, Y., 1975. *Publ. Astron. Soc. Japan*, 27, 581.
Berthomieu, G., Provost, J. & Schatzman, E., 1984. *Nature*, 308, 254.

Brown, T.M., Mihalas, B.W. & Rhodes, E.J., 1986. *Physics of the Sun*, (ed. Sturrock, P.A., Holzer, T.E., Mihalas, D. & Ulrich, R.K., Reidel) (Part 1).

Böhm-Vitense, E., 1958. *Z. Astrophys.*, **46**, 108.

Chandrasekhar, S., 1964. *Astrophys. J.*, **139**, 664.

Christensen-Dalsgaard, J., 1980. *Mon. Not. R. astr. Soc.*, **190**, 765.

Christensen-Dalsgaard, J., 1981. *Mon. Not. R. astr. Soc.*, **194**, 229.

Christensen-Dalsgaard, J., 1982. *Mon. Not. R. astr. Soc.*, **199**, 735.

Christensen-Dalsgaard, J., 1984a. *Space Research Prospects in Stellar Activity and Variability*, (ed. Mangeney, A. & Praderie, F., Observatoire de Paris), p. 11.

Christensen-Dalsgaard, J., 1984b. *Theoretical Problems in Stellar Stability and Oscillations*, (Institute d'Astrophysique, Liège), p.155.

Christensen-Dalsgaard, J., 1984c. *The Hydromagnetics of the Sun* (ESA SP-220, ESTEC), p. 3.

Christensen-Dalsgaard, J. & Frandsen, S., 1983a. *Solar Phys.*, **82**, 165.

Christensen-Dalsgaard, J. & Frandsen, S., 1983b. *Solar Phys.* **82**, 469.

Christensen-Dalsgaard, J. & Gough, D.O., 1982. *Mon. Not. R. astr. Soc.*, **198**, 141.

Christensen-Dalsgaard, J., Dziembowski, W. & Gough, D.O., 1980. *Lecture Notes in Physics*, **125**, 313 (ed. Hill, H.A. & Dziembowski, W., Springer-Verlag).

Christensen-Dalsgaard, J., Gough, D.O. & Toomre, J., 1985a. *Science*, **229**, 923.

Christensen-Dalsgaard, J., Duvall, T.L., Gough, D.O., Harvey, J.W. & Rhodes, E.J., 1985b. *Nature*, **315**, 378.

Claverie, A., Isaak, G.R., McLeod, C.P. & van der Raay, H.B., 1981. *Nature*, **293**, 443.

Cox, A.N. & Kidman, R.B., 1984. *Theoretical Problems in Stellar Stability and Oscillations* (Institute d'Astrophysique, Liège) p. 259.

Delache, P., 1984. *Mem. della Societa Astronomica Italiana*, **55**, 75.

Delache, P. & Scherrer, P.H., 1983. *Nature*, **306**, 651

Deubner, F.-L. & Gough, D.O., 1984. *Ann. Rev. Astron. Astrophys.*, **22**, 593.

Duvall, T.L., 1982. *Nature*, **300**, 242.

Duvall, T.L. & Harvey, J.W., 1983. *Nature*, **302**, 24.

Dziembowski, W. & Pamjatnykh, A.A., 1978. *Pleins feux sur la physique solaire*, (ed. Rösch, J., CNRS, Paris), p. 135.

Dziembowski, W., Paternò, L. & Ventura, R., 1985. *Astron. Astrophys.*, in press.

Däppen, W., 1983. *Astron. Astrophys.*, **124**, 11.

Eisenfeld, J., 1969. *J. Math. Anal. Applic.*, **26**, 357.

Fossat, E., 1985. *Future missions in solar, heliospheric and space plasma physics* (ed. Rolfe, E. & Battrick, B., ESA SP-235, ESTEC, Noordwijk), p. 209.

Goldreich, P. & Keeley, D.A., 1977. *Astrophys. J.*, **212**, 243.

Gough, D.O., 1984a. *Adv. Space. Res.*, **4**, 85.

Gough, D.O., 1984b. *Mem. della Societa Astronomica Italiana*, **55**, 12.

Gough, D.O., 1984c. *Phil. Trans. R. Soc. London*, A **313**, 27.

Gough, D.O., 1985a. *Future missions in solar, heliospheric and space plasma physics* (ed. Rolfe, E. & Battrick, B., ESA SP-235, ESTEC, Noordwijk), p. 183.

Gough, D.O., 1985b. *Proc. Kun Ming Workshop on Solar Physics* (ed. Chen, B. & de Jager, C.; Yunnan Observatory) in press.

Gough, D.O., 1985. *Solar Phys.*, p. 65-99.

Gough, D.O. & Weiss, N.O., 1976. *Mon. Not. R. astr. Soc.*, 176, 589.

Grec, G., Fossat, E. & Pomerantz, M., 1983. *Solar Phys.*, 82, 55.

Harvey, J., 1985. *Future missions in solar, heliospheric and space plasma physics* (ed. Rolfe, E. & Battrick, B., ESA SP-235, ESTEC, Noordwijk), p. 199.

Kidman, R.B. & Cox, A.N., 1985. *Solar Seismology from Space* (ed. Ulrich, R.K., JPL Publ. 84-84), p. 335.

Lamb, H., 1932. *Hydrodynamics*, 6th ed. (Cambridge University Press).

Ledoux, P. & Pekeris, C.L., 1941. *Astrophys. J.*, 94, 124.

Provost, J., 1984. *Proc. IAU Symposium No 105: Observational Tests of Stellar Evolution Theory*, (ed. Maeder, A. & Renzini, A., Reidel), p. 47.

Schatzman, E. & Maeder, A., Angrand, F. & Glowinski, R., 1981. *Astron. Astrophys.*, 96, 1.

Scherrer, P.H., Wilcox, J.M., Christensen-Dalsgaard, J. & Gough, D.O., 1983. *Solar Phys.*, 82, 75.

Tassoul, M., 1980. *Astrophys. J. Suppl.*, 43, 469.

Ulrich, R.K. & Rhodes, E.J., 1977. *Astrophys. J.*, 218, 521.

Unno, W., Osaki, Y., Ando, H. & Shibahashi, H., 1979. *Nonradial Oscillations of Stars* (University of Tokyo Press).

Vandakurov, Yu. V., 1967. *Astron. Zh.*, 44, 786.

OBSERVATIONS OF LOW-DEGREE P-MODE OSCILLATIONS IN 1984

Harald M. Henning and Philip H. Scherrer
Stanford University
CSSA, ERL 328
Stanford, CA 94305
USA

ABSTRACT. Analysis of Stanford differential velocity observations has been extended through the 1984 observing season. Excellent quality observations were obtained in 1984 on 38 days in a 49 day interval from June 20th through August 7th. The power spectrum of this data has been examined and improved frequency determinations have been made for p-modes of degree 2 through 5 and order 5 through 34. Of special interest are the modes of the lower orders, n ranging from 5 to 10, which have not been identified previously.

1. ANALYSIS

Differential velocity observations of the sun hace been made at the Wilcox Solar Observatory at Stanford since 1977. The method of observation and sensitivity to p-modes of different degrees l have been described previously (Scherrer et al 1983). The year 1984 has presented us with the best set of data yet obtained at Stanford. We have used a span of 49 days for this analysis, with good data available for 38 of those days. To enable conversion of the daily data into one 49 day string, a 35 minute high-pass filter was first applied to each day's data. Then each day was normalized to have a variance of 1 and a mean of 0. Finally, the data was synchronized on 30 second time steps and concatenated. The full power spectrum is shown in figure 1. It should be noted that figure 1 is plotted at a much lower resolution than 1/49 days, the full resolution of the power spectrum.

Even though the data coverage during the interval is excellent, the average length of one day's usable observations is still only around 10 hours. Thus the filling factor is quite low, around a third. The resulting window function causes many difficulties in isolating and identifying solar oscillation modes, introducing large sidelobes for each peak at 1/day splittings from the central frequency. A plot of the window function, obtained by putting a single sine wave into our data window and computing a power spectrum, is shown in figure 2. The function looks quite nice, and its simplicity has greatly helped in the analysis. However, autocorrelations reveal that the smaller peaks close

55

D. O. Gough (ed.), Seismology of the Sun and the Distant Stars, 55–62.
© 1986 by D. Reidel Publishing Company.

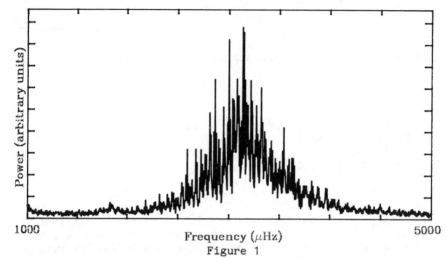

Figure 1

The power spectrum of the full data, from 1000 μHz to 5000 μHz.

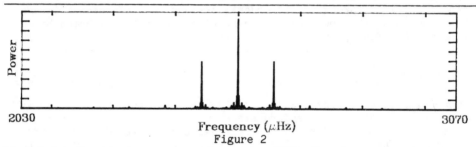

Figure 2

The window function: power spectrum of a 3000 μHz sine wave put into the actual data window.

to the main peaks greatly affect the fine scale structure of the spectrum and make the identification of individual peaks difficult.

In addition to the complications introduced by the window, the effects of splitting due to rotation and lifetimes, each producing multiple peaks with sidelobes, as well as noise make the spectrum very complex. Several computerized deconvolution and peak-subtracting techniques failed to make significant inroads on the problem (Scherrer 1986). A scheme combining computed power spectra and manual pattern recognition was decided upon as the best method to identify real solar peaks. Figures 3 and 4 are typical examples of the graphs used. These figures each show a section of the spectrum computed with several resolutions. Frequency runs along the abscissa, with each panel covering the same span, from 1432 μHz to 1572 μHz in figure 3, and from 3200 μHz to 3340 μHz in figure 4. We realized that the full resolution spectrum (panel E) is too complex to enable recognition of any given peak as a

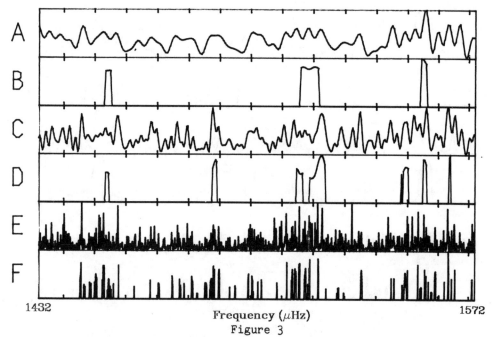

Figure 3

Plot of data in the frequency range from 1432 µHz to 1572 µHz :
A) average power spectrum of 4 day chunks of data, B) peak search on A,
C) average power spectrum of 8 day chunks of data, D) peak search on C,
E) full resolution power spectrum, F) peak search on E.

---

solar oscillation mode. At lower resolution, however, it becomes much
clearer to see where there is power in the spectrum. Thus, the average
of 6 spectra of 4-day segments of the data is plotted in panel A. Panel
C shows the average of 3 spectra of 8-day segments of the data. To
further aid in locating modes a simple computer program was used to scan
the spectra, looking for structures which correspond to the main
features of the window function, i.e. a central peak with two sidelobes
at 1/day distance and of about 50% magnitude. The scan allowed the
sidelobes to be as large as 90% or as small as 20% of the central peak.
The program took the value of the spectrum at a frequency if it
corresponded to a possible central peak and zero otherwise. The results
of this search are plotted below the corresponding spectrum, in panels
B, D and F. This composite graph proves to be a very useful tool in
determining the frequencies of the modes. The first four panels, A-D,
serve to locate and emphasize small frequency ranges where there is
power. Using this information and a transparency with the window function plotted on it, one can then pinpoint the group of peaks (and its
sidelobes) in the high-resolution full spectrum, panels E and F, that
correspond to a solar p-mode. The frequency recorded for each mode was

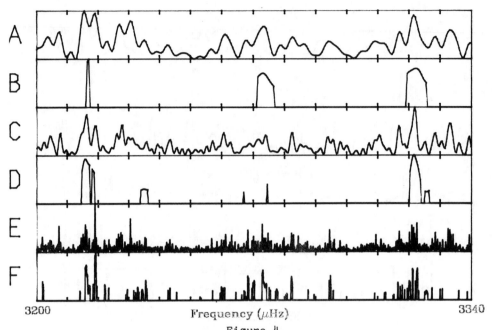

Figure 4

As figure 3, but for the range from 3200 µHz to 3340 µHz.

---

that of the central peak or approximate center of the group of peaks. A
conservative estimate of the error in the frequency of any given mode is
2 µHz.

Analyzing the spectrum, we located modes ranging from 1035 µHz up
to 5008 µHz. Identification of the modes was accomplished using previously published frequencies (Duvall and Harvey 1983, Duvall - private
communication) for modes in the range from about 2000 to 4000 µHz as
"anchors" for the determination of order n and degree l. The fact that
the Stanford instrument is most sensitive to modes of degree 2 through 5
prevents confusion with higher degrees, although those modes certainly
contribute to the "noise". Modes with frequencies outside of this range
were then identified by making an echelle diagram (figure 5) and following the curve described by different n values of a given l. It is
important to note that although there certainly is some bias in the
location and identification of the modes with previously published frequencies, there is no such bias in the frequencies of modes of lower and
higher order, since there was no preconception of where they should be.
Table I lists the frequencies of the identified modes.

In addition to identification of the modes, we have also examined
the data for evidence of rotational splitting. As mentioned before, the
power of each mode is split into a multitude of peaks resulting from a
combination of mode lifetimes, window function, and noise, as well as
rotational splitting. This confusion effectively prevents one from

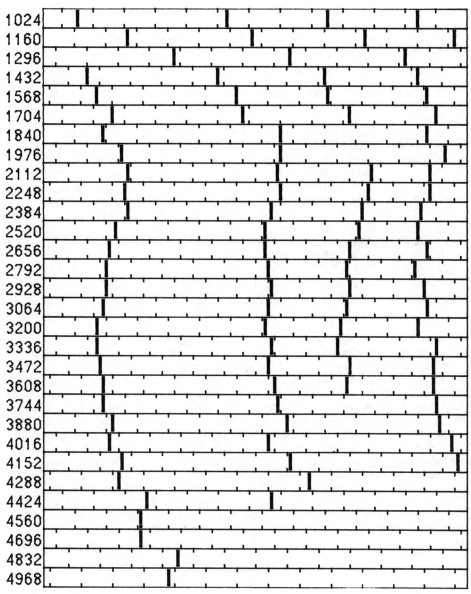

Figure 5
Echelle diagram of the identified modes, each strip covers 136 μHz.

n	frequency				n	frequency			
	l=2	l=3	l=4	l=5		l=2	l=3	l=4	l=5
5			1035	1115	20	3026	3083	3136	3187
6	1083	1144	1187	1263	21	3161	3217	3271	3320
7	1227	1292	1338	1412	22	3295	3353	3409	3462
8	1375	1446	1488	1552	23	3430	3490	3544	3597
9	1522	1585	1630	1691	24	3570	3627	3682	3733
10	1659	1726	1768	1830	25	3705	3763	3819	3870
11	1802	1859	1916	1963	26		3902	3958	4007
12		2001	2052	2105	27		4037	4088	4147
13		2139	2187	2236	28		4177	4231	4285
14	2217	2274	2052	2372	29		4312	4373	
15	2352	2411	2457	2505	30		4457	4497	
16	2486	2543	2591	2640	31		4591		
17	2621	2677	2727	2779	32		4727		
18	2754	2812	2864	2911	33		4875		
19	2889	2948	3001	3050	34		5008		

Table I - Frequencies of identified modes

assigning m-values to given peaks and directly measuring the splitting.
Rather, we have analyzed the spectrum using autocorrelations. Using the
determined frequencies we calculated the autocorrelation of a small section of the full spectrum around each mode. All autocorrelations with a
given degree l, but different order n, were then averaged together. The
assumption is that any rotational splitting will not change much with
order n. The results are plotted in the first four panels of figure 6.
An autocorrelation of the window function is shown in the last panel of
figure 6. It can be seen that the window is responsible for much of the
systematic splitting, as evidenced by the peaks in its autocorrelation
which are repeated and actually dominate the autocorrelations of the
modes. However, all degrees show a peak around 0.8 µHz, which is close
to twice the rotational splitting as measured by Duvall and Harvey
(1984). Since the Stanford instrument is sensitive only to modes with
l+m even, this peak can be identified as the result of rotational splitting.

2. CONCLUSION

We have determined the frequencies of over 20 modes which have not been
identified before. The majority of these modes lies in the frequency
range from 1035 µHz to 1800 µHz. The echelle diagram of the modes shows
clearly the change in slope at low n, as the modes become spread further
apart than 136 µHz. Also, we have been able to see a characteristic
splitting which is consistent with the rotational splittings measured by

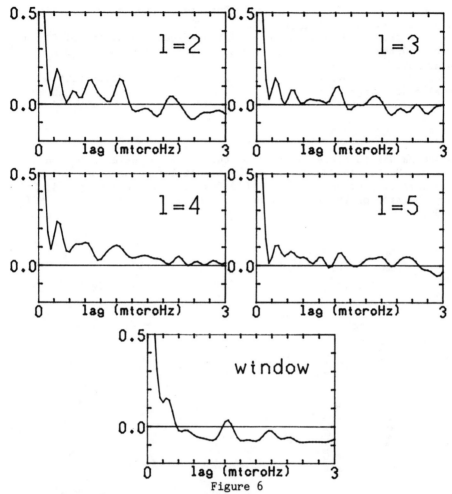

Figure 6

Autocorrelations of the power spectrum around different degrees l and the window function.

Duvall and Harvey (1984). Future examination of other Stanford data will hopefully refine the values, although no other year has had data as complete as 1984.

3. REFERENCES

Duvall Jr., T. L., and J. W. Harvey, 'Frequency spectrum of solar 5-minute oscillations with degree less than 140', Nature, 302, 24, 1983.

Duvall Jr., T. L., and J. W. Harvey, 'Rotational frequency splitting of
    solar oscillations', Nature, 310, 19, 1984.
Scherrer, P. H., J. M. Wilcox, J. Christensen-Dalsgaard, and D. O.
    Gough, 'Detection of solar five-minute oscillations of low degree',
    Solar Physics, 82, 75, 1983.
Scherrer, P. H., 'Comments on techniques for spectral deconvolution',
    these proceedings, 1986.

4. ACKNOWLEDGEMENTS

This work was supported in part by the Office of Naval Research under
Contract N00014-76-C-0207, by the National Aeronautics and Space
Administration under Grant NGR5-020-559, and by the Atmospheric Sciences Section of the National Science Foundation under Grant ATM77-
20580.

ANOTHER REASON TO SEARCH FOR SOLAR g-MODES AND NEW LIMITS FROM SOLAR
ELLIPTICITY MEASUREMENTS

J. R. Kuhn
Joseph Henry Laboratories
Physics Department
Princeton University
Princeton, N. J. 08544 U.S.A.

ABSTRACT. Observations of solar g-modes will teach us some gravitational physics. The present sensitivity of the Princeton Solar
Distortion Telescope and other recent claims of solar g-modes suggest
that these low frequency modes should be observable in shape measurements. From about 250 days and nearly 1000 hours of observations we
find no evidence for significant spectral power that can be associated
with g-modes and no sign of the elusive 160.01 min period solar
oscillation.

1. NEW GRAVITATIONAL PHYSICS INTERESTS

We will undoubtedly learn a great deal about the solar interior
from the long period g-modes if they are reliably observed, but there's
another perspective from which their observation is interesting. I'm
impressed that we (almost!) understand the sun to the extent that as we
learn more about the oscillations we can treat the sun as a "detector."
In particular, a good example of this are potential observations of low
degree modes with periods near one hour. Such observations may yield
interesting limits to a stochastic gravitational wave background
(Boughn and Kuhn, 1984). It's also interesting that the low degree
modes with periods of order an hour have a relatively large gravitational
energy associated with their oscillation. In light of recent geophysical
evidence (Holding and Tuck, 1984) that the effective gravitational
constant may be larger on distance scales much larger than laboratory
scales, a solar determination of G becomes interesting. Some hope that
such a measurement can be disentangled from the mass uncertainty of the
sun is gained by noting how the eigenfrequencies of the low order modes
are affected by perturbations in the gravitational constant. Figure 1
shows the relative frequency shift, in terms of the fractional perturbation in G, expected for some low order $\ell = 2$ and $\ell = 3$ g- and p-modes.
The figure suggests that a possible 1-percent effect seen in the geophysical data might be detected from the solar oscillation spectrum. At
present such predictions are premature and more work must be done to
constrain the density-gravitational constant product from other solar
observational constraints.

63

D. O. Gough (ed.), Seismology of the Sun and the Distant Stars, 63–72.
© *1986 by D. Reidel Publishing Company.;*

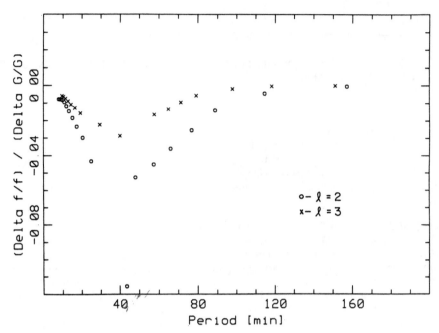

Figure 1. Modal frequency shifts due to gravitation constant perturbations.

2. LONG PERIOD SOLAR OSCILLATIONS

Several claims for detection of solar g-modes with periods between 2 and 10 hours have been made (e.g. Scherrer, 1984; Isaak, et al. 1984; Frohlich and Delache, 1984). The surface velocity amplitudes of these modes appear to be between 0.1 and 1 m/s but there is poor agreement on the excitation spectrum between different observers. On the other hand several groups have consistently claimed detection of a 160.01 min period solar oscillation (cf. Kotov, et al., 1984) also with a velocity amplitude near 1m/s. All of these modes have been observed with low spatial resolution and are most likely of low spherical harmonic degree ($\ell \leq 2$).

It's notable that the surface amplitudes of these oscillations are comparable to 5-min p-mode amplitudes of about 15 cm/s (cf. Kuhn, 1984a). If the observations are correct then the energy per g-mode is roughly 10^8 times larger than typical p-mode energies and of considerable curiosity.

3. SHAPE OBSERVATIONS

Except for the ACRIM data (cf. Frohlich and Delache, 1984) and

Hill's data (1985) most of the low frequency data are due to Doppler velocity observations. Since the period of the modes discussed in this paper are long compared to a radiative relaxation time at the photosphere ($\sim 10^3$ s) the pressure and density perturbations due to the oscillation will be nearly isothermal. It is therefore reasonable to calculate the shape change by following photospheric brightness contours as they move due to mass motion of the oscillation displacement field below the photosphere.

The Distortion Telescope and oblateness data have been described in detail elsewhere (Kuhn, et al., 1985; Libbrecht and Kuhn, 1984, Dicke and Goldenberg, 1974). In brief, measurements of the deviation from a circular solar limb shape are obtained by measuring the flux from a solar image occulted by a slightly undersize circular occulting disk. The solar flux is measured in two broadband colors (.5µ and .8µ) and at 256 positions around the limb. Between 9 and 20 arcseconds of the limb extends beyond the occulting disk. The data are obtained with a specialized telescope now operating at the Mount Wilson Observatory.

A data set at one of 3 possible limb exposures and two simultaneous colors is obtained approximately once every 5 min. The flux measurements in 128 distinct position angles (only symmetric distortions about the image center may be observed) around the image are reduced to a displacement in arcseconds using a geometrical calibration. Corrections for optical imperfections of the telescope, mirror distortions, atmospheric transparency and refraction, and facular contamination are applied to the data before obtaining the two coefficients that describe the oblateness of the solar image. For the discussion below the oblateness is described by the coefficients D and V of a fit of D sin 2θ + V cos 2θ to the binned shape data. Here θ measures the angle from the solar rotation axis projected onto the image plane. Thus each observation at a given limb exposure yields 4 coefficients -- a V and D term for each of the two color bands. Except for additional atmospheric corrections described by Dicke, et al. (1985) the analysis sketched above essentially follows the description in the references.

The two summers of observations yielded about 254 useful days of data. This amounted to about 11,000 5 min observations distributed between the three possible limb exposures and the two summers. After subtracting least-squares fit daily quadratic polynomials, to remove daily trends, the standard deviation of the residuals was approximately 24 milliarcseconds in D or V.

4. SEEING SOLAR OSCILLATIONS IN SHAPE DATA

The oblateness data will not be sensitive to $\ell = 0$ or 1 oscillations. We may expect g-modes with $\ell = 2, 3, \ldots$ to contribute to the observed power in the V or D coefficients. If the 160 min mode is, for example, an $\ell = 2$ g-mode it will also contribute to the shape data.

For definiteness we will estimate the expected shape signal amplitude from the assumption that the 160 min mode is an $n = 9$, $\ell = 2$ g-mode (the closest predicted frequency for low order modes). The velocity amplitude of an $m = 0$ mode can be written

$$\vec{V}_n(r, \theta) = V_n(4) \, P_2(\cos \theta) \, \hat{r} + \frac{U_n(r)}{r} \frac{\partial P_2(\cos \theta)}{\partial \theta} \, \hat{\theta} \quad . \tag{1}$$

A model calculation (cf. Kuhn and Boughn, 1984) gives for the n = 9 mode $V_9(R_{sun})/U_9(R_{sun}) = 1.14$. The line-of-sight velocity V_{On} can be related to the modal amplitude V_n from

$$V_{On} = S_n \cdot V_n \quad , \tag{2}$$

where V_{On} is the intensity weighted line-of-sight velocity, $\vec{V}_n \cdot \hat{z}$, integrated over the solar disk. The factor S_n is easily evaluated for full disk observations and has been tabulated by Christensen-Daalsgard and Gough (1982). For example $S_9 = 1.3$ for the radial-transverse velocity ratio given above for the n = 9, ℓ = 2 mode. Other long period ℓ = 2 modes have values of S_n within about 50% of this value. Projecting the $P_2(\cos \theta)$ term against $\sin 2\theta$ or $\cos 2\theta$ gives an approximate relation between the coefficient V and the observed full disk velocity amplitude, V_0 of a g-mode of frequency ω,

$$V \cong 1.3 \, V_0/\omega \tag{3}$$

Thus, from the 160 min mode with an observed amplitude of about 1 m/s we expect an oblateness coefficient of $V \sim 2 \times 10^3$ m = 2.7 milliarcsec. If the oblateness residuals were uncorrelated such an oscillating shape signal would be observable with high signal-to-noise.

5.   IS THE 160 MIN OSCILLATION VISIBLE?

     In the following sections we address two questions: 1) Is there evidence of a 160.01 min solar oscillation in these data? and 2) Is there evidence of any other solar oscillating power in the oblateness residuals? These questions are complicated by the very uneven sampling in time of the data. For example in 1984 there were about 5500 observations between day number 107 and 274, spaced at about 5 min intervals. This represents only about 11% coverage of the full interval with only an approximately even spaced 5 min sample interval domain. It's clear from the power spectrum of the window function for one of the data sets, plotted in Figure 2, that any mode structure will be highly aliased in the observed power estimates.

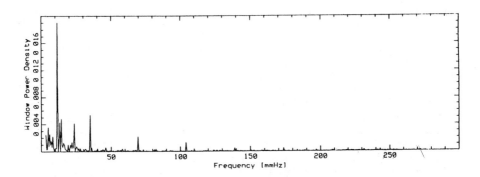

Figure 2. Power spectrum of the window function for a 1983 dataset.

   Data from the two color channels are obtained simultaneously and
are highly correlated. Thus to reduce the noise the colors are averaged
before computing power spectra. By averaging the power spectrum from 3
possible limb exposures and both V and D coefficients we may hope to
find a 160 min signal. This has been done in Figure 3 for 1983 and 1984
data separately, and combined. "Power spectra" are computed from the
summed squared amplitude of least squares fits of sin ωt and cos ωt to
the color averaged residuals.

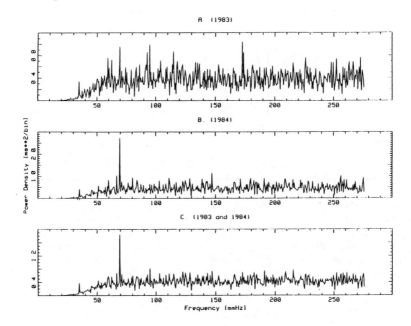

Figure 3(a). Average of the least-squares spectra from the six 1983
datasets.   3(b). Same as 3(a) but for 1984 datasets.   3(c). All 12
spectra averaged from 1983 and 1984.

Figure 3c is an average of 12 "pseudo-spectra" and is our first approximation to the actual mean power spectrum. We note that at frequencies
near 104.16 μHz (T = 160.01 min) all peaks are smaller than twice the
mean power level. To the extent that this approximates the actual mean
spectrum the probability distributions of signal in the presence of
noise calculated by Groth (1975) are applicable here. We find that
there is less than a 1% chance that the signal power at f = 104.16 μHz
is larger than the mean noise power of about 0.4 milliarc^2. It's
notable that there are prominent peaks in the averaged and yearly
spectra at periods of 1/10, 1/9, 1/6, 1/5 and 1/3 days. Some 1 day^{-1}
frequency sidelobes of the stronger peaks are also prominent.

Since we know the frequency and phase of the 160.01 min signal we
can also fit for it directly in the data. To simplify the analysis the
data were again averaged by color and limb exposure and binned in half
hour intervals to obtain a single time series for each V and D coefficient for each year. The phase and period described by Grec, et al.
(1980) has been used to find the 160 min amplitude and fit error for
1983, 1984 and combined datasets, displayed in Table I. The results of
a non-linear fit with frequency treated as an adjustable parameter are
also displayed.

TABLE I. Least Squares Fits for 160.01 min Oscillation

		1983	1984	Combined
f = 104.16 μHz Amplitude	V	0.4 + .5	1.7 + .8	0.3 + .4
[milliarcsec]	D	0.3 + .5	0.6 + .8	0.3 + .4
Frequency [μHz]	V	103.92 + .02	104.23 + .03	104.235 + .003
	D	104.18 + .01	104.00 + .05	104.171 + .003
Amplitude	V	1.3 + 1.3	2.1 + 2.0	1.7 + .8
[milliarcsec]	D	1.8 + 1.3	1.3 + 1.9	1.4 + .8

Only the V coefficient of the 1984 data shows a marginally
significant amplitude at about the 2σ level, but even this is probably
due to the nearby strong peak at a period of 159.90 min. The results in
Table I provide no evidence for the 160.01 min oscillation with an
amplitude greater than, approximately, the fit uncertainty of 0.4
milliarcsec.

6.   EVIDENCE OF OTHER g-MODES?

Several approaches have been considered to look for other discrete
frequency power contributions to the data.
Scherrer (1984) claimed detection of g-modes by searching for peaks

in the Stanford data using an iterative least-squares fitting technique
(the "Clean" algorithm) and then noted that the peaks fit the asymptotic
expression for the mode periods, $T(n, \ell)$:

$$T = T_0(n + \frac{\ell}{2} - 1/4)/\sqrt{\ell(\ell + 1)} \quad , \tag{4}$$

where n and ℓ are chosen separately for each mode to best fit the
observed periods using a single value of T_0. With 14 power spectrum
peaks with periods between 172 and 329 min Scherrer found a good fit
with T_0 = 38.6 min and ℓ and n values between 1 and 2, and 6 and 20. As
a test of the uniqueness of such a fit we've simulated this procedure on
100 sets of random numbers evenly distributed in frequency between the
corresponding limiting periods in Scherrer's data. Each set of 14
"peaks" is used to estimate a value of T_0 which minimizes the summed
squared deviation of the period calculated from (4) and random peaks.
The integer n and ℓ values are constrained to be less than or equal to
25 and 2 respectively and are adjusted to minimize the error for each
trial T_0 value. A 1% resolution search between 30 and 50 minutes finds
a value of T_0 which minimizes the summed error. The 100 trials yielded
the T_0 and mean error distributions shown in Figure 4. Notice that
such a fit is not very conclusive since an RMS error of about 3 minutes
is often possible given the range of available parameters. From
Scherrer's n, ℓ and T_0 values we calculate an RMS value of 3.0 minutes,
from his data. Thus his T_0 provides no supporting evidence that the
power spectrum peaks are of solar origin.

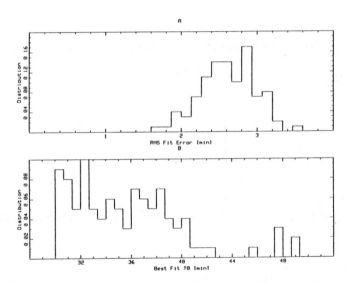

Figure 4(a). Distribution of the mean fit errors to the g-mode asymptotic period relation from 100 synthetic peaks lists. 4(b). Distribution of the minimizing parameter T_0 from the synthetic datasets.

A second approach for finding discrete power contributions is to
reduce the aliased structure in the spectrum by deconvolving the window
function from the observed power spectra. The data have been averaged
over the 3 limb exposures and 2 colors and binned in half hour intervals
for this purpose. Even so the "fill factor" on the evenly spaced half-
hour domain is only 23% in 1983 and less in 1984. The technique we've
applied to work around the gaps is described in Kuhn (1982, 1984b). By
choosing a particular set of evenly spaced frequencies the window
function effects can be minimized (the "swap" technique). Unfortunately
the domain is too sparse for this method (and probably any other).
Dummy data have been generated by adding unit Gaussian noise to a unit
amplitude sinusoid of frequency 101 µHz. Figure 5 compares the least-
squares and swap technique spectra for the 1983 domain and the above
synthetic signal. It's notable that the swap algorithm helps some but
that the sidelobes still dominate the observed "spectrum." We must rely
on statistical techniques to deduce evidence of g-modes in the oblateness data.

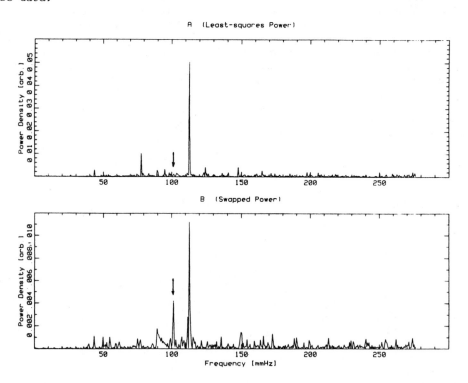

Figure 5(a). Least squares power spectrum of a 101 µHz frequency signal
plus Gaussian noise using the 1983 domain. 5(b). Same as 5(a) but using
the "swap" technique to evaluate the power spectrum.

We take as a null hypothesis that the 1983 and 1984 power spectra

(Fig. 3) are the result of random uncorrelated normally distributed
noise. If this is correct then the distribution of power from both
years should be consistent with the power distribution generated from
noise analyzed in the same way as the data. Thus we generate fake datasets on the same domain as the actual data. The values are distributed
normally from a distribution whose variance is scaled so that the mean
fake power and actual power agree. The resulting cumulative power
probability distributions are plotted in Figure 6. The agreement is
quite good if the anamolous peak at 1/6 day period in the 1984 data is
ignored. The Kolmogorov-Smirnov test yields a useful quantitative
description of the probability that both samples come from the same
parent distribution. The K-S test is essentially a measure of the
largest deviation between the two distributions (cf. Hollander and Wolfe,
1973). Applying the test to the 1983 and 1984 data separately yields
statistic values of 0.34 and .86 respectively, indicating that we should
accept the null hypothesis. In short we find no evidence of solar
g-modes in oblateness power spectra of mean noise 0.4 milliarcsec^2.

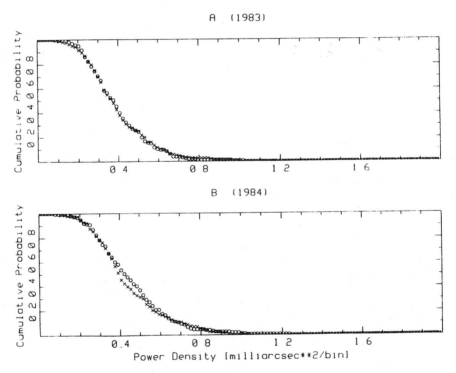

Figure 6(a). Cumulative power distribution of the data in Fig. 3(a)
(crosses) and the spectra of normally distributed white noise (circles).
6(b). Same as 6(a) except for 1984 data and time domain.

7. SUMMARY

There is no evidence of a 160.01 min solar periodicity in the
oblateness data at a level corresponding to a g-mode velocity amplitude
of about 0.2 m/s. Assuming the validity of previous 160.01 min signal
claims this suggests that the oscillation is not an $\ell = 2$ mode but
probably has $\ell = 1$. Further, we see no evidence of other g-modes to
approximately the same level.

I'm grateful to K. G. Libbrecht who operated the telescope in 1983
and has been involved in the data analysis and to R. H. Dicke who has
also contributed much to the analysis and numerous fruitful conversations.

This research was supported in part by the AFGL under contract no.
F19628-84-K-0024, and in part by the National Science Foundation.

REFERENCES

Boughn, S. B. and Kuhn, J. R. 1984, Ap. J. 286, 387.
Christensen-Daalsgard, C. D. and Gough, D. O. 1982, Mon. Not. Roy.
Astron. Soc. 198, 141.
Dicke, R. H. and Goldenberg, H. M. 1974, Ap. J. Supp. 27, 131.
Dicke, R. H., Kuhn, J. R. and Libbrecht, K. G. 1985, Nature, in press.
Frolich, C. and Delache, P. 1984, Mem. Soc. Astron. Ital. 55, 75.
Grec, C., Fossat, E. and Pomeratz, M. 1980, Nature 288, 541.
Groth, E. J. 1975, Ap. J. Supp. 29, 285.
Hill, H. A., Bos, R. J. and Goode, P. R. 1982, Phys. Rev. Lett. 49, 1794.
Holding, S. C. and Tuck, G. J. 1984, Nature, 307, 714.
Hollander, M. and Wolfe, D. 1973, Non-parametric Statistical Methods,
(John Wiley, New York).
Isaak, G. R., Van der Raay, H. B., Palk, P. L., Roca Cortes, T. and
Delache, P. 1984, Mem. Soc. Astron. Ital. 55, 91.
Kotov, V. A. Seveny, A. B. and Tsap, T. T. 1984, Mem. Soc. Astron. Ital.
55, 117.
Kuhn, J. R. 1982, Astron. J. 87, 196.
_____ 1984a, Mem. Soc. Astron. Ital. 55, 69.
_____ 1984b, Solar Seismology from Space, (ed. R. K. Ulrich; JPL:
Pasadena), p. 283.
Kuhn, J. R. and Boughn, S. P. 1984, Nature 308, 164.
Kuhn, J. R., Libbrecht, K. G. and Dicke, R. H. 1985, Ap. J. 290, 758.
Libbrecht, K. G. and Kuhn, J. R. 1984, Ap. J. 277, 889.
Scherrer, P. H. 1984, Mem. Soc. Astron. Ital. 55, 83.

PHASE STUDIES OF SOLAR 5 MIN OSCILLATIONS

Søren Frandsen
Institute of Astronomy
University of Aarhus
DK-8000 Aarhus C
Denmark

ABSTRACT

Observations of time series of spectral lines contain information about
the phase (and the amplitude) of the evanescent waves at many heights in
the solar atmosphere. The calculated phases turn out to agree with the
observed phases in the middle part of the atmosphere, but in the chromosphere, where the theory is only a crude description of the actual environment, differences occur. More surprisingly the calculations predict
wrong phases at the deeper layers around unit optical depth. The reason
is not known, but the coupling between convection and pulsation, which
is ignored in the model used, might be the responsible agent.

1. INTRODUCTION

In the solar atmosphere we have the possibility to study the 5 min acoustic modes in detail. By analysing time series of spectral line profiles
we can obtain information about the amplitude and the phase of the velocity and the temperature (and pressure) variation through the atmosphere
from the chromosphere, using the centers of strong lines, to the bottom
of the photosphere, observing the continuum fluctuations. If we choose
low l-value modes, the waves are predominantly vertically propagating in
the atmosphere. The horizontal wavelength for low l-values is much
larger than the scale of the inhomogeneities in the photosphere, but
maybe not so much in the chromosphere. Consequently the results we obtain
can be thought of as a description of a mean solar atmosphere.

As shown by Christensen-Dalsgaard (1984) computed and observed pulsational frequencies disagree in a way, that suggests theory fails at
the top of the solar model and at the lower boundary of the convection
region. Now the top of the model is precisely the atmosphere whereas the
bottom of the convection region is irrelevant in the context of this
paper. The idea one could pursue, is whether the detailed observations
in the atmosphere could be used to empirically reconstruct the eigenfunctions at the top of the model, thus removing the uncertainty in the
boundary condition at the very top of the model. But before trying to do

D. O. Gough (ed.), Seismology of the Sun and the Distant Stars, 73–80.
© 1986 by D. Reidel Publishing Company.

this reconstruction, the failure of the present theory to predict the
detailed form of the eigenfunction, describing the waves in the atmosphere, must be shown. Otherwise we already have the correct eigenfunctions in the atmosphere, and there is nothing to gain.

The behaviour of the amplitudes of the intensity fluctuations in a
spectral line observed at the center of the solar disk was observed by
Andersen (1984). At the present level of accuracy the observed amplitudes
do not deviate significantly from the predicted amplitudes as calculated
by Frandsen (1984), but there are small systematic differences, that
might contain new information. Here the attention is directed towards
the phase delays of velocities and intensity variations in various combinations. As there would be no phase changes in an adiabatic atmosphere,
the phase delays tell us about the radiative damping and other non-
adiabatic terms, which complicates the problem in the atmosphere. Staiger et al. (1984) have published extensive observations of phase delays.
Earlier some analysis of phases has been initiated especially by Schmieder (1979).

2. THE THEORETICAL METHODS

Radial eigenfunctions are used. The coupling between the pulsation and
the convection and other mechanical forms of energy is ignored. Quite a
detailed treatment of the continuum radiation and the line blanketing is
included as described by Christensen-Dalsgaard and Frandsen (1984). The
numerical technique used to obtain the eigenfunctions and the eigenfrequencies iterates on the eigenfrequency. The model is split into an
inner and an outer part. In the inner part the problem is a boundary
value exercise with 4 complex variables. In the outer part a Rybicki
elimination technique is used to solve a large number of equations simultaneously. This solution for a fixed frequency provides an equation
for the boundary for the inner part of the model. An iterative approach
can be used also for obtaining solutions in the outer part to speed up
the computations. The iterations compute successive approximations to
the radiative terms in the energy equation.

Given the eigenfunctions the oscillatory spectral line profile is
calculated straightaway using a numerical quadrature formula for weak
lines, where LTE is a valid assumption. For stronger lines an equivalent two level atom is introduced. Otherwise artificial emission appears
in the line cores, because we are using a VAL solar model (similar to
model C of Vernazza et al., 1981) including the chromospheric temperature
rise. This appears in the static profile as well as in the oscillating
signal. A consistent solution is found for the populations and the radiation field in the line, again using Rybicki type elimination schemes to
solve linear equations. The calculations have been done on CRAY1 computers at NCAR, Boulder and at IPP, Garching by Munich.

3. RESULTS

The calculations presented here all refer to a pulsational frequency of

3 mHz corresponding to the peak power of the 5 min models. The immediate result of the computations is tables of the complex function $\delta I(\Delta\lambda)/I(\Delta\lambda)$ for various lines. The lines studied until now are Fe I 5434Å and 5576Å and the KI line 7688Å. The scintillation makes it impossible to observe $\delta I(\Delta\lambda)/I(\Delta\lambda)$ directly. Immediate observable quantities are obtained by combining the intensity observed on both sides of the line center at equal distances into a sum

$$S = \frac{\delta I(\Delta\lambda)}{I(\Delta\lambda)} + \frac{\delta I(-\Delta\lambda)}{I(-\Delta\lambda)} - 2\frac{\delta I_c}{I_c} \qquad (3.1)$$

relative to the continuum variation $\delta I_c/I_c$, and the difference

$$D = \frac{\delta I(\Delta\lambda)}{I(\Delta\lambda)} - \frac{\delta I(-\Delta\lambda)}{I(-\Delta\lambda)} . \qquad (3.2)$$

Another observable is the phase delay of $\delta I(\Delta\lambda)/I(\Delta\lambda)+\delta I(-\Delta\lambda)/I(-\Delta\lambda)$ to the difference D.

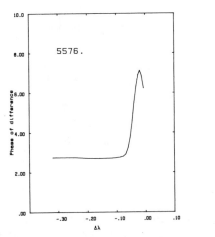
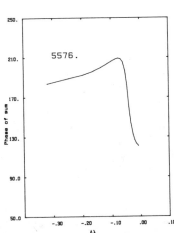

Figure 1 and 2. Phase plot of D and S for the FeI 5576 line. The horizontal axis is in Å and the vertical axis in degrees. The zero point of the vertical axis can be chosen freely.

The first two Figures 1 and 2 show the computed phase of D and S for the FeI 5576 line. Large changes occur in the center of the line. Before a comparison with observed phases is done, the connection between D and S with the eigenfunctions v, the velocity, and θ, the temperature perturbation, is discussed. For each line center distance $\Delta\lambda$, we define a mean optical depth

$$\tau(\Delta\lambda) = \int C(\Delta\lambda,\tau)\tau d\tau / \int C(\Delta\lambda,\tau) d\tau \qquad\qquad (3.3)$$

where τ is a mean optical depth in the atmosphere (Rosseland mean or τ_{5000}). The contribution function is the contribution from the interval $\Delta\tau$ to the integral expression used to calculate $\delta I(\Delta\lambda)$. Note that $\tau(\Delta\lambda) \neq \tau(-\Delta\lambda)$. We now expect the phase of D to be the same as the phase of v at an optical depth $\sim \tau(\Delta\lambda)$. And we expect the phase of S to be the same as the phase of $\theta(\tau(\Delta\lambda))$. One can be more accurate and define separate formation depths for D and S (Schmieder, 1979), but for the present purpose this refinement is not necessary. A plot of a set of contribution functions running from the line center to the far wing is shown in the Figures 3 and 4.

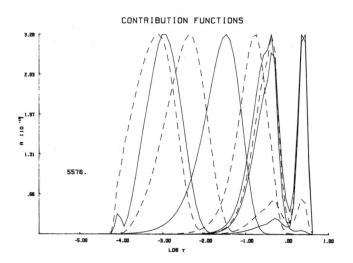

Figure 3. Normalized contribution functions for the FeI 5576 line. Every second curve is dashed to improve readability. The optical depth τ is closely related to the Rosseland mean.

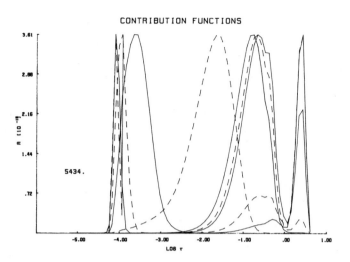

Figure 4. Normalized contribution functions for the FeI 5434 line.
This line is stronger than 5576.

The reason for the double peaks seen in some cases is not yet known.
Something like this is found also by Schmieder (1979) in the contribution
function for the velocity. The Figures 5 and 6 summarize the result. In
Figure 5 the phase has been scaled to give zero phase at the bottom of
the atmosphere (τ= 5). The expected relation between v and D is confirmed
although a small offset is seen. Observational material has been collected
from Schmieder (1979) from observations of the MgI 5172Å line and to
less extent from Staiger et al. (1984). There are some problems with the
construction of the observed points, due to the different choice of
reference model for the Sun, so the error bars might be a little underestimated. The observations differ from the computed phase by having a
much steeper slope in the photosphere and a smaller slope at small optical
depths. Too much emphasis should not be attached to this, because the data
are not homogeneous, but the difference of the slope in the photosphere
is probably real. Schmieder predicts a steep slope, but uses a much simpler theoretical description of the eigenfunctions.

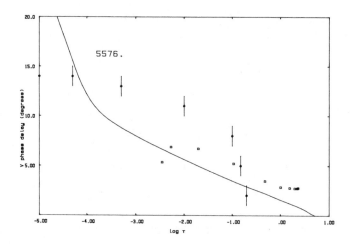

Figure 5. A combined plot of computed phases of D (squares, see text) compared to the phase of the velocity eigenfunction (curve) and observed phases (circles with error bars). The observations have been scaled to match the computed curve at log τ = 0. τ is closely related to the Rosseland mean optical depth.

For the temperature there is good agreement with the phase of S in the line centers. But as one approaches the continuum, the phase of S ceases to reflect the phase of θ = δT/T anymore. It seems to be a real effect in the sense, that different methods for computing δI are consistent. The points with the error bars are observed phase delays from Staiger et al. (1984) with some additional values not appearing in the paper but provided privately. The error bars are not provided by Staiger et al. They are an estimated common upper limit to all lines. A continuum value is given by Frohlich and Van der Raay (1984) with a large error bar, which is from their paper. Once more the line centers agree with the phase of the temperature perturbation. For the strongest lines, where the line centers are formed in the chromosphere, some differences are seen, but at these levels the assumptions going into the computations are dubious anyway. For weak lines and the continuum the situations could not be much worse. The computed phases differ by as much as 180 degrees from the observations, and in between one finds the phase of the temperature perturbation. The observations seem to agree fairly well. From the figures of Lites and Chipman (1979) one can deduce values for the line centers: FeI 5576Å 60 degrees, MgI 5173Å 120 deg. and CaII 8498 120 deg. This is the opposite behaviour of the observations of Staiger et al. where the phase shift of 5576 is larger

than the two other lines, but the values are in the same range. Schmieder gets for the I-v phase at the edge of the core of MgI 5172 (point 9)
110 deg. decreasing to 80 deg. in the line center more in line with
Staiger et al.

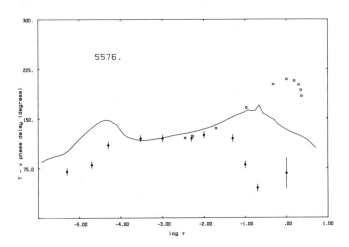

Figure 6. The phase delay of the intensity to the velocity as computed
and observed compared to the delay of the temperature eigenfunction to
the velocity eigenfunction. Same signature as in Fig. 5.

4. CONCLUSION

Beginning with the positive remarks one can conclude, that the qualitative description of the variation of the phase of the evanescent 5 min
modes is correct in a large part of the atmosphere. This is true for
such features as the direction of the phase changes with height in the
atmosphere. But from the earlier results on amplitudes (Andersen, 1984,
and Frandsen, 1984) and particularly the present results on phases there
is surely evidence for systematic problems with the fundamental assumptions going into the calculations of the eigenfunctions in the photosphere. The first thing to do is to check, whether the computed eigenfunctions can be understood from simple principles and the calculated
phases explained as physically reasonable. Next the influence of
convection has to be checked, but that is a difficult problem. Instead
one might attempt to deduce the eigenfunctions from the observations
directly. Once a fit has been obtained to the observations by an empirical model, one can then attempt to improve the theoretical picture, so
that the empirical eigenfunctions can be calculated from first principles.

Until this is possible one can use the empirical eigenfunction as
boundary conditions for interior calculations, and perhaps then the
difference between observed and calculated periods described by Christensen-Dalsgaard (1984) is diminished. Alternatively one can seek
for an explanation of the discrepancies seen above in terms of an
ill-posed problem, where the results are extremely sensitive to the
details of the physics (and possibly also to the numerical techniques
used). I consider this possibility less probable in the current circumstances.

References

Andersen, B., 1984, *Proc. of the 25th Liege Int. Astroph. Coll.*, 220
Christensen-Dalsgaard, J., 1984, *Proc. of the 25th Liege Int. Astroph.
    Coll.*, 155
Christensen-Dalsgaard, J. and Frandsen, S., 1984,
    Mem. Soc. Astr. Ital. $\underline{55}$, 285
Frandsen, S., 1984, *Proc. of the 25th Liege Int. Astroph. Coll.*, 303
Frohlich, C. and van der Raay, H.B., 1984, *Proc. of the Fourth Euro-
    pean Meeting on Solar Phys*. 'The Hydromagnetics of the Sun', 17
Lites, B.W. and Chipman, E.G., 1979, *Astroph. J.* $\underline{231}$, 570
Schmieder, B., 1979, *Astron. Astrophys.* $\underline{74}$, 273
Staiger, J., Schmieder, B., Deubner, F.-L. and Mattig, W., 1984,
    Mem. Soc. Astr. Ital. $\underline{55}$, 147
Vernazza, J.E., Avrett, E.H. and Loeser, R., 1981,
    Astrophys. J. Suppl. $\underline{45}$, 635

OBSERVATIONS OF GRAVITY WAVES IN THE SOLAR ATMOSPHERE

Franz-Ludwig Deubner
Institut für Astronomie und Astrophysik
Universität Würzburg
D-8700 Würzburg, Germany F.R.

EXTENDED ABSTRACT:

In direct analogy to the techniques of seismic measurements on the Earth, information on the vertical structure of the solar atmosphere can be obtained by monitoring the differences of arrival times of periodic signals at various "stations" (= successive line forming layers) or the time lag between signals of different type arriving at the same position. In contrast to Helioseismology of the interior progressive wave modes are studied in the atmosphere as well, and the essential information is displayed in phase diagrams.

There have been various attempts in recent years to prove the existence of internal gravity waves in the solar atmosphere by direct observation. Two examples are given which demonstrate, how the phase information can be used to single out gravity waves among other possible candidates, even if various wave modes are active in the same spatial and spectral region.

1)   The heating source of the bright chromospheric network elements has not been determined yet. Damé et al. (1984) have shown that the brightness of the network elements is correlated with the amplitude of intensity oscillations in the Ca II K-line at frequencies $\nu \approx$ 1-2 mHz, but anticorrelated at frequencies $\nu > 4$ mHz. New measurements with the Multi Diode Array system at the Vacuum Solar Tower (NSO, Sac Peak) including the infrared Ca II lines (8498Å, 8542Å) and a photospheric Fe I line (8497Å) have been used to confirm this result and to demonstrate that the low frequency oscillations are very likely downward propagating (magneto-) gravity waves.

This conclusion is supported by the following findings:

*   The photospheric line fails to show any indication of anamplitude-brightness correlation similar to the one observed in the Ca II lines (i.e. the source of the brightening cannot be traced to the photosphere.
*   Acoustic waves cannot propagate at 1 mHz.
*   Gravity waves with $l > 600$ are propagating at 1 mHZ.
*   The velocity-intensity (V-I) phase difference diagram indicates

D. O. Gough (ed.), Seismology of the Sun and the Distant Stars, 81–82.

the transition from evanescent p-modes to a different type of
wave motions at frequencies below $\nu \approx 3$ mHz.
*   The velocity-velocity (V-V) phases between the two CaII lines
    are distinctly positive, indicating downward propagation in the
    case of gravity waves, in the bright network regions (see also
    Staiger et al., 1984, Fig.6).

Generally speaking, and with regard to the three-dimensional
network geometry, it appears much easier to concentrate the required
mechanical flux by collecting it from a wide uniformly excited um-
brella "above" by means of magnetic funnels, than to feed this energy
from "below" selectively through the narrow supergranular borders.

2)   V-V phase spectra of photospheric lines exhibit a reversal of
sign near $\nu \approx 2.5$ mHz which was clearly noted first by Schmieder
(1976) and interpreted by herself as the signature of gravity waves
dominating the wave motions in this range of frequencies.

Schmieder also pointed out, that convective motions (granulation) strongly bias the V-I phase spectra, causing there a reversal
of sign as well, in the same frequency range which in addition is
haunted by the low-l extensions of the lowest-n p-mode ridges. This
rather complex mix of modes evidently calls for a decomposition of the
frequency spectra with respect to wavenumber l.

Observations were obtained with the MDA system at the Vacuum
Solar Tower (NSO, Sac Peak) in two photospheric Fe I lines (5929.7Å,
5930.2Å) along with the Na I D$_1$ line. For reasons of statistical
stability the l scale has been only partially resolved in five contiguous wavenumber intervals.

Results:
*   The influence of convection pulls the V-I phases only in the
    lowest two intervals (200"> λ > 4".6) whereas gravity modes appear
    to be more dominant in the higher wavenumber range (4".6> λ > 1".8).
*   This V-I phase difference remains below 50° at the lowest atmo-
    spheric levels (upward velocity lagging), but it jumps to much
    higher values at the level of the Na I D$_1$ line. This is conceiv-
    ably a consequence of overshoot.
*   The effects of radiation relaxation and of seeing (see Deubner
    et al., 1984) on the observed phase spectra can be clearly re-
    cognized.

References

Damé, L., Gouttebroze, P., Malherbe, J.-M.: 1984, Astron. Astrophys.
    130, 331.

Deubner, F.-L., Endler, F., Staiger, J.: 1984, Mem.S.A.It. 55, 135.

Schmieder, B.: 1976, Solar Phys. 47, 435.

Staiger, J., Schmieder, B., Deubner, F.-L., Mattig, W.: 1984,
    Mem.S.A.It. 55, 147.

IS THERE AN ANISOTROPY OF p-MODES ?

F.-L. Deubner
Institut für Astronomie und Astrophysik
Universität Würzburg
D-8700 Würzburg, Germany F.R.

EXTENDED ABSTRACT:

The question of the isotropy of the p-modes with respect to orientation on th solar disk was recently addressed by Kuhn (1984). He found significantly greater power at low wavenumbers ($l \lesssim 30$) in his measurements with the slit oriented in E-W direction as compared to the N-S orientation.

A similar experiment was conducted at the Capri station of the Kiepenheuer Institute, including the intermediate and the high wavenumber range from $l = 12$ to $l = 3000$. In view of the question of isotropy the purpose of this experiment was both to provide an update of the older (Deubner, 1975) measurements which suffered from rather low statistical stability, and to have a second look at the lower wavenumbers, if feasible.

The one-dimensional scanning technique, using a long slit to average the data in the other dimension, was applied. For this reason, the N-S and the E-W recordings had to be obtained on two different days. (This certainly poses some problems with respect to changing seeing quality and instrument stability).

After correcting the E-W data for solar rotation two-dimensional FFT transforms were performed in order to compare the power in the two sets of data as a function of ν and l. The preliminary results shown at this workshop unfortunately were largely mutilated by a trivial error in the data reduction code. The trends in the diagrams which seemed to support Kuhn's results do not exist.

In the correct results, any excess power in the E-W scans is certainly less than 30% in amplitude, it is not peaked at low l-values, and it is not restricted to frequencies with high p-mode power. It is expected that taking instrumental and seeing effects fully into account the upper limit of 30% can be lowered further substantially. This work is still in progress.

A full account of this study is in preparation for Astronomy and Astrophysics.

D. O. Gough (ed.), Seismology of the Sun and the Distant Stars, 83–84.
© 1986 by D. Reidel Publishing Company.

References

Deubner, F.-L.: 1975, *Astron. Astrophys.* 44, 371.

Kuhn, J.R.: 1984, *Mem.S.A.It.* 55, 69.

INFLUENCE OF SPATIAL FILTERING ON
POSSIBLE ANISOTROPIES IN SOLAR OSCILLATIONS

Frank Hill

National Solar Observatory
National Optical Astronomy Observatories
Tucson, Arizona 85726 USA

Deborah A. Haber and Juri Toomre

Joint Institute for Laboratory Astrophysics
Department of Astrophysical, Planetary and Atmospheric Sciences
University of Colorado
Boulder, Colorado 80309 USA

and

Laurence J. November

Air Force Geophysical Laboratory
National Solar Observatory
National Optical Astronomy Observatories
Sunspot, New Mexico 88349 USA

ABSTRACT. We have used full disk Doppler observations of solar oscillations to compare the amplitudes of sectoral modes propagating along the equator with those of similar modes propagating along a great circle aligned with the poles. We find that the amplitudes are generally not equal for the two classes of modes, but the results are sensitive to analysis procedures attempting to isolate the different modes of oscillation. Spatial filtering of the data using spherical harmonics suggests that greater amplitudes are associated with "polar" sectoral modes than with "equatorial" sectoral modes.

1. INTRODUCTION

Determination of the amplitudes of solar oscillations, in addition to their frequencies, is essential if we are to understand how these oscillations are excited. Several mechanisms have been proposed for driving the oscillations. Ando & Osaki (1975) deduced that the κ-mechanism could serve to excite many solar modes within the observed range of periods and horizontal wavenumbers. Yet Goldreich & Keeley (1977a, b) found such driving to be marginal when dissipative effects of turbulent convection were taken roughly into account, and favored stochastic excitation by the turbulence itself as a possible means of driving the modes. Antia et al. (1984)

D. O. Gough (ed.), Seismology of the Sun and the Distant Stars, 85–92.

have reconsidered the combined effects of the κ-mechanism and turbulent convection. Ando (1985) has examined the possible role of differential rotation in the generation of the waves. Each of these models predicts a different structure for the amplitudes of the oscillations as a function of ℓ and ν, where ℓ is the spherical harmonic degree of the mode and ν is its temporal frequency.

Measurement of the amplitudes is relatively simple when individual modes can be temporally or spatially resolved. This has almost been accomplished with observations of Doppler shifts in integrated sunlight which provide information on low-degree modes ($\ell < 5$). Turning to data of this type, Christensen-Dalsgaard & Gough (1982) concluded that the observed amplitudes are consistent with excitation of the modes by stochastic convection. However, determination of amplitudes for modes with larger ℓ has been challenging.

At high ℓ it is the sectoral modes (with $m = \ell$, where m is the azimuthal order of the oscillation) that have been studied most extensively. These sectoral modes are confined to the equator of the chosen spherical coordinate system. The degree of concentration increases with increasing ℓ, modes with $\ell > 500$ being completely confined within a narrow $10°$ band centered on the equator. Although the propagation path of sectoral modes needs to be a great circle, any such circle on a sphere is admissible, and modes on any such path are expressible by a matrix multiplication that transforms the set of spherical harmonics from one choice of axes to another. There is, however, a natural physical equator for the Sun due to its rotation axis. Since one of the proposed sources of excitation is differential rotation, one might expect that modes with various orientations with respect to the rotation axis might show differences in amplitude. For instance, Hansen *et al.* (1978) argued that rotation enhances the instability of prograde modes, while it diminishes the instability of similar retrograde modes. Thus prograde sectoral modes propagating with the rotation might possess higher amplitudes than similar retrograde modes. Likewise, should differential rotation have a role in exciting the waves, then sectoral modes propagating in the east-west equatorial direction (parallel to rotational velocities) might well have different amplitudes than sectoral modes propagating in the north-south polar direction (perpendicular to rotation).

A number of studies have searched for such anisotropies in the oscillations by observing the Sun in both the north-south and east-west directions. Deubner (1975) and Rhodes (1977), whose observations were sensitive mainly to sectoral modes with $\ell > 100$, found no significant differences in power spectra obtained from observations with detectors oriented along these two directions on the Sun. Kuhn & O'Hanlon (1983) reported that for $\ell < 30$ the east-west propagating modes had more power than modes propagating north-south, and the converse for $\ell > 60$. Deubner (1986) has seen similar behavior, except the turnover point occurred at $\ell \approx 900$. In order to try to isolate these two classes of sectoral modes, the two-dimensional spatial Doppler data in these studies had been averaged over straight lines (perpendicular to the propagation path of the modes) in the plane of the sky. Since this is only one of several possible filtering techniques, we have analyzed full disk Doppler with three different spatial filtering procedures to assess their effects upon possible anisotropies in "equatorial" and "polar" sectoral modes.

2. OBSERVATIONAL METHODS

The data discussed here were obtained in December 1984 and April 1985 using the Universal Birefringent Filter (UBF) and the Vacuum Tower Telescope at the

Sacramento Peak site of the National Solar Observatory. The UBF is a promising instrument because of its ability to image the full disk of the Sun with potentially high spatial resolution at many wavelengths. Full disk Doppler measurements are desirable because they allow comparison of different spatial filtering techniques using a single data set, as well as the possibility of determining the amplitudes of waves traveling in different directions.

Intensity images in the red and blue wing of the Fe I λ 5576 spectral line were recorded simultaneously on 35 mm film. Typical exposure times for the images were 3 seconds to allow some averaging of the atmospheric seeing distortions as suggested by the simulations reported by Toomre et al. (1985). The film images were digitized with the fast microphotometer at Sac Peak with a spot size of 51 μm. Registration of the pairs of intensity images was accomplished during the digitizing process by fitting a circle to four limbs of each image. Each frame was digitized on a grid of 512 $\times$ 512 points, with the solar diameter typically covering 480 pixels. The digitized densities were converted to intensities using a measured calibration curve for the film; velocities were computed from the ratio of the difference of the intensity images to their sum. The temporal sampling interval of the data was 60 seconds. Observations were carried out over a total of twenty days, though only two individual days are considered here. Observations on the first of these days, 10 December 1984, consisted of 512 time samples spanning 8.5 hours in time, and the other, 11 April 1985, had 700 time samples spanning 11.7 hours. The observational setup was still in a preliminary state when this data was taken. It appears that using film as a detector caused inhomogeneities in the images that precludes consideration of modes with $\ell < 30$. Observations now under way use CCD detector arrays.

3. SPATIAL FILTERING ANALYSIS

The spatial pattern of the oscillation modes can be expressed in terms of spherical harmonics as

$$P_\ell^m(\cos\theta)e^{im\phi},$$

where P_ℓ^m is the associated Legendre function, θ is the colatitude, and ϕ is the longitude in the chosen spherical coordinate system. For sectoral modes, with $m = \ell$, the function P_m^m is symmetric about the equator, and decays to zero towards the poles.

To assess the effects of spatial filtering on possible anisotropies of equatorial and polar sectoral modes, we perform the analysis of the data in three different ways. In the first method, we simply average each image on straight lines (either east-west or north-south) extending over the entire solar disk, in a manner similar to the earlier studies. The resulting two-dimensional array of velocities as a function of space and time is interpolated onto an equally spaced longitude grid of 512 points with a spacing of 0.270°, and is then Fourier transformed to produce a $m - \nu$ power diagram. This is the filtering technique applied to the April data. Since sectoral modes with relatively high m values $(m > 500)$ are confined to regions close to the equator where the lines of constant longitude are nearly straight, this method is probably adequate to isolate such modes. However, the strategy is not optimal for lower values of m, as the modes extend to higher latitudes where the lines of constant ϕ become curved. Consequently, averaging along straight lines rather than lines of constant longitude will pick up contributions from tesseral modes with $\ell > m$.

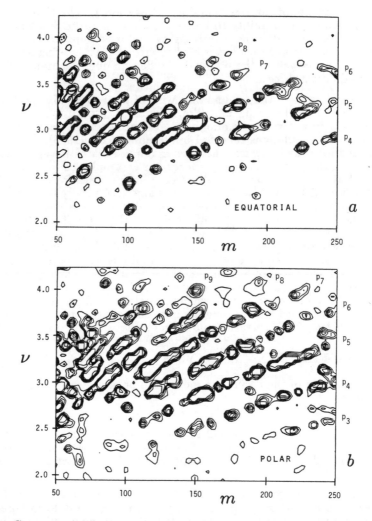

Figure 1. Contour plots of power as a function of m and ν in (a) for equatorial sectoral modes and in (b) for polar sectoral modes. Here the spherical harmonic projections of method 3 were applied to full disk Doppler data obtained on 10 December 1984 with the UBF instrument. (Frequency units are mHz.)

In the second analysis method, each full disk image is interpolated onto an equally spaced longitude grid of 256 points covering 180°, and then simply averaged over lines of constant ϕ. For the December data considered here, the interpolation is simplified because the inclination angle B of the rotation axis of the Sun to the plane of the sky is less than $1/2°$ and can be ignored. The averaging of the data is typically carried out over 31° in latitude (128 points) centered on the chosen equator. However, such an unweighted average probably does not isolate a mode as well as a weighting based on the appropriate Legendre function. The data are then Fourier transformed in ϕ and in time.

In the third method, the spherical harmonic properties of the modes are utilized fully. This is probably the best method for isolating sectoral modes. We begin by again interpolating a given Doppler image in time onto a uniform $(\phi, \cos\theta)$ grid, followed by a Fourier transform in ϕ to produce a data array in m and $\cos\theta$. The data are then averaged in the θ direction for each m using the appropriate P_m^m as a weighting function. These operations correspond to projecting the velocities upon the spherical harmonic basis functions. The resulting data arrays are then Fourier transformed in time to produce a diagram of power as a function of m and ν. The second and third methods are shown here as applied to the December data.

All three of these methods were performed with the Fourier transform in m along both the east-west and north-south directions in the image, yielding $m - \nu$ diagrams which show the power in "equatorial" and "polar" sectoral modes. Figure 1 displays power diagrams obtained for both classes of modes using method 3. These $m - \nu$ diagrams show the familiar ridge structure of the oscillations, with each ridge of power corresponding to different radial orders n of the modes.

4. COMPARING EQUATORIAL AND POLAR AMPLITUDES

It is evident from Figure 1 that the ridges of the equatorial modes are displaced from the ridges for the corresponding polar modes. This is due to the advection of the wave fronts of the equatorial modes by solar rotation. In order to directly compare the power of the equatorial and polar modes, we have registered the two sets of ridges by using a two-dimensional cubic spline to interpolate each equatorial $m - \nu$ diagram onto a new grid in ν. The relationship between the new grid of frequencies ν' and the apparent frequencies ν is

$$\nu' = \nu + \frac{m}{2\pi R_\circ}V,$$

where $R_\circ$ is the solar radius and V is a solar rotational velocity. The value of V that maximized the cross-correlation between the polar and equatorial $m - \nu$ diagrams was used to produce an equatorial $m - \nu'$ diagram with the effects of rotation removed.

We have summed the power in each $m - \nu$ diagram over 8×8 pixels in order to try to compare our results with the studies of Kuhn & O'Hanlon (1983) and Deubner (1985), both of which considered smoothed power spectra. This smoothing results in decreased noise levels but reduces resolution and mixes the background power with the oscillatory power in the ridges. Further, in the region where the ridges are converging, this summing mixes power from more than one ridge. We then proceeded to compute R, the logarithm of the ratio of the power in the polar $m - \nu$ diagram to that in the equatorial modes for each of the three spatial filtering methods. Computing a ratio of power R for each pair of diagrams allows us to avoid difficulties in the choice of normalization constants in the power spectra.

The results are shown in Figure 2, which displays contour plots of R in the region of the ridges for the three spatial filtering methods. In the hatched areas where $R < 0$, the power in equatorial modes is greater than that in polar modes; the unhatched areas are regions where the polar modes have the greater power.

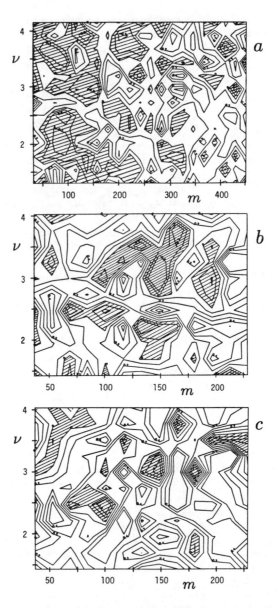

Figure 2. Contour plots of R, the logarithm of the ratio of power in the polar sectoral modes to the power in the equatorial sectoral modes, as function of m and ν. The contour levels range from -0.2 to +0.5 in increments of +0.1, with areas of negative R hatched. (a) R for method 1 of spatial averaging over straight lines, (b) for method 2 of averaging over lines of constant longitude, and (c) for met' od 3 based on projections onto spherical harmonic functions. Method 1 was here applied to data from 11 April 1985, methods 2 and 3 to data from 10 December 1984.

Figure 2a displays R from method 1 of averaging over straight lines; the results are consistent with both Kuhn & O'Hanlon (1983) and Deubner (1986) in that the equatorial modes have greater power than the polar modes below some value of m, and that the converse is true above that value of m. If we collapse our R data by summing in ν, we can identify this turnover point as being at $m \approx 230$; Kuhn & O'Hanlon found a turnover at $m \approx 50$, and Deubner at $m \approx 900$. It is interesting that the turnover value of m seems to be about $1/2$ of the number of pixels across the solar diameter for each of the three studies, implying that the effect may be due to the spatial filtering characteristics of the observations rather than some property of the solar oscillations.

This suggestion is further strengthened by the other two contour plots in Figure 2, which show R for method 2 involving simple averaging over lines of constant ϕ and for method 3 of spherical harmonic projection. Figure 2b shows that with method 2 the region of equatorial mode dominance ($R < 0$) is restricted to somewhat smaller values of m than with method 1. However, analysis based on method 3 and shown in Figure 2c indicates that the $R < 0$ regions have largely vanished, implying that polar modes have typically greater power than the equatorial modes at nearly all values of m. We have also examined grey scale images of R computed from unsmoothed $m - \nu$ diagrams. Such images are very noisy, but the gross distribution of R displayed in Figure 2 can still be discerned. It thus seems that the choice of spatial filtering has an important effect in influencing what may appear as anisotropies in solar oscillations. Of the three methods of analysis, method 3 with projection onto spherical harmonics should be the most reliable at isolating sectoral modes from tesseral modes, and thus we would place greater credence on the results for R shown in Figure 2c.

5. CONCLUSIONS

We have found that various spatial filtering methods used to isolate sectoral modes can lead to different conclusions about the apparent amplitudes of these oscillations. That in itself is not too surprising, since spatial filtering by averaging over straight lines as in method 1 clearly must become less effective for isolating the sectoral modes at low m. Yet such procedures are very appropriate for detecting zonal modes when combined with Legendre transforms. Our results with method 3, which we believe to be the most reliable of the analysis procedures for sectoral modes, indicate that the polar sectoral modes generally possess greater power than the equatorial ones. If this result is indeed of solar origin, we then have a challenge to understand it theoretically.

The comparison of power levels using the ratio R based on smoothed $m - \nu$ power diagrams could certainly be improved upon by just concentrating on power in individual ridges, using centroiding methods discussed in Hill, Gough & Toomre (1984) to locate the ridges. Further, we need to determine how our measurement of power levels in the polar modes is affected by the lateral advection of their great circle of propagation by solar rotation in the course of an observing day. We also recognize that mode beating effects upon power levels, due to many unresolved modes within each resolution bin in ν and m, restrict us at best to estimates of amplitudes for individual modes. There may also be differences in the sensitivity of the instrumentation to modes oriented in the two directions, though we have attempted to minimize such anisotropies. There are indeed a number of issues to be pursued before we can reliably state that certain excitation mechanisms are more effective, or damping and dispersive mechanisms more damaging, for one orientation of sectoral modes over another. Our results so far hint that polar sectoral modes may be somewhat favored.

Scientific discussions with Drs. Jørgen Christensen-Dalsgaard, Tom Duvall, Douglas Gough and Jack Harvey have been very helpful. The instrumental expertise of Dr. Dick Dunn made it possible to explore a number of unusual optical configurations necessary to image the entire Sun through the UBF with the Tower Telescope; the observing and technical staff at Sacramento Peak have made vital contributions in implementing these observations. The National Optical Astronomy Observatories are operated by the Association of Universities for Research in Astronomy, Inc. under contract with the National Science Foundation. Studies at the University of Colorado were supported in part by the National Aeronautics and Space Administration through grants NSG-7511 and NAGW-91 and by the Air Force Geophysics Laboratory through contract F19628-82-K-0008.

6. REFERENCES

Ando, H. 1985, *Pub. Astron. Soc. Japan* **37**, 47.

Ando, H., and Osaki, Y. 1975, *Pub. Astron. Soc. Japan* **27**, 581.

Antia, H.M., Chitre, S.M., and Narashima, D. 1984, *Mem. Soc. Astron. Ital.* **55**, 175.

Christensen-Dalsgaard, J., and Gough, D.O. 1982, *Mon. Not. R. Astron. Soc.* **198**, 141.

Deubner, F.-L. 1975, *Astron. Astrophys.* **44**, 371.

Deubner, F.-L. 1986, these proceedings.

Goldreich, P., and Keeley, D.A. 1977a, *Astrophys. J.* **211**, 934.

Goldreich, P., and Keeley, D.A. 1977b, *Astrophys. J.* **212**, 243.

Hansen, C.J., Cox, J.P., and Carroll, B.W. 1978, *Astrophys. J.* **226**, 210.

Hill, F., Gough, D.O., and Toomre,J. 1984, *Mem. Soc. Astron. Ital.* **55**, 153.

Kuhn, J., and O'Hanlon, M. 1983, *Solar Phys.* **87**, 207.

Rhodes, E.J., Jr. 1977, Ph.D. Thesis, Univ. California, Los Angeles.

Toomre, J., Hill, F., Merryfield, W.J., and Gough, D.O. 1985, Astrophys. J., to be submitted; see also 1984, *Probing the Depths of a Star: The Study of Solar Oscillations from Space* (eds. R.W. Noyes and E.J. Rhodes, Jr.; NASA Jet Propulsion Laboratory Publ. *400-237*), pp. 37-42.

THE DETECTION OF GLOBAL CONVECTIVE WAVE FLOWS ON THE SUN

Philip H. Scherrer,
Richard Bogart,
J. Todd Hoeksema[†] and,
Hirokazu Yoshimura[‡]
†Center for Space Science and Astrophysics, Stanford University
‡Department of Astronomy, University of Tokyo

ABSTRACT. Global convective flows in the solar convection zone have been predicted by theoretical interpretations of the global-scale ordering of magnetic fields and activity centers and by theoretical analyses of rotating convection zones. Direct evidence of these flows in the photosphere has not previously been found despite several long-term efforts. The signatures of such flows have now been detected by analyzing the daily series of low-resolution Dopplergrams obtained at the Wilcox Solar Observatory at Stanford University. The signatures are patterns of alternating east and west flows with amplitudes on the order of 25 m/s and longitudinal extent of about 30 degrees. The patterns move across the disc at approximately the solar rotation rate and have lifetimes of at least several rotations. Boundaries of the fast and slow flows are often associated with large magnetic active regions.

The detection and study of global convective wave flows on the sun is of great importance for understanding the basic mechanisms of the generation of magnetic fields and magnetic activity on the sun (Yoshimura, 1971). They have been considered as the most likely kind of fluid motions to drive the solar dynamo which generates magnetic fields on the sun (Yoshimura, 1972, Yoshimura, 1983). This global convection has escaped detection despite a number of long-term observing and analysis programs (Durney et al., 1985, Howard and Yoshimura, 1976, Howard and LaBonte, 1980, LaBonte et al., 1981, Snodgrass and Howard, 1984).

Observations made at the Wilcox Solar Observatory at Stanford University have been examined for the signature of global scale convective motions. This series of observations began in 1976 and has been found to have lower noise in the detection of solar velocities than the instruments used in previous analyses (Scherrer et al., 1980). We now have a sufficient time span of observations to study the evidence for global-scale horizontal motions superimposed on solar rotation. The observations are Doppler-shift measurements which yield low-resolution full-disc daily velocity maps. Clear velocity patterns with amplitudes

93

D. O. Gough (ed.), Seismology of the Sun and the Distant Stars, 93–102.
© 1986 by D. Reidel Publishing Company.

of at least 25 m/s have been found in zones near the equator. These
patterns can be seen to move across the disc with solar rotation. They
have a large latitudinal extent, crossing the equator, but are not
aligned on meridians. There are from 4 to 7 alternating patterns around
the sun (Scherrer and Yoshimura, 1985, Yoshimura et al., 3-15).

There are several fundamental difficulties in detecting global-
scale small-amplitude horizontal motions on the sun. One problem is
caused by the large contributions of solar rotation (about 2000 m/s at
the equator) and the limb shift (a "W"-shaped red-shift profile across
the disc with an amplitude of 600 m/s at the limb and 30 m/s at the disk
center). Other sources of solar "noise" are the motions of 5-minute
oscillations and supergranulation. These noise velocities are larger in
amplitude but smaller in scale than the postulated horizontal flows.
The large aperture used for the Stanford observations greatly attenuates
these signals. A third solar source of velocity noise comes from the
red-shifts associated with regions of intense magnetic fields. These
"downflows" have been examined by a number of authors (Cavallini et al.,
1985, Labonte and Howard, 1982) and are generally believed to be due to
a local reduction in small scale convective motions in the vicinity of
magnetic fields.

A final difficulty is that we are restricted to measuring the
line-of-sight component of the motion. In order to infer individual
velocity components, one must assume either spatial continuity or temporal continuity during the disc passage of individual features. In the
present work, we have used two different methods in an attempt to understand and avoid the weaknesses of each. In both approaches, we assume
the rotation and limb shift signals are constant over several rotations,
and that the remaining horizontal motions are organized on a scale much
larger than supergranulation. In the first approach (Scherrer and
Yoshimura, 1985) we further restrict ourselves to horizontal velocity
structures with a large east-west extent. We assume that by averaging
over a large window (up to a solar radius) the components other than
east-west motions will cancel. In the second approach (Yoshimura et
al., 3-15), we do not assume such a large spatial extent, but we assume
that the structures are stable during their disc passage. The two
methods have different sensitivities to noise and changing solar velocity fields. We refer to the two methods as the slope method and the
decomposition method.

The Stanford velocity maps are a product of a 9-year synoptic
series of observations. The observations and instrument have been
described previously (Duvall, 1979, Scherrer et al., 1980) and will be
only briefly described here. The instrument is a scanning magnetograph
attached to a high dispersion spectrograph. The raw data consist of
intensity, magnetic field, spectral line position, position on the solar
image, and time. Daily scans are made with a three-arc-minute square
aperture. The solar image is scanned in 21 steps of 90 arc-sec in the
solar east-west direction and stepped to 11 scan lines in the north-
south direction. In this way we obtain simultaneous magnetograms and
Dopplergrams on every suitable day. The observations are made in the
absorption line of neutral iron near 525.02nm. The calibration of the
line shift signal in m/s and removal of the earth's motion is done as

described previously. After excluding observations contaminated by clouds we have 1494 daily Dopplergrams for the present study.

Previous analyses have removed the large rotation and limb-shift signal by fitting and subtracting a particular functional form of differential rotation and limb-shift from each Dopplergram. To avoid any spurious signal from an improper fit and to avoid the possibility of removing some of the desired signal as rotation, we have adopted a new procedure. The first step is to subtract the 54-day running mean of the observed velocity at each point on the disc. This procedure removes any signal dependent on disc position, such as rotation and limb shift, without making any assumptions regarding its functional form. Next, we subtract from each scan point residual the mean of all residuals within the central 80% of that Dopplergram. This step is necessary because we have no absolute velocity reference for our measurements. To reduce the contamination of our horizontal velocity signal by active-region red-shifts, we next discard all scan points which have a velocity residual amplitude exceeding 30 m/s or a magnetic field exceeding 20 gauss. An average of 27 percent of the scan points were discarded from each Dopplergram. Discarding these points may remove evidence of the larger amplitude flows, but we believe it is necessary to remove most of the contamination from the active regions.

The resulting series of residual Dopplergrams is then analyzed for horizontal velocity components by fitting the east-west component of motion. In the first approach (slope method) we fit the east-west component in a restricted portion of the solar disc with a large longitudinal window at a fixed latitude. This is done by finding (by least squares) the east-west component of velocity (V_{west}) such that: $V_{obs} = V_{west}\sin(L)\cos(B_0)$ where L is the longitude measured from disc center and B_0 is the latitude of disc center. The regions included in each fit were east-west strips 810 arc-seconds long by 180 arc-seconds wide. These fits were computed for the nine central scan lines and all possible east-west positions for each Dopplergram. From these fits we can display daily maps of east-west motions. The relatively large area used in these fits greatly attenuates the noise from oscillations and supergranulation. It also acts as a spatial filter allowing only very large-scale structures to produce a signal. The method is not sensitive to day-to-day variability in instrumental zero or noise from changing solar velocities, but it is sensitive to contamination from line shifts not associated with east-west motion.

We also derived east-west motions by a second approach (decomposition method) in which we examine the Doppler-shift variation of each location on the sun as it crosses the disc. This is accomplished by using all Dopplergrams having a given Carrington coordinate within 0.8 R_0 of disc center to find the inferred components U, V, and W that best fit (least squares) the observed velocity residual V_{obs}:

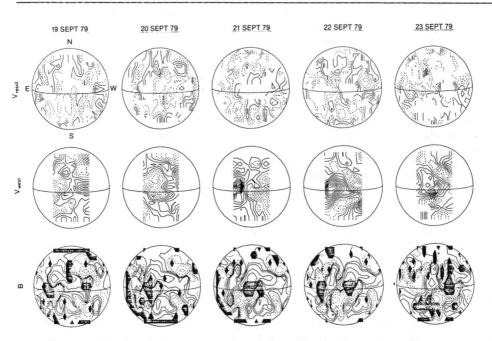

Figure 1: Velocity residual maps for September 19 to 23, 1979. The top
row is the line-of-sight "velocity" with limb-shift and rotation removed
(positive away from observer). The middle row if the computed east-west
velocity (positive to the west) and the bottom row is the magnetic
field. The contour levels are 10 m/s for the velocity maps and 0, ±1,
2, 5, 10 ... gaus for the matnetograms. The regions excluded from the
analysis are shaded on the magnetograms.

$$V_{obs} = U\cos(B_O)\sin(L)$$

$$+V(\cos(B_O)\sin(B)\cos(L) - \sin(B_O)\sin(B))$$

$$-W(\cos(B_O)\cos(B)\cos(L) + \sin(B_O)\cos(B))$$

where B is the latitude and U, V, W are the westward, northward, and
upward components respectively. This method has the advantage that it
cleanly separates the westward velocity from active-region "downdrafts"

but has the disadvantage that it assumes a static velocity structure, no day-to-day zero level errors, and the Carrington rotation rate.

Using the slope method, Scherrer and Yoshimura (1985) have shown that there are global-scale velocity features that are stable for their transit across the disk. This is the first requirement for detecting large scale flows. Figure 1 shows data from 5 consecutive days, September 19 to 23, 1981. The top row shows residual line-of-sight signals (before excluding points with large residuals or fields), the center row shows V_{west} as described above, and the bottom row shows the corresponding magnetograms for comparison. The V_{west} signal can easily be seen to move across the disc as the sun rotates. The V_{west} signal is shown as positive values with solid contours in regions of westward motion (i.e. motion in the same direction as rotation).

To study the morphology of the horizontal flows we have constructed synoptic charts of both V_{west} and U in the same way as is usually done for magnetic fields. These charts show a map of the solar surface made by averaging the signals from each day's observations for each latitude-longitude bin. The synoptic charts are computed on a grid of 5-degrees in longitude by 30 equal steps in sine-latitude. Some additional temporal and spatial averaging results from this process. To reduce the small-scale noise a further averaging with a 40-degree-in-longitude low-pass filter was applied to each latitude strip.

Figure 2a shows V_{west} for three consecutive rotations from the fall of 1979, Figure 2b shows three rotations from the fall of 1981, and Figure 2c shows the fall of 1984. The velocity pattern shows 4 or 5 structures of alternating eastward and westward motion per rotation. These structures are most clearly seen in low northern latitudes. They can be followed for several rotations.

Synoptic maps computed with the decomposition method have roughly the same character but differ in detail. They differ most in the vicinity of active regions, where the slope method could be contaminated by "downflows" if their spatial extend is large, and the decomposition method would be contaminated if they are changing.

This signal appears to persist for more than one rotation. Flows of about the same magnitude are generally present during the past six years during periods of both intense and low magnetic activity. For instance, the signal is present in the fall of 1984 when activity was quite low indicating that the while there is a correlation with the magnetic structures, we really are measuring different physical quantities. A full understanding of the relation between the flow patterns and magnetic activity will require more analysis and more complete observations sequences than are presently available.

Figure 2: Synoptic maps of east-west velocity V_{west} for three consecutive rotations in 1979 (part a), in 1981 (part b), and in 1984 (part c). Westward moving regions are shown in unshaded. Eastward moving regions are shaded dark. The contour levels are ± 10, 20, 50, ... m/s.

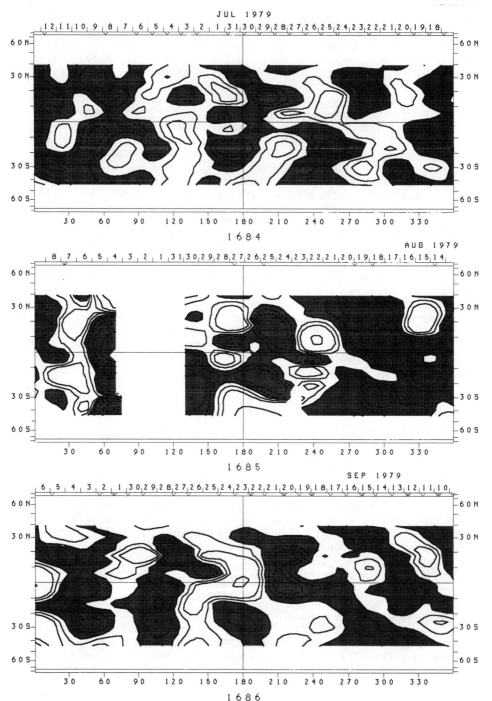

Figure 2a

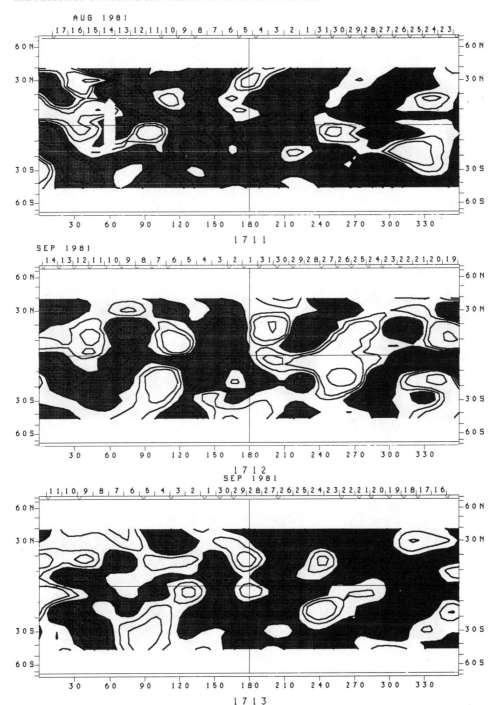

Figure 2b

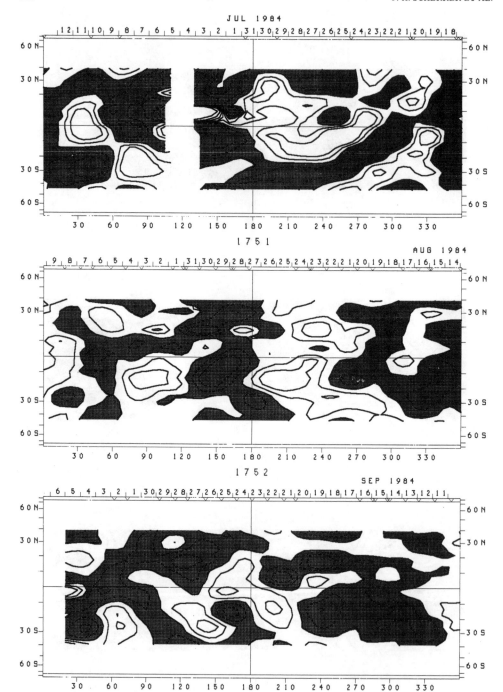

Figure 2c

The intervals presented in Figure 2 were chosen both because of sufficient data coverage to show nearly complete maps and because an initial survey had identified these intervals as times of simple modal structure. This survey was done by computing power spectra of V_{west} for intervals of nearly continuous data coverage. It is interesting to note that the day-to-day variation in the whole-disc fit equatorial rotation rate (Scherrer et al., 1980) is very similar to the meridional average of V_{west} for the intervals examined. This was examined both by comparing power spectra and by inspection of the raw data. Essentially all of the day-to-day variation in the full disc fit of rotation comes from the velocity structures described here.

These velocity patterns have a large amplitude compared to the upper limits reported in previous analyses of observations from Mt. Wilson (Snodgrass and Howard, 1984). However, we note that the present analysis of Stanford observations finds horizontal flow patterns which are different in structure, and larger in scale than those searched for in the Mt. Wilson studies.

The complete understanding of the flow patterns reported here will require an extended set of high quality continuous observations throughout a solar cycle. The present results have been obtained with the instrument at Stanford. This instrument at present is limited to low resolution measurements and seldom has more than 30 or 40 days without interruption of a day or two. We are taking steps to increase both the resolution and the Doppler zero stability for future observations. We are also examining the comparable data of improved quality available from Mt. Wilson since 1982. We hope to be able to test our interpretation of the Stanford observations as signatures of the global convection and to begin the process of examining the morphology of these flows by comparisons with other data and further observations.

Acknowledgements: We thank Harald Henning for useful comments. This work was supported in part by the Office of Naval Research under Contract N00014-76-C-0207, by the National Aeronautics and Space Administration under Grant NGR5-020-559, and by the Atmospheric Sciences Section of the National Science Foundation under Grant ATM77-20580.

References

F. Cavallini, G. Ceppatelli, and A. Righini, Meridional and Equatorial Center to Limb Variation of the Asymmetry and Shift of three Fe I Solar Photospheric Lines Around 6300A, Astron. and Astrophys. in press(1985).

B. R. Durney, L. E. Cram, D. B. Guenther, S. L. Keil, and D. M. Lytle, A Search for Long-Lived Velocity Fields at the Solar Poles, Ap. J. 292(1985), 752-762.

Thomas L. Duvall, Jr., Large-Scale Solar Velocity Fields, Sol. Phys. 63(1979), 3-15.

Robert Howard and Hirokazu Yoshimura, Differential Rotation and Global-

Scale Velocity Fields, in Basic Mechanisms of Solar Activity, Bumba and Kleczek (editor), 1976.

R. Howard and B.J. LaBonte, A Search for Large-Scale Convection Cells in the Solar Atmosphere, Ap. J. 239(1980), 738-745.

B.J. LaBonte, R. Howard, and P.A. Gilman, An Improved Search for Large-Scale Convection Cells in the Solar Atmosphere, Ap. J. 250(1981), 796-798.

Barry J. Labonte and Robert Howard, Solar Rotation Measurements at Mount Wilson III. Meridional Flow and Limbshift, Sol. Phys. 80,2 (1982), 361-372.

Philip H. Scherrer, John M. Wilcox, and Leif Svalgaard, The Rotation of the Sun: Observations at Stanford, Ap. J. 241(1980), 811-819.

P.H. Scherrer and H. Yoshimura, The Detection of Global Convective Wave Flows on the Sun, submitted to Nature, 1985.

Hershel B. Snodgrass and Robert Howard, Limits of Photospheric Doppler Signatures for Solar Giant Cells, Ap. J. 284,2 (1984), 848-855.

H. Yoshimura, Philip H. Scherrer, J. Todd Hoeksema, and Richard Bogart, The Detection of Global (Giant Cell) Convective Wave Flows on the Sun, Presented at the 1985 Meeting of the Solar Physics Division, American Astronomical Society, Tucson, Arizona, 1985 May 13-15.

Hirokazu Yoshimura, Complexes of Activity of the Solar Cycle and Very Large Scale Convection, Sol. Phys. 18(1971), 417-433.

Hirokazu Yoshimura, On the Dynamo Action of the Global Convection in the Solar Convection Zone, Ap. J. 178,3 (1972), 863-886.

Hirokazu Yoshimura, Dynamo Generation of Magnetic Fields in Three-Dimensional Space: Solar Cycle Min Flux Tube Formation and Reversals, Ap. J. Supplement Series 52,4 (1983), 363-385.

FIRST EVIDENCE OF A LARGE-SCALE CIRCULATION IN THE SOLAR CONVECTION ZONE

E. Ribes and P. Mein
Observatoire de Meudon
France

ABSTRACT. A large-scale circulation has been detected in the solar convection zone, for the first time. Tracers used to probe the convective layers are young sunspots and long-lived H alpha filaments. On one hand, we found that sunspots trace a four-zonal meridional circulation which varies through the solar cycle. On the other hand, the meridional circulation pattern fits in nicely with the magnetic pattern drawn by the long-lived H alpha filaments. These observations suggest that we are looking at azimuthal rolls. The direction of rotation seems to be associated with the magnetic polarity which characterizes each roll.
    The properties of the rolls are as follows:
    - width: 20 to 40 degrees
    - velocity fields: 30 to 50 metres per second, comparable with predicted convective velocities
    - stability: marginal; dislocation of a roll and appearance of a new one every 2 to 3 years
    - connection with sunspot activity.
We suggest that the large-scale circulation is of convective origin. The properties of this large-scale convection are very different from what was predicted, and shed a completely new light on problems dealing with the differential rotation, the transport of angular momentum and the dynamo action.
    A detailed report of this work is to appear in <u>Nature</u>.

D. O. Gough (ed.), Seismology of the Sun and the Distant Stars, 103.
© *1986 by D. Reidel Publishing Company.*

SOLAR DOPPLER SHIFTS: SOURCES OF CONTINUOUS SPECTRA

T. L. Duvall, Jr.
Laboratory for Astronomy and Solar Physics
NASA/Goddard Space Flight Center
Greenbelt, Maryland 20771 USA

and

J. W. Harvey
National Solar Observatory
National Optical Astronomy Observatories[*]
Tucson, Arizona 85726 USA

ABSTRACT. Oscillation observations can be used to study non-oscillatory solar phenomena that exhibit Doppler shifts. In this paper we discuss several effects of these phenomena and their associated temporal and spatial power spectra: 1) They limit the signal-to-noise ratio and sometimes detectability of oscillation modes. 2) There is the potential for better understanding and/or detection of solar phenomena: surface rotation, supergranulation, granulation, active regions, giant cells, and mesogranulation. 3) Large-scale convection may spatially modulate oscillation modes, leading to a continuous background spectrum. 4) In regions of the spectrum where we lack the resolution to separate modes, we can determine upper limits for the integrated effects of modes.

1. INTRODUCTION

Observations of solar oscillations are made in the presence of noise from instrumental, atmospheric and solar sources. This noise limits the accuracy of oscillation mode frequency and amplitude measurements. In this paper we demonstrate how accurate estimates of the precision of mode frequency measurements can be made and outline a method for modelling observed spectra. The solar component of the "noise" provides important information about many solar phenomena. We discuss some of the results from available observations of the background spectrum. The inhomogeneous structure of the convection zone may alter the modal structure of the oscillations in such a way to produce sidebands around modal features in k,ω spectra. We conjecture that such a process may contribute to the rise of background power level in such spectra.

---

[*]Operated by the Association of Universities for Research in Astronomy, Inc. under contract with the National Science Foundation.

D. O. Gough (ed.), Seismology of the Sun and the Distant Stars, 105–116.
© 1986 by D. Reidel Publishing Company.

Finally, we use results from a 17-day observation sequence to set limits on the integrated strength of unresolved g-modes of degree 2.

2. THE EFFECT OF BACKGROUND POWER ON THE S/N RATIO OF OSCILLATION FREQUENCY MEASUREMENTS

It is obvious that sources of background power will limit the accuracy of estimates of oscillation mode frequencies. Here we address the following issues: 1. Quantitatively, how much are the frequencies affected as a function of signal to noise ratio? 2. What is an optimum method for estimating frequencies and their uncertainties? The need for reliable estimates of frequency uncertainties is particularly evident in the inversion calculations, where the uncertainties in the final product (say solar rotation versus depth) are critically dependent on the estimates of internal error in the frequency measurements (Christensen-Dalsgaard and Gough, 1984; Gough, 1984). What we assume in this section is that there are two sources of uncertainty in frequency measurements, one component due to background power and the other due to the stochastic nature of the modes themselves. It is this second source, due basically to the finite lifetimes of the modes, that gives the fundamental limitation of mode frequency measurement. In practice, however, we feel that the observations to date have not approached this fundamental limit but instead are limited by background sources of power.

2.1. A Model of Oscillations and Noise

We will begin the discussion (following Jenkins and Watts, 1968) by considering a simple example of a stochastic process: a harmonic oscillator with damping driven by a random forcing function. The differential equation describing the system is:

$$\frac{1}{\omega_o^2} \frac{d^2y}{dt^2} + \frac{1}{\omega_o Q} \frac{dy}{dt} + y(t) = x(t) \tag{1}$$

where $y(t)$ is the displacement, ω_o is the undamped oscillator frequency, Q is a constant describing the damping, and $x(t)$ is the forcing function. This is an example of an equation (linear differential equation with constant coefficients) whose solution can be expressed as a convolution:

$$y(t) = \int_{-\infty}^{\infty} h(t')x(t-t')dt' \tag{2}$$

The function $h(t)$, called the impulse response of the system, is independent of the forcing function $x(t)$ and so only depends on the left side of equation (1). $h(t)$ must satisfy a condition of physical realizability ($h(t)=0, t<0$) and must be integrable. Using the convolution theorem, the Fourier transform of equation (2) is:

$$\tilde{y}(f) = H(f)\tilde{x}(f) \tag{3}$$

where $\tilde{y}$, $\tilde{x}$ are the transforms of $y(t)$, $x(t)$, and $H(f)$ is the transform of $h(t)$ and is termed the frequency response function of the system. To derive the frequency response function for our example of the harmonic oscillator, one would do the Fourier transform of equation (1). And, in fact, $|H(f)|^2$ is what gives our normal Lorentz line shape. It is the parameters describing $|H(f)|^2$ that we would ultimately wish to estimate.

If one computes the power spectrum of a finite realization (of length T) of the $y(t)$ process (we will call this a sample spectrum), one does not obtain $|H(f)|^2$ directly because of the random nature of $\tilde{x}(f)$. Considering the sample spectrum of the $x(t)$ process separately (calling it $C_x^T(f)$), we find that at frequencies separated by $1/T$ the values are independent and distributed as χ^2 with two degrees of freedom. This leads to a definition of the "spectrum" of the $x(t)$ process as:

$$\Gamma_x(f) = \lim_{T\to\infty} E[C_x^T(f)] \tag{4}$$

where the expectation operator, E, implies an average over many independent realizations. For a white noise process, this is just a constant, independent of f (and hence a flat spectrum from white noise). Similarly, the spectrum of the $y(t)$ process is:

$$\Gamma_y(f) = |H(f)|^2 \, \Gamma_x. \tag{5}$$

As for the white noise case, the ratio (at a given frequency) of the sample spectrum divided by the true spectrum is distributed as χ^2 with two degrees of freedom. This is used as the basis for a maximum likelihood method for estimation of $\Gamma_y(f)$ given a sample spectrum. That the solar data actually possess this property has been demonstrated by Woodard (1984).

The maximum likelihood technique consists of constructing the joint probability density for the outcome of an experiment in terms of some model parameters. Then maximizing this probability density (or likelihood function) for a given experimental outcome yields the maximum likelihood estimates of the parameters. In the present example, we have at a given frequency f_i, the probability density

$$p(C_i) = \frac{1}{\Gamma_i} \exp(-C_i/\Gamma_i) \tag{6}$$

where we have adopted the simplified notation that C_i, Γ_i are respectively the sample spectrum and the spectrum at frequency f_i. The joint probability density for the experimental outcome at all the independent frequencies is just the product of the individual densities:

$$L = \exp[-\sum_i(\ln\Gamma_i + C_i/\Gamma_i)] \tag{7}$$

The model might be, for example, a sum of Lorentzian plus background power:

$$\Gamma_i = \frac{A}{1 + \left[\dfrac{f_i - f_o}{w}\right]^2} + r \tag{8}$$

where A is the maximum signal power, w is the half-width and r is the background power. For this case, one takes a given sample spectrum C_i and maximizes equation (7) in terms of the model parameters f_o, A, w, r.

2.2. Fitting Observations and Estimating Uncertainty

The analysis of the uncertainties is most easily done if the likelihood function is approximately Gaussian for variations of the fitting parameters. If it is Gaussian, the likelihood function can be considered equivalent to a probability

density function for the parameters and the error analysis can proceed similarly to that for a normal multiple regression. If it is not Gaussian, it is sometimes possible to change the desired parameters to make it Gaussian. For
example, in the case of Lorentzian plus background we found in simulations that
if the likelihood function was expressed in terms of the logarithms of the parameters A,w,r, that it was approximately Gaussian.

To test the fitting procedure and error estimation, a series of simulations
was performed on the model of Lorentzian plus background. In all cases, the
resolution used was that of the 1980 ACRIM data: $1/290$ days $= 0.04$ μHz and the
FWHM of the modes was 1.6 μHz.

Sample spectra were generated at this resolution over a range of ±10 μHz
from a mode position for four different values of the peak signal to background
ratio (A/r). Fifty realizations were computed for each value of A/r under the
assumption that the ratio of sample spectrum to model should be distributed as
χ^2 with two degrees of freedom. To check the error estimation procedure, the
internal error estimates were averaged over the 50 cases and compared to the
external scatter found in the parameters. This comparison showed that the
internal error estimates are in fact good estimates of the parameter uncertainties.

The critical parameter is the frequency estimation error and the results of
the simulations are shown in Fig. 1 and Table 1. The results are presented in the
figure as σ^2 (variance of frequency estimate) versus r/A in order to show the
linear relationship between these two variables.

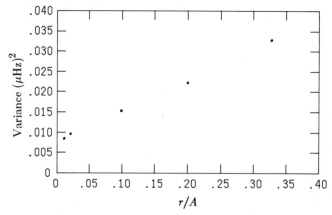

Figure 1. The mean uncertainty squared (variance) of frequency estimates as a
function of background noise level divided by peak signal power for the simulations discussed in the text. Note the linear trend.

One sees that measurement uncertainty decreases as S/N ratio increases, as
expected. However, even for infinite S/N ratio, a finite measurement error
remains. This is partly due to the finite mode lifetime and also partly due to
only using the data $\pm10\mu$Hz around the mode. One rather disappointing aspect
of this plot is the rather slow decrease of measurement error for substantial
increases of S/N ratio.

TABLE I. Simulation results

A/r	3	5	10	50	100
$\sigma\ (\mu\text{Hz})$	0.181	0.149	0.124	0.098	0.092

3. DETECTION OF SURFACE PHENOMENA -- THE SOLAR BACKGROUND SPECTRUM

The other solar surface phenomena (besides oscillations) that contribute to Doppler-shift observations are rotation, limb effect, gravitational redshift, supergranulation, active regions, granulation, giant cells and mesogranulation. All of these phenomena are impediments to observing oscillations, but we may be able to learn something about these phenomena from oscillation observations. For example, one might like to know what is the spatio-temporal power spectrum of supergranulation. Phrased in this way, the question is very difficult to answer: all of the phenomena are lumped together and it will be nontrivial to separate them. A more reasonable question might be: what is the spatio-temporal power spectrum of solar non-oscillatory Doppler shifts? As we shall see, this also is not an easy question to answer.

What do we expect for the spatio-temporal power spectra? We can get some guidance from a simple model given by Harvey (1985) for the lowest spatial frequencies. In this model each solar phenomenon is described by an exponentially decaying in time, autocovariance function. This leads to a Lorentzian-shaped temporal power spectrum (centered at $\nu=0$). Each of the different phenomena has its own lifetime and amplitude. The proposed temporal power spectrum for integrated sunlight is shown in Fig. 2. The steep falloff with frequency is due to the ν^{-2} wings of the Lorentz profile. The spectrum is not expected to change significantly up to spherical harmonic degree of roughly 50.

There are a number of observational difficulties with measuring the spatio-temporal power spectrum of the non-oscillatory Doppler shifts. The power spectrum of the temporal window function for a completely sampled interval falls off as ν^{-2}. Thus, even if signals were only present at low frequencies, the resulting spectrum would appear similar to Fig. 2 because of leakage. This could be cured by apodization for a completely sampled sequence. However, for a sequence with significant data gaps the problem is more severe. In general, the spectrum of the window function falls off more slowly than ν^{-2} and is not as easily corrected by apodization.

Another severe problem at least at higher spatial frequencies is heterodyning of the low frequency signal (mostly supergranulation and active regions) by the atmospheric seeing (Ulrich *et al.* 1984). As the seeing is generally at a higher frequency than the sampling, the result is a roughly flat background spectrum (Hill, 1984). This background may be the dominant signal at higher spatial frequencies and many temporal frequencies. This would prevent the measurement of the solar background spectrum in these areas. It is also a severe impediment for oscillation observations. The amount of "homogenizing" of the temporal spectrum will change as a function of spatial frequency, making the variation of solar power with spatial frequency uncertain. Another source of uncertainty of variations with spatial frequency is the instrumental point-spread function, which needs to be measured. Potentially severe problems at all spatial frequencies are purely instrumental sources of noise increasing at low frequencies. The

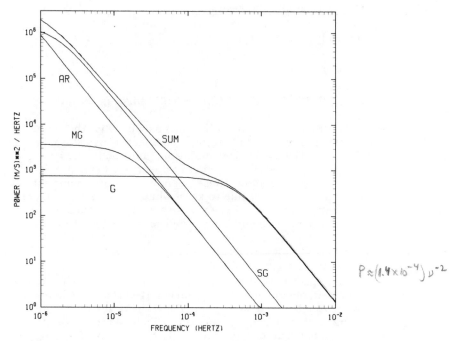

Figure 2. The model power spectrum for integrated sun observations. The curve labelled SUM is the sum of the spectra for the individual physical processes (shown separately) : AR (active regions), SG (supergranulation), G (granulation), and MG (mesogranulation).

spectra of such noise sources are almost never known accurately and can be difficult to separate from solar phenomena, although sometimes solar rotation can be used to provide a means of discrimination.

Given these uncertainties, what can be learned about the background spectrum from present observations? One good data set to examine is the 17-day sequence of sectoral mode observations described by Duvall and Harvey (1984). In Fig. 3 is shown a low-pass (temporally) filtered picture of the data. The dominant signal is solar, as evidenced by the rotation across the disk of almost all visible features. The picture would appear to be dominated by active region signals and a signal showing horizontal motion. The active region signal is manifested as apparent recession (white) most strongly visible near disk center. The most obvious of these features have been identified with particular active regions rotating across the disk. The horizontal motion signal, probably supergranulation but possibly also giant cells, is of generally higher spatial frequency and visible mostly away from the disk center.

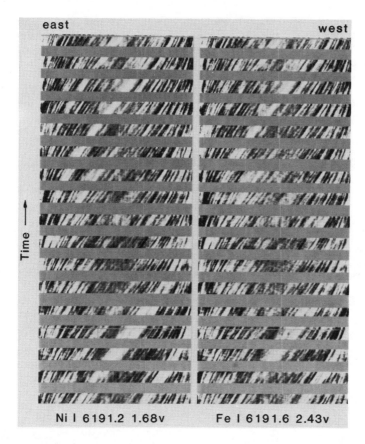

Figure 3. Low-pass filtered Doppler shift versus time for the data of Duvall and Harvey (1984). The two pictures are for the two spectrum lines used. Time proceeds upwards for 17 days. Space goes horizontally, east limb to west limb. White corresponds to apparent recession. Features do not move in straight lines because we are observing the surface of a sphere projected onto a disk.

The spatial-temporal power spectrum of the low frequency data is shown in Fig. 4. This picture is dominated by the rotating pattern of Fig. 3, which appears as an apparent prograde sectoral ridge with frequency $l\Omega$, l being the spherical harmonic degree, Ω the cyclic rotation frequency. The sideband structure of this feature associated with the nightly gaps is generally the strongest thing seen at frequencies separated by ≥ 10 μHz from $l\Omega$. So we conclude that, without a more detailed analysis, the only definitely solar signal is within a few μHz of $l\Omega$. A marked feature is the strong falloff of the power at high degree. This could be attributed to the instrumental point spread function, as we do not have a good measure of it for these observations. We notice that at higher degrees, the rotational feature is somewhat broader, suggesting a positive correlation of lifetime with feature sizes. Such a correlation was also seen in the supergranulation observations of Duvall (1980). Another noticeable feature is an enhancement of power near degree = 40. This relatively narrow feature in l is at a size scale about three times larger than supergranulation and is of unknown origin. We unsuccessfully sought a corresponding feature in the sectoral harmonic spectrum of magnetic field patterns observed at the same time as the Doppler measurements. This feature is better seen in Fig. 5, a plot of the integrated power ±5 μHz about the frequency of $l\Omega$. The low frequency results shown in Figs. 4 and 5 are about the only solid information we can currently get from these data about the solar background.

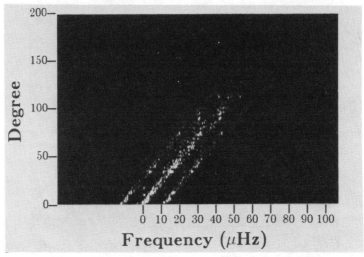

Figure 4. The spatio-temporal power spectrum of Fig. 3. The ordinate is spherical harmonic degree and the abscissa is temporal frequency with waves propagating in the direction of rotation having positive frequency. The central ridge is due to the approximately steady features moving across Fig. 3. The sideband structure is due to the nightly gaps.

A plot of observed power at higher frequencies is shown in Fig. 6. For the reasons discussed above we cannot be certain that the background power observed at higher frequencies is entirely of solar origin. Nevertheless, a comparison of the results of Fig. 6 with the model estimate in Fig. 2 is interesting. The agreement is very good which suggests that either the observations are revealing the true solar background noise or that the simple model estimate seriously overestimates the amount of background power to be expected.

4. A PROPOSED EFFECT OF LARGE-SCALE CONVECTION ON OSCILLATIONS

Helioseismology observations of the inhomogeneities in the Sun due to convection promise the first opportunity for a comprehensive study of astrophysical convection. Convection's effect on the oscillation modes should provide a means for detection of the convective inhomogeneities. The initial attempts to exploit this possibility (Hill, Toomre, and November, 1982, 1983; Gough and Toomre, 1983; Hill, Gough, and Toomre, 1984) have focused on the largest scales of

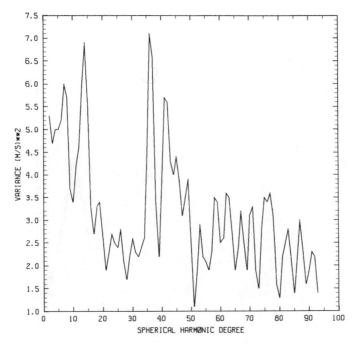

Figure 5. The integrated power over a band ±5 μHz centered on the frequency $l\Omega$ as a function of spherical harmonic degree.

convection. Searches have been made for variations (from day to day) in advection and frequency for high-degree sectoral modes observed over a large longitude range ($\sim 1/6$ of the solar circumference). The limitation to the largest scales comes from the large longitude range observed: a convective inhomogeneity with more than one wavelength within the observing aperture will effectively be filtered out by the above procedures. The present contribution is concerned with the observation of these higher wavenumber convective inhomogeneities.

The eigenfunction of an oscillation mode for a spherically symmetric system is of the form:

$$E_S = V(r)P_{lm}(\theta)\exp[i(m\phi+\omega t)] \qquad (9)$$

For a nonspherical system with a convective perturbation which is sinusoidal in longitude and long-lived, modes will still exist with a well-defined frequency. However, their spatial eigenfunction will be modified. One possibility for this modification is that the horizontal spatial frequency will be modulated, an analogy being an FM radio signal. The form of the eigenfunction in this case would be:

$$E_N = V(r)P_{lm}(\theta)\exp[i(m\phi+a\sin n\phi+\omega t)] \qquad (10)$$

where n is the wavenumber of the convection and a is a constant measuring the strength of interaction. This type of relation is normally considered only in the case $n<<m$. Another possibility is that the mode would be amplitude modulated

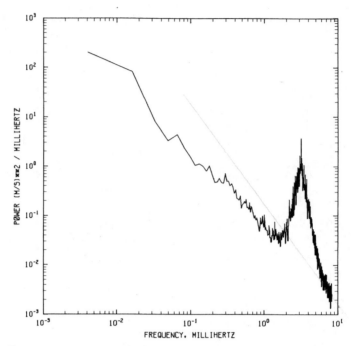

Figure 6. Power spectrum of Doppler shifts for spherical harmonic degree 15. The resolution has been reduced by piece-wise integration. Comparison with the model of Fig. 2 shows good agreement.

in longitude, with the result in the spectrum being qualitatively similar for weak FM modulation.

For the case in which the convective perturbation is resolved in longitude, the spectrum of eq. (10) consists of a carrier and sideband structure. The carrier appears at (m, ω) and for weak modulation, a single pair of sidebands appear at $(m \pm n, \omega)$. Except for edge effects, this spectrum would be constant from day to day, in contrast to the case in which the convective mode is not resolved spatially. For fairly high wavenumber convection (e.g. $n=40$), the effect of the convection (the sidebands) appears far from the normal mode frequency. And, in fact, if we expect a continuous spectrum of convection, there should be generated a continuous background in the oscillation spectrum which has a higher amplitude at temporal frequencies where the oscillation power is high.

This excess background is one potentially observable effect of the convection. To look for this effect, we have examined the data of Duvall and Harvey (1984). Fig. 7 is a plot of the degree = 150 spectrum at reduced temporal frequency resolution. There is an excess of background power between the mode peaks, amounting to about 2% of the peak oscillation power. However, until a more detailed analysis can be done to decide whether this could be due to the finite extent of the spectrum of the window function, this can only be considered an upper limit to the magnitude of the effect.

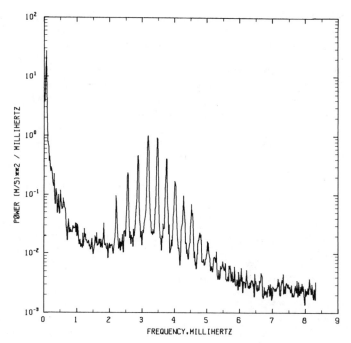

Figure 7. Power spectrum for degree 150. Note the rise in background power between the peaks.

5. UPPER LIMITS TO MODE AMPLITUDES IN UNRESOLVED REGIONS OF THE SPECTRUM

There has been much interest recently in the detection of g-modes because of the diagnostic capabilities of these modes. One contribution that low temporal-resolution observations can make is to set upper limits on the integrated modal power. For example, from Fig. 5, we can state that the upper limit of the integrated power of g-modes of degree = 2 in the frequency band $\pm 5\mu$Hz centered on $l\Omega$ is ~ 5 (m/s)^2. At first sight this might not seem to be a stringent upper limit, but there are a lot of modes in this frequency band and so the mean power per mode must be small. The asymptotic formulas suggest that there would be an infinite number of modes in this frequency band or about 1700 in the band 1-5μHz.

6. CONCLUSIONS

A study of a reasonable model of a solar oscillation mode in the presence of background noise shows that at low signal-to-noise ratio, background noise is a serious source of uncertainty in making estimates of the frequencies of oscillation modes. The reduction of this uncertainty with increasing signal-to-noise ratio is disappointingly slow because of the intrinsic finite lifetime of the modes.

Comparison of a model of solar non-oscillatory power with observations gives good results at high frequencies but there are many uncertainties that prevent us from learning more about solar phenomena from this comparison.

Future observations with better equipment and smaller window-function sidelobes should allow considerable information about surface velocity patterns to be reliably obtained. In the meantime, available observations of the spatial spectrum below a frequency of 5μHz suggest that a velocity pattern having a degree of about 40 is a significant solar feature. The same observations indicate that the integrated power from g modes at low frequency is small.

We conjecture that small-scale convection may modulate the pattern of oscillation modes so as to spread power rather far from its origin in k,ω spectra. If proven to be true, this phenomenon is both a nuisance because it adds to the background noise but is also a potential diagnostic of some of the properties of solar convection.

7. REFERENCES

Christensen-Dalsgaard, J. and Gough, D. O.: 1984, in R. K. Ulrich (ed.) *Solar Seismology from Space* (Jet Propulsion Laboratory Publication 84-84, Pasadena) p. 79.

Duvall, T. L., Jr.: 1980, *Solar Physics* **66**, 213.

Duvall, T. L., Jr., and Harvey, J. W.: 1984, *Nature* **310**, 19.

Gough, D. O.: 1984, in R. K. Ulrich (ed.) *Solar Seismology from Space* (Jet Propulsion Laboratory Publication 84-84, Pasadena) p. 49.

Gough, D. and Toomre, J.: 1983, *Solar Physics* **82**, 401.

Harvey, J.: 1985, in E. Rolfe and B. Battrick (eds.) *Future Missions in Solar , Heliospheric, & Space Plasma Physics* (ESA SP-233, European Space Agency, Noordwijk) p.199.

Hill, F.: 1984, in R. K. Ulrich (ed.) *Solar Seismology from Space* (Jet Propulsion Laboratory Publication 84-84, Pasadena) p. 255.

Hill, F., Toomre, J. and November, L. J.: 1982, in J. P. Cox and C. J. Hansen (eds.) *Pulsations in Classical and Cataclysmic Variable Stars* (Joint Institute for Laboratory Astrophysics, Boulder) p. 139.

Hill, F., Gough, D. and Toomre, J.: 1984, *Mem. Soc. Astron. Ital.* **55**, 153.

Hill, F., Toomre, J. and November, L. J.: 1983, *Solar Physics* **82**, 411.

Jenkins, G. M. and Watts, D. G.: 1968, *Spectral Analysis and its Applications*, Holden-Day, San Francisco.

Ulrich, R. K., Rhodes, E. J., Jr., Cacciani, A. and Tomczyk, S.: 1984, in R. K. Ulrich (ed.) *Solar Seismology from Space* (Jet Propulsion Laboratory Publication 84-84, Pasadena) p. 263.

Woodard, M.: 1984, Ph.D. Thesis, University of California, San Diego, Dept. of Physics.

COMMENTS ON TECHNIQUES FOR SPECTRAL DECONVOLUTION

Philip H. Scherrer,
Center for Space Science and Astrophysics
Stanford University
Stanford, California
U.S.A.

ABSTRACT. Current observational questions in asteroseismology require
high spectral resolution that can only be obtained with observations
spanning many days or months. The primary constraint in the full utili-
zation of single mid-latitude observing sites is the presence of diurnal
data gaps. Several methods for removing the effect of these gaps in the
spectra obtained from velocity observations have been suggested. In the
case of data coverage of less than 50%, none of the methods considered
has been successful at unambiguously recovering the true spectrum from
test spectra with realistic complexity. The limitations of these
methods and their applicability to the helioseismology problem is discussed.

The presence of data gaps is particularly troublesome when computing
power spectra of multi-day observations of solar or stellar oscillations. If the expected lifetime of the observed modes is longer than
individual uninterrupted observing sequences, one naturally wants to
combine many days of data to obtain higher spectral resolution. The
gaps from nights or poor weather produce a complicated structure of
sidebands in the power spectrum unless the gaps are filled with
estimated data. Since the power spectrum is computed from the Fourier
transform, and the observed transform can be considered as the convolution of the true spectrum and the transform of the observing window,
filling the gaps is essentially a deconvolution process. Since there
are an unlimited number of ways to fill the gaps with data, most of them
non-physical, we must have some guidance from our a priori knowledge of
the sun or star in question and we must have some method for incorporating this knowledge in an analytical procedure. Thus the whole problem
of deconvolution or gap-filling is one of explicitly recognizing the
assumptions about the star and finding objective ways to incorporate
those assumptions in the analysis.

This problem is not unique to asteroseismology. In fact, there is
a significant body of literature dealing with the deconvolution problem.
The specific problems of asteroseismology are different from many of the
previous applications however, so care must be exercised in applying
previously developed methods. Some of the particular problems of solar

117

D. O. Gough (ed.), Seismology of the Sun and the Distant Stars, 117–120.
© 1986 by D. Reidel Publishing Company.

and stellar observations are low filling factors (i.e. large gaps relative to intervals of continuous data), very complex spectra with
thousands of modes allowed over several decades in frequency, line
separation on the same order as the main contributions from typical
observing windows, and low signal to noise in frequency bands that contain important oscillation modes. A number of gap-filling methods have
been suggested for the asteroseismology problem. These include maximum
entropy methods of "predicting" the data in the gaps (Fahlman and
Ulrych, 1982), variations of the CLEAN algorithm of iterative peak subtraction (Scherrer, 1984), rearranging the data to fill the gaps at the
expense of resolution (Kuhn, 1982), and iterative multiplicative deconvolution (Connes and Connes, 1984).

Brown and Christensen-Dalsgaard (1985) have completed a thorough
study of the applicability of the maximum entropy method to the problem.
They found that complex spectra can be correctly recovered provided the
filling factor is high (greater than 0.8) and the signal-to-noise ratio
is high (greater than 100). Fortunately, these conditions are expected
to be met for p-mode observations from an earth-bound network They found
that as the filling factor dropped below 0.5 or the s/n dropped below
about 30 the error rate increased rapidly. Kuhn found similar results
when applying the data-swapping method to data with a low filling factor
(Kuhn et al., 1985).

Delache and Scherrer (1983) successfully applied the CLEAN-like
iterative peak subtraction method to low-filling factor observations in
the long-period spectral region but that method was not successful when
applied to p-mode spectra (Henning and Scherrer, 1986). Experiments
with linear iterative deconvolution methods have not been as successful
as the CLEAN technique. There is a tendency for noise peaks to grow
rather than to be simply extracted unchanged as in the CLEAN method.
This is a characteristic of linear iterative methods (Jansson, 1984).

In order to push these procedures to the regime of low-filling
factor, the methodology of inserting a priori knowledge into the
analysis must be explicitly examined. There are different assumptions
implicitly made by the various methods. The maximum entropy method
assumes that the signal is locally stationary. This is a fairly weak
assumption and leads to failure when the gaps are large. The stronger
assumption of completely stationary signals made in the CLEAN method can
do a better job of recovering the true spectrum with low filling factors, but only in cases where the assumption is valid. Thus, the long
period success and the p-mode failure of the method.

Particularly strong claims have been made for the advantages of
the non-linear spectral deconvolution method (Connes and Connes, 1984).
This has lead to a careful examination of the applicability of that
method to the problem. The method can be summarized as the iterative
task of finding a spectrum that when convolved with the window function
yields the observed spectrum. The iteration is done under some constraints that provide the mechanism for including prior knowledge. This
method has the advantage of explicitly specifying the assumptions. It
has been applied for some time to the deconvolution of instrumental profiles from spectra and has been described by (Blass and Halsey, 1981)
for that application. The general method has been examined by a number

of authors in other applications and it has been shown that convergence can be guaranteed in many cases (Sanz and Huang, 1983, Schafer et al., 1981). A general discussions of both subtractive and multiplicative deconvolution techniques have been presented by Bates and colleagues (Bates et al., 1982a, 1982b, 1982c) and by Jansson (1984).

Connes and Connes suggested applying the multiplicative deconvolution method independently to the real and imaginary parts of the Fourier transform of oscillation observations. It turns out that constraints suggested do not quite match the problem. In particular, the method assumes that the true signal is band-limited, and that power outside the known band comes from the gaps. Unfortunately, it is not possible to arrange the observations of the sun (or many stars) to match this criterion. The p-mode frequency band, for example, extends from possibly 300 micro-Hz through 4500 micro-Hz. The sidebands from nightly gaps are spaced at about 11.6 micro-Hz. Thus most of the sidebands are within the desired frequency band. The method probably works but is expected to be rather inefficient. Thus, this method at present does not solve our problem (J. Connes, private communication, 1985).

What has yet to be done, is to find a methodology for tuning an analysis to specific prior knowledge or assumptions. For example, if we were asking what is the rotational splitting for solar p-modes, then since we know the central frequencies for each multiplet (to a few micro-Hz) we should be able to tell our analysis procedure just where power is from gaps and where it is probably of solar origin. If we believe that the lifetime of low-frequency p-modes is several months but the lifetime of high-frequency modes may be only a few days, we should be able to find peak centroids using varying resolution across the p-mode band. Perhaps the iterative methods can be adapted to such more problem-specific constraints. In any case, we must always be aware of the explicit and implicit assumptions in our procedures. In summary, two things are clear: First, if we do not attempt to account for the sidebands introduced by the data gaps we are making the implicit assumption that the sun does not oscillate at night and second, we do not yet have a reliable, objective method for filling the gaps in the case of low filling factor.

Acknowledgements: This work was supported in part by the Office of Naval Research under Contract N00014-76-C-0207, by the National Aeronautics and Space Administration under Grant NGR5-020-559, and by the Atmospheric Sciences Section of the National Science Foundation under Grant ATM77-20580.

References

J.H.T. Bates, W.R. Fright, R.P. Millane, A.D. Seagar, G.T.H. Bates, W.A. Norton, A.E. McKinnon, and R.H.T. Bates, Subtractive Image restoration. III. Some practical Applications, Optik 62(1982), 333-346.

J.H.T. Bates, A.E. McKinnon, and R.H.T. Bates, Subtractive Image Restoration. II): Comparison with Multiplicative Deconvolution, Optik 62(1982), 1-14.

J.H.T. Bates, A.E. McKinnon, and R.H.T. Bates, Subtractive Image Restoration. I: Basic Theory, Optik 61(1982), 349-364.

William E. Blass and George W. Halsey, Deconvolution of Absorption Spectra, Academic Press, Inc., 1981.

Timothy M. Brown and Jorgen Christensen-Dalsgaard, A Technique for Filling Gaps in Time Series with Complicated Power Spectra, preprint, High Altitude Observatory, NCAR, 1985.

Janine Connes and Pierre Connes, A Numerical Solution of the Data-Gaps Problem, Space Research Prospects in Stellar Activity and Variability, Meudon Observatory, February 1984.

Philippe Delache and Philip H. Scherrer, Detection of Solar Gravity Mode Oscillations, Nature 306(December 15, 1983), 651-653.

G.G. Fahlman and T.J. Ulrych, A New Method for Estimating the Power Spectrum of Gapped Data, Mon. Not. R. astr. Soc. 199(1982), 53-65.

Harald M. Henning and Philip H. Scherrer, Observations of Low-Degree P-Mode Oscillations in 1984, These Proceedings, 1986.

Peter A. Jansson, ed., Deconvolution with Applications is Spectroscopy, Academic Press, Inc., 1984.

J.R. Kuhn, Recovering Spactral Information from Unevenly Sampled Data: Two Machine-Efficient Solutions, The Astronomical Journal 87,1 (January 1982), 196-202.

J.R. Kuhn, K.G. Libbrecht, and R.H. Dicke, Solar Ellipticity Fluctuations Yield No Evidence of g-Modes, submitted to Nature, 1985.

Jorge L.C. Sanz and Thomas S. Huang, Iterative Time-Limited Signal Restoration, IEEE Transactions on Acoustics, Speech, and Signal Processing ASSP-31,3 (June 1983), 643-650.

Ronald W. Schafer, Russell M. Mersereau, and Mark A. Richards, Constrained Iterative Restoration Algorithms, Proceedings of the IEEE 69,4 (April 1981), 432-450.

Philip H. Scherrer, Detection of Solar Gravity Mode Oscillations, in Solar Seismology From Space, A Conference at Snowmass, Colorado, R. K. Ulrich, J. Harvey, E. J. Rhodes, Jr., and J. Toomre (editor), NASA, December 15, 1984, 173-182.

SOUND SPEED IN THE INTERIOR OF SOLAR MODELS

Ian W. Roxburgh,
Theoretical Astronomy Unit,
School of Mathematical Sciences,
Queen Mary College, London E1 4NS.

The analysis of Christensen-Dalsgaard et al (1985) yields the run of the sound speed c(r) as a function of radial distance r inside the sun. This note presents theoretical calculations of this variable for four solar models, all having the present solar luminosity and solar radius: a standard inhomogeneous evolved model with an age of $4.6 \ 10^9$ years, and an outer convective zone whose base is at 0.73 $R_\odot$ (Fig.1) a similar model but including convective overshooting beneath the base of the unstable convective zone (c.f. Roxburgh 1978, 1985) which gives the base of the almost adiabatic zone at 0.7 $R_\odot$ (Fig.2), a fully homogeneous model (Fig.3), and a homogeneous model with overshooting to give the base of the outer convective zone at 0.7 $R_\odot$ (Fig.4).

The point I wish to make is that the difference between c(r) for these models is quite small except for r < 0.2 $R_\odot$ - for example the inhomogeneous model with overshooting differs from the standard model by less than 5% for 0.2 < $r/R_\odot$ < 1.

REFERENCES

Christensen-Dalsgaard J., Duvall T.L., Gough D.O., Harvey J.W. and
    Rhodes E.J., 1985, Nature 315, 378.

Roxburgh I.W. 1978, Astron. and Astrophys., 65, 281.

Roxburgh I.W. 1985, Solar Physics 100, p. 221.

D. O. Gough (ed.), Seismology of the Sun and the Distant Stars, 121–123.
© *1986 by D. Reidel Publishing Company.*

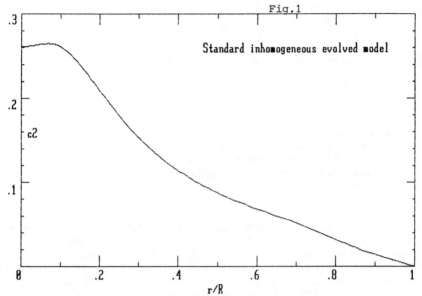

Variation of the square of the sound speed with radial distance
in Solar models (units of 10^16 cm^2/sec^2)

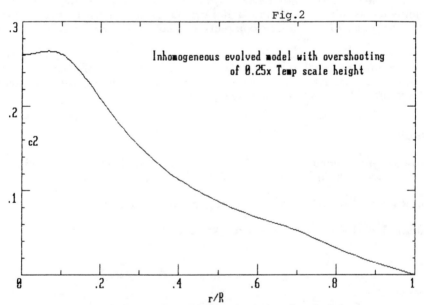

Variation of the square of the sound speed with radial distance
in Solar models (units of 10^16 cm^2/sec^2)

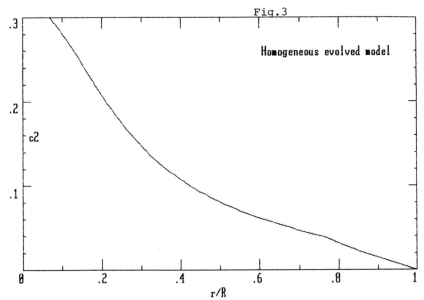

Variation of the square of the sound speed with radial distance
in Solar models (units of 10^16 cm^2/sec^2)

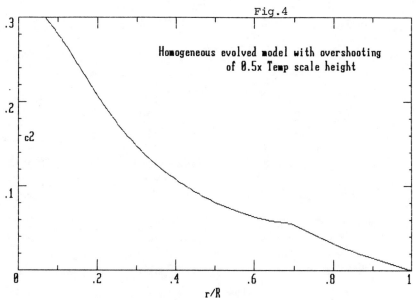

Variation of the square of the sound speed with radial distance
in Solar models (units of 10^16 cm^2/sec^2)

ASYMPTOTIC SOUND-SPEED INVERSIONS

D.O. Gough
Institute of Astronomy, and Department of Applied
Mathematics and Theoretical Physics,
University of Cambridge

ABSTRACT.   An asymptotic analysis of stellar p modes in the short-wavelength limit is summarized. The formula can be used as a means of estimating the sound speed in the solar interior from a knowledge of identified eigenfrequencies. The accuracy of the procedure is discussed, and it is shown how the results can be used to measure properties such as the location of the base of the convection zone.

1. INTRODUCTION

Although asymptotic formulae are imprecise, for the purposes of inverting solar oscillation data they have two obvious advantages over totally numerical integrations. The first is that they are perhaps simpler to implement, and certainly require less computer time to execute. The second, and possibly the more important, is that because the formulae are quite simple they appear to be relatively easy to interpret, and readily give one an appreciation both of what kind of information is contained in the data and of how it might be extracted. It is important, however, to have some idea of how accurate the inversions are. Only then can one judge how seriously to regard the details of the results.

   In this paper I report some results of an investigation to assess the reliability of inverting an asymptotic formula for the sound speed in terms of five-minute p-mode eigenfrequencies. It appears that useful, albeit approximate answers to some important questions might be emerging, even though a direct formal numerical inversion will undoubtedly be required to extract all the information the data contain.

2. ASYMPTOTIC ANALYSIS OF ADIABATIC MODES

In the short-wavelength limit the equation describing adiabatic

125

D. O. Gough (ed.), Seismology of the Sun and the Distant Stars, 125–140.

acoustic-gravity waves of frequency ω can be written

$$\nabla^2\psi - \frac{N^2}{\omega^2}\nabla_h^2\psi + \frac{\omega^2-\omega_c^2}{c^2}\psi = 0, \tag{1}$$

where $\psi = \rho^{\frac{1}{2}}c^2\text{div}\xi$, ρ and c being the density and adiabatic sound speed of the equilibrium state and ξ the displacement eigenfunction, and ∇_h^2 is the horizontal Laplacian. Also

$$N^2 = g(\frac{1}{H} - \frac{g}{c^2}) \tag{2}$$

is the square of the buoyancy frequency, where g is the acceleration due to gravity and H is the density scale height of the equilibrium state, and

$$\omega_c^2 = \frac{c^2}{4H^2}(1 - 2H'), \tag{3}$$

where a prime denotes differentiation with respect to the radial coordinate r. The condition for equation (1) to be satisfied is that the characteristic scale κ^{-1} of radial variation of ψ is both much less than r and small enough for perturbations in the gravitational potential to be negligible. If, furthermore, κ^{-1} is much less than the scale of variation of the equilibrium stratification, which is likely to be the case throughout most of the region within which the modes are concentrated if the order n or the degree ℓ is large, then asymptotic theory may be applied. One may either separate ψ into the product of a spherical harmonic of degree ℓ and a function of r alone, the equation for which is solved by JWKB theory, or apply ray theory directly to equation (1). In both cases one obtains

$$\int_{r_1}^{r_2} \kappa dr \simeq (n + \varepsilon)\pi, \tag{4}$$

which determines ω in terms of n and ℓ. The quantity κ, which is the vertical component of the local wave number vector, is defined by the local disperson relation

$$\kappa^2 = \frac{\omega^2-\omega_c^2}{c^2} + \frac{L^2}{r^2}(\frac{N^2}{\omega^2} - 1). \tag{5}$$

The quantity L/r is the horizontal component of the wave number, where L is given by

$$L^2 = \ell(\ell+1) \text{ or } (\ell+\tfrac{1}{2})^2, \tag{6}$$

depending on whether a spherical harmonic decomposition or ray theory is used. The limits of integration r_1 and r_2 are the turning points, where κ^2 vanishes and between which $\kappa^2 > 0$. These are the radii of caustic surfaces of the rays. For five-minute p modes, r_1 occurs where $Lc/\omega r \simeq 0$ and thus depends on ℓ and n in the combination $w = \omega/L$ alone, and r_2 is roughly where $\omega_c = \omega$, which is situated at the top of the convection zone and depends on ω alone.

Finally, the quantity ε is a phase factor which one would expect to depend principally on conditions in the outer layers of the star. Formally $\varepsilon = -\frac{1}{2}$ {this is $-1+2\times1/4$, the term -1 arising because counting starts at n = 1 and not n = 0 (cf. Deubner and Gough, 1984), and the contributions 1/4 coming from each distinct caustic through which a 'complete' ray passes (cf. Keller, 1958; Keller and Rubinow, 1960)} , but near the top of the convection zone the equilibrium state varies on a scale comparable with or less than κ^{-1}, and the asymptotic approximation breaks down. This breakdown is in a layer near $r = r_2$ which is thin compared to the thickness r_2-r_1 of the region of propagation. It is for this reason that the errors in the approximation can be accommodated as a phase change near the upper turning point, and thus absorbed into ε. It is also the case that for most of the modes I shall be considering the thickness of this layer is also much less than the horizontal wavelength of the oscillations; this leads to the conclusion that the dynamics of acoustic waves in this layer is insensitive to the value of L, and that for a given equilibrium model ε is a function of ω alone. I must point out, however, that if $L/n \ll 1$, the waves penetrate close to the centre of the star where the geometrical terms neglected in equation (1) become noticeable. In that case there is an additional contribution to the phase near $r = r_1$; this can also be absorbed into ε, but it is not independent of L.

3. AN ASYMPTOTIC FORMULA FOR THE SOUND SPEED

The main thrust of the work on the diagnostic aspects of the asymptotic formula (equation 4) for solar p-mode eigenfrequencies has been to approximate it to the point where it is simple enough to be inverted analytically (Christensen-Dalsgaard, 1984; Gough, 1984a,b, 1985; Christensen-Dalsgaard et al., 1985). This has been achieved by neglecting N^2, which for five-minute oscillations is everywhere much less than ω^2, and adding the contribution to the integral from ω_c^2 to ε to give a new phase factor α. One also replaces r_2 by the solar radius R, to obtain

$$\frac{\pi(n+\alpha)}{\omega} \simeq F \equiv \int_{r_1}^{R} a^{-1}(1-a^2/w^2)^{\frac{1}{2}} \, d\ln r, \tag{7}$$

where $w = \omega/L$ and $a = c/r$.

The justification for absorbing the contribution from ω_c^2 into the phase factor first came from noticing that ω_c^2 is comparable with ω^2 only near the upper turning point $r = r_2$, and that therefore its influence can be considered as being substantial only in a thin layer which can be regarded as a reflecting boundary. Thus one expects α to be independent of L. This idea was supported by noticing that in the case of an adiabatically stratified plane-parallel polytropic layer of index μ under constant gravity g, in which c^2 increases linearly with depth, equation (7) can be integrated to yield

$$\frac{n + \alpha}{\omega} = \frac{\mu Rw}{2g} \quad , \tag{8}$$

which is identical to the exact result if $\alpha = \mu/2$.

In a complete polytropic atmosphere ω_c^2 is inversely proportional to depth, and so actually is not really very strongly concentrated near the surface. However, its influence in equation (4) is such as merely to add a constant to ε, even when the polytrope is not adiabatically stratified, leading to

$$(\mu+1)\omega^4 - 2\gamma gk(n+\alpha)\omega^2 - (\mu+1-\gamma\mu)g^2k^2 = 0, \tag{9}$$

with $\alpha = \varepsilon + \sqrt{\mu(\mu+2)} /2$ (cf. Christensen-Dalsgaard, 1984) and $k = L/r$.

As with equation (8), this equation would be exact if α were replaced by $\mu/2$ (e.g. Gough, 1977; Christensen-Dalsgaard, 1980). Thus we have an indication of the accuracy of equation (4), at least in the polytropic case: observations suggest that for the sun the effective value of μ is about 3 (Duvall, 1982; Christensen-Dalsgaard et al., 1985), and for that value the relative differences between $\mu/2$ and $(-1 + \sqrt{\mu(\mu+2)})/2$, using the asymptotic value of $-\frac{1}{2}$ for ε, is about 4 per cent. This implies that for n = 10, say, a value typical of five-minute modes that penetrate just to the base of the convection zone, the error in $n+\alpha$ is about $\frac{1}{2}$ per cent.

According to theoretical models the stratification of the sun is not as smooth as in a polytrope, so one might expect to encounter errors rather greater than $\frac{1}{2}$ per cent. Nevertheless, since most of the rapid variation occurs near the surface, one might also hope to be able to absorb part of those errors into a potentially frequency-dependent α. At present, the degree to which that can be accomplished can be assessed only by performing numerical experiments.

It should be noticed that since r_1 is the lower turning point, which is determined in the current approximation by the condition $a(r_1) = w$, the right-hand side of equation (7) is a function F of w alone. This property, together with the special functional form of the left-hand side, permits the simultaneous determination of F(w) and $\alpha(\omega)$ from the data. It was first carried out by Duvall (1982) using the approximation that α is constant. Finally one can differentiate equation (7) with respect to w and make the substitution $\xi = a^{-2}$, $\eta = w^{-2}$ to cast it into Abel's integral equation, which can be inverted to give ξ as a function of η. After restoring the original variables one obtains

$$r = R \exp\{ -\frac{2}{\pi} \int_{a_s}^{a} (w^{-2} - a^{-2})^{-\frac{1}{2}} \frac{dF}{dw} dw \}, \tag{10}$$

where $a_s = a(R)$. This can be regarded as giving c as a function of r.

4.   DETERMINATION OF F(w)

The determination of F is accomplished by finding that function $\alpha(\omega)$
which renders $\pi(n+\alpha)/\omega$ as close as possible to a function of w alone.
Several measures of closeness have been tried in the past (Duvall, 1982;
Christensen-Dalsgaard et al., 1985), all yielding quite similar results.
In the procedure described here I have ordered the modes sequentially
in order of increasing w, labelling each with an integer i. The deviation of the values F_i from a single function of w I have chosen is
defined by

$$\Delta = \sum_{i=2}^{I-1} (F_{i-1} - 2F_i + F_{i+1})^2 ,   \tag{11}$$

where I is the number of modes and $F_i \equiv F(w_i)$.
    The functions F and α were determined by iteration. Equation (7)
was written

$$F = \Phi + \phi ,   \tag{12}$$

where

$$\Phi(n,\omega;\alpha_o) \equiv \pi(n+\alpha_o)/\omega   \tag{13}$$

for some constant α_o, and $\phi(\omega;\alpha_o)$ is such that

$$\alpha(\omega) = \alpha_o + \pi^{-1}\omega\phi .   \tag{14}$$

Then, using ϕ from a previous iteration, α_o was determined by minimizing Δ using $\Phi+\phi$ in place of F . For the first iteration ϕ was taken
to be zero. The function F was then determined from $\Phi+\phi$ by smoothing
the latter with a cosine-weighted running mean. Next, since ϕ is
independent of w at fixed ω, it could be obtained at frequency $\bar{\omega}j$, say,
from an average of equation (12), namely

$$\phi(\bar{\omega}_j) = \Sigma(1 - F_i^{-1} \Phi_i)/\Sigma F_i^{-1},   \tag{15}$$

$$\bar{\omega}_j = J^{-1}\Sigma\omega_i ,   \tag{16}$$

where the sums are over the J modes with frequencies ω_i within some
prescribed interval. The process was repeated until the iterations con-
verged, after which $\alpha(\omega)$ was computed from equation (14). Finally, F
was transferred to a computational grid in w, using either linear
regression on all the modes having w within half a mesh interval from
any mesh point to determine the value of F at that point, if suitable
modes existed, or linear interpolation between the closest modes
otherwise.

5. EVALUATION OF THE SOUND SPEED: FIRST APPROXIMATION

When evaluating the integral in equation (10), care must be taken to
account for the singularity in the integrand.   In evaluating integrals
of the type

$$I_i(G;a) \equiv \int_{w_{i-1}}^{w_i} (w^{-2} - a^{-2})^{-\frac{1}{2}} G(w) dw , \qquad (17)$$

in the mesh interval (w_{i-1}, w_i), the function G was approximated by the
linear function

$$G_{i-1} + h_i^{-1} (G_i - G_{i-1})(w - w_{i-1}),$$

where $h_i = w_i - w_{i-1}$, and the kernel $(w^{-2} - a^{-2})^{-\frac{1}{2}}$ was treated exactly.

This yields

$$I_i \simeq (A_i - B_i)G_{i-1} + B_i G_i, \qquad (18)$$

where

$$A_i = a\left[\sqrt{(a^2 - w_{i-1}^2)} - \sqrt{(a^2 - w_i^2)}\right], \qquad (19)$$

$$B_i = a\{a^2[(\cos^{-1}(w_{i-1}/a) - \cos^{-1}(w_i/a)] - w_{i-1}\sqrt{(a^2-w_{i-1}^2)} +$$

$$+ (2w_{i-1}-w_i)\sqrt{(a^2-w_i^2)}\} /2h_i . \qquad (20)$$

The function G = dF/dw was calculated from F by second-order accuracy
differencing:

$$G_i \simeq [(F_{i+1}-F_i)h_i/h_{i+1} + (F_i-F_{i-1})h_{i+1}/h_i]/(h_{i+1}+h_i) . \qquad (21)$$

In practice the integral in equation (10) cannot be evaluated
completely because the lowest value w_0 of w available from data is
substantially greater than a_s.   Consequently the structure of the superadiabatic boundary layer at the top of the convection zone cannot be
probed by this simple method.   To overcome this deficiency the function
F(w), which is approximately a power law at the lower values of w for
which there are data (e.g. Gough 1984a; Christensen-Dalsgaard et al.,
1985), was extrapolated to a_s with the power law $F_0 w^\nu$ that best fits
the data at the lower end of w, the constants F_0 and ν being determined
by linear regression in logarithmic variables.   This implies a functional form for the sound speed c that approaches the power law

$c_0 z^{1/(1+\nu)}$ at great depths z beneath the photosphere, provided that z remains small compared with the radius R of the sun. The power of z so obtained from solar data is about 0.5 (Christensen–Dalsgaard et al., 1985), which agrees with the value for a plane–parallel polytrope. Of course that does not describe the upper layers of the sun adequately, and therefore introduces an error into the value of $c(r)$ inferred. What form does this error take?

The approximation actually used to compute the sound speed is

$$r = R \exp\{-\lambda - \frac{2}{\pi} \int_{w_0}^{a} (w^{-2} - a^{-2})^{-\frac{1}{2}} \frac{dF}{dw} dw\} , \tag{22}$$

where

$$\lambda \simeq \lambda_1(w_0,a) \equiv \nu F_0 \int_{a_s}^{w_0} (w^{-2} - a^{-2})^{-\frac{1}{2}} w^{\nu-1} dw \tag{23}$$

$$\sim \frac{\nu F_0 w_0^{\nu+1}}{\nu+1} \left[1 - \left(\frac{a_s}{w_0}\right)^{\nu+1}\right] \quad \text{as} \ \left(\frac{a}{w_0}\right)^2 \to \infty . \tag{24}$$

Sufficiently deep in the star where $(a/w_0)^2 \gg 1$, the factor λ can be regarded as providing a constant scaling factor for the radius variable r. An inversion of equation (7) in the region $w > w_0$, where there are data, also leads to equation (22), but with

$$\lambda = \lambda_2(w_0,a) \equiv \ln(R/R_0) - \frac{2}{\pi} \int_{a'=w_0}^{a'=a_s} \sin^{-1}\left[\left(\frac{w_0^{-2}-a^{-2}}{a'^{-2}-a^{-2}}\right)^{\frac{1}{2}}\right] d\ln r' \tag{25}$$

$$\sim \ln(R/R_0) - \frac{2}{\pi} \int_{a'=w_0}^{a'=a_s} \sin^{-1}(\frac{a'}{w_0}) d\ln r' \quad \text{as} \ \left(\frac{a}{w_0}\right)^2 \to \infty , \tag{26}$$

where $a'=a(r')$ and R_0 is the radius at which $a = w_0$. Thus the error introduced into the asymptotic formula by the power–law extrapolation of F to the surface is, far below the lower turning point of the shallowest mode, equivalent to multiplying the radius scale by the constant factor $\Lambda = \exp\left[\lambda_2(w_0,\infty) - \lambda_1(w_0,\infty)\right]$. For Christensen–Dalsgaard (1982) solar Model 1, for example $\Lambda = 0.9989$ when $w_0 = 2.7 \times 10^{-5}$ s^{-1} (the value corresponding to the data analysed by Christensen–Dalsgaard et al., 1985).

It is likely that the uncertainty in the outer layers of the sun leads to an error in the sound speed inferred from real solar data of the same kind, and with a similar magnitude. Moreover, there is additional uncertainty arising from both the errors in the asymptotic formula (4) and the neglected terms in the approximation (7). This is more difficult to assess. Of course the accuracy of equation (7) can be determined by comparing the eigenfrequencies it predicts with numerical computations. But this does not immediately provide an estimate of the reliability of the asymptotic procedure that is used to

determine c(r). That is probably best judged by inverting artificial
data computed from a theoretical model of the sun, and comparing the
sound speed inferred from those data with that of the model. Some
results of such comparisons are summarized in the next section.

6. EVALUATION OF THE SOUND SPEED: SECOND APPROXIMATION

In Figure 1 the sound speed inferred from 2820 theoretical eigenfrequencies of Christensen-Dalsgaard's (1982)solar Model 1 is compared
with the actual sound speed. (The inferred curve was actually computed
from equation (10) with F having been determined for constant α by
minimizing a measure of the deviation from a single curve that is
somewhat different from Δ (Christensen-Dalsgaard et al., 1985).) The
two curves appear to be very similar for r ≳ 0.4R, and suggest that
the asymptotic method provides at least a rough idea of the sound speed
in the outer 60 per cent or so by radius of the sun. Agreement can be
improved somewhat by permitting α to vary with ω in the manner described above. This leads to a reduction in Δ by more than a factor 10.
The function α so obtained is illustrated in Figure 2.

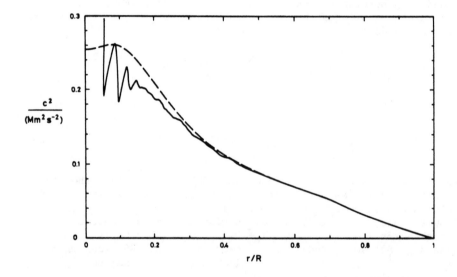

Figure 1. The continuous line is the square of the sound speed of
Model 1 of Christensen-Dalsgaard (1982) inferred from 2820 adiabatic
oscillation eigenfrequencies, using equation (10) with F having been
estimated with α constant. The dashed line is the square of the actual
sound speed (from Christensen-Dalsgaard et al., 1985).

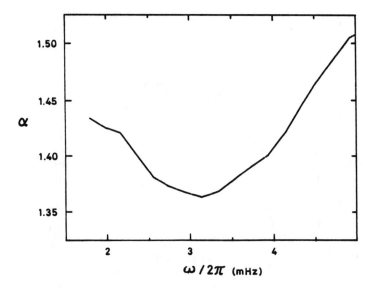

Figure 2. The function $\alpha(\omega)$ computed from the eigenfrequencies of
Christensen-Dalsgaard's (1982) Model 1.

At the larger radii the errors in the asymptotic method are too
small to be visible in Figure 1. Therefore it is necessary to expand
the scale, which is conveniently accomplished by plotting the relative
difference between the inferred and the actual sound speeds. This is
illustrated in Figure 3, where plotted also is the difference between
the sound speed inferred from the real solar data and the sound speed
of Model 1. In addition to the large errors in the central regions
of the sun, large relative errors that are not evident in Figure 1 are
revealed near the surface. This is not surprising, since it is in the
surface layers (and very close to the centre) that the asymptotic
analysis is expected to fail.

It was noticed by Christensen-Dalsgaard et al. (1985) that both
the small-scale fluctuations and many of the systematic trends exhibi-
ted in Figure 3 are similar. In the case at least of the inversion of
the theoretical frequencies, we know that these are due to relatively
sudden variations in the smoothed representation of F, the details of
which depend on which modes are included in the analysis, the degree
of smoothing adopted in the computation of F, and the inaccuracy of
the asymptotic formula. I must stress that the same modes were used
in the inversion of the real and the artificial data, and that the
smoothing and the mesh distributions were identical for the two
inversions. It was hoped that by so doing the systematic errors would
be very similar. Thus by subtracting the sound speed inferred from
the artificial data from that inferred from the solar data the systematic errors should be largely eliminated, and the result should be a
good estimate of the difference between the solar sound speed and the

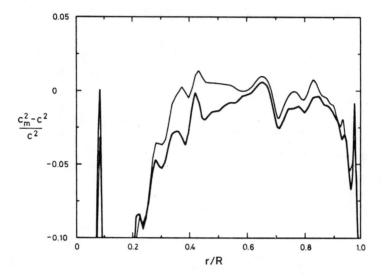

Figure 3. The thick line is the relative difference $(c_m^2 - c^2)/c^2$ between the square of the sound speed c_m^2 inferred from oscillation eigenfrequencies and the square of the actual sound speed c^2 of Model 1 of Christensen-Dalsgaard (1982). The thin line is the relative deviation from c^2 of the square of the sound speed inferred from solar data (from Christensen-Dalsgaard et al., 1985).

actual sound speed in the theoretical model. This constitutes the second approximation.

The procedure was tested by Christensen-Dalsgaard et al. (1985), using a different theoretical solar model, Model A of Christensen-Dalsgaard, Gough and Morgan (1979), in place of the sun. Thus the difference between the inferred sound speeds could be compared with the difference between the actual sound speeds. The result is shown in figure 4. Only 363 corresponding five-minute modes of oscillation were used for the comparison, so, disregarding observational error, one would expect the errors in the result to be greater than for the 2820 solar modes that were analysed. The systematic errors are likely to have cancelled less completely too in this case, because the difference between the two theoretical models is rather larger than the difference between Model 1 and the sun. The errors in this case are also likely to be greater than necessary, because the analysis was carried out with α constant. By permitting α to vary with ω, the errors can be reduced by nearly a factor 2 throughout much of the domain.

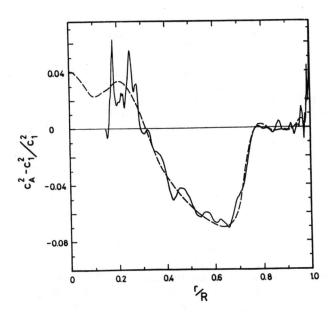

Figure 4. The dashed line is the relative difference $(c_A^2 - c_1^2)/c_1^2$ between the squares of the sound speeds in Model A of Christensen-Dalsgaard, Gough and Morgan (1979) and Model 1 of Christensen-Dalsgaard (1982). The continuous line is the relative difference inferred from corresponding sets of 363 frequencies of five-minute p modes (from Christensen-Dalsgaard et al., 1985).

7. LOCATING THE BASE OF THE CONVECTION ZONE

Throughout most of the solar convection zone the stratification is extremely close to being adiabatic, whereas in the radiative region beneath one expects the temperature gradient to be equal to the so-called radiative gradient. In most theoretical models of the sun a local relationship between heat flux and temperature gradient is used, which leads to a discontinuity in the second derivative of the sound speed at the base of the convection zone. This is discernible in Figure 1, where a relatively rapid variation in the gradient of the sound speed is evident particularly near $r = 0.7R$; the base of the convection zone of Model 1 is at $r = 0.717R$. The sound speed inferred from the real solar data looks similar in this region, suggesting that the base of the solar convection zone is also located near $r = 0.7R$.

A convenient way to exhibit the base of the convection zone is with a plot of

$$\tilde{W} = \frac{r^2}{GM} \frac{dc^2}{dr} \, , \tag{27}$$

where M is the mass of the sun and G is the gravitational constant.
A plot of d^2c^2/dr^2 inferred by the technique in its present state is
much too noisy to be useful, and the noise in only the first derivative
is great enough to conceal the appropriate abrupt change in slope.
However, multiplying the gradient by r^2 renders it approximately constant in the lower layers of the convection zone, which makes the
departure from the convective value in the radiative zone easier to
detect. The reason for this property is that because the convection
zone contains only about 3 per cent of the sun's mass, $\tilde{W}$ is close to
W, which is defined as in equation (27) save that M is replaced by the
mass enclosed by the sphere at radius r. In an adiabatically stratified region

$$W = \Theta \equiv (1 - \gamma_{c^2})^{-1}(1 - \gamma_\rho - \gamma) , \tag{28}$$

where γ is the adiabatic exponent and γ_{c^2} and γ_ρ are the partial
logarithmic derivatives of γ with respect to c^2 and ρ at constant ρ
and c^2 respectively (Gough, 1984a). Near the base of the convection
zone ionization of hydrogen and helium is essentially complete,
$\gamma \simeq 5/3$ and therefore $\Theta \simeq -2/3$. On the convective side of the
boundary W shares that value, but in the radiative region beneath it
rises abruptly (Däppen and Gough, 1984), indicating quite distinctly
where the boundary is located.

The extent to which the function $\tilde{W}$ might be resolved by oscillation
data can be judged from Figure 5, where the value inferred from the
asymptotic inversion is compared with the actual value for Christensen-
Dalsgaard's Model 1. The 2820 modes used by Christensen-Dalsgaard
et al. (1985) were employed, and α was considered to vary with ω in
the manner illustrated in Figure 2. The horizontal line is drawn at
$-2/3$; the deviations* of the actual values of $\tilde{W}$ from $-2/3$ result from
ionization of carbon and oxygen. It is interesting that the largest
fluctuation in the inferred value of $\tilde{W}$ interior to $r \simeq 0.97R$ in the
convection zone roughly coincides with the largest excursion of the
actual $\tilde{W}$ from $-2/3$, though it has a substantially larger amplitude.
This provides hope that a direct measurement of the absolute abundances
of carbon and oxygen might eventually be possible with helioseismic
data. The hump at $r \simeq 0.98R$, which is also overestimated by the
inversion, results from the second ionization of helium. It is illustrated on an expanded scale by Däppen and Gough (these proceedings).

* in the lower half of the convection zone.

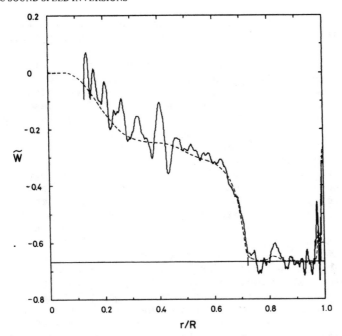

Figure 5. The dashed curve is the function $\tilde{W}$ defined by equation (27)
for Model 1 of Christensen-Dalsgaard (1982). The continuous line is
the value inferred from the sound speed obtained from the 2820 adiabatic eigenfrequencies used to construct Figure 1, except that here
the variation of $\alpha(\omega)$ illustrated in Figure 2 was used to compute F.
The thin horizontal line is drawn at $-2/3$, the value of Θ for a
perfect fully ionized gas; it is intersected by a tick at the position
of the base of the convection zone.

8.   ON THE ROLE OF THE OUTER LAYERS

The extent to which it is necessary to know the structure of the very
outer layers of the solar envelope in order to be able to infer conditions deeper down is of considerable importance, particularly for
deciding on observational strategy. The experiment illustrated by
Figure 4 shows that at least in one particular case one can obtain the
value of c^2 when $r \gtrsim 0.3R$ with an error of less than one per cent,
except very close to the surface. Because in the very outer layers the
eigenfunctions of the deeply penetrating modes of low degree are hardly
distinguishable from those of intermediate degree with the same frequency, it seems unlikely that those modes are any more susceptible to
the superficial uncertainties. Therefore it should be possible to
probe the solar core. The failure to have done so by the asymptotic
method appears to have resulted from a deficiency of the method itself,
at least in its present form. In the light of current evidence it seems
likely that an inversion procedure that is capable of penetrating more
deeply into the sun will not be severely impeded by our lack of knowledge of the surface layers, at least at the one-per-cent level.

I must emphasize that the case is not proven, for the argument, though superficially plausible, has so far been substantiated by but one example. Although the calculations discussed here show little indication that it is necessary to measure modes of very high degree in order to determine conditions in the solar core, it would be unwise for observers to conclude that measurements of high-degree modes are not useful. This has been stressed, for example, by E.J. Rhodes Jr (private communication) and by R.K. Ulrich at this meeting, who have argued that the appreciable contribution from the outer layers to the frequencies of all the five-minute modes must surely be substantially deleterious to our view of the deeper interior.

As a further step towards clarifying the issue I report the results of a blind experiment performed a few days before this meeting. Three sets of   oscillation frequencies were provided by Ulrich for the purposes of distinguishing between the sound speeds of the theoretical models from which they were computed. The data were inverted in the manner outlined in this paper. The theoretical models were evidently different, because their frequencies were different, but no significant difference in the sound speeds for $r \leq 0.95R$ was apparent from the asymptotic inversions. Therefore it was concluded that the sound speeds of the models differed by no more than one per cent except in the very outer layers, where differences were manifest by the different values of α that were found. The three models were compared with the sun, by comparing the inversions with a similar inversion using only the subset of the observed solar modes that corresponded to the theoretical frequencies that had been provided. The magnitude of the greatest differences between the sound speeds in the sun and in Ulrich's models were found to be comparable with the greatest difference between Christensen-Dalsgaard's Model 1 and the sun, though the region of substantial difference extended over a broader range of r. This is probably due in part to the rather shallow convection zones in Ulrich's models. The radii of the bases of the convection zones were estimated separately from $\tilde{W}$ computed from the inferred sound speeds, yielding a mean value of 0.747R with standard deviation 0.005R.

At the meeting Ulrich confirmed that the models did not differ by more than one per cent in the region in which the inversions detected no significant difference, but he emphasized that an accuracy of one per cent is insufficient to answer many of the scientific questions we need to ask. He subsequently reported (private communication) that the bases of the convection zones were at radii 0.749R, 0.749R and 0.750R.*

We hope to perform more extensive experiments of this kind in the future.

9.   FUTURE PROSPECTS

Undoubtedly there will be further improvements to the asymptotic results as additional refinements are made. For example, the influence of ω_c^2 and N^2 on the resonance condition (4), partially neglected in the discussion above, can be taken into account.

* Just before going to press, Ulrich revised the third radius to 0.749R.

It was pointed out above that in a plane parallel polytrope under constant gravity the inclusion of ω_c^2 is equivalent to adding a constant to α in equation (7). In the solar convection zone the depth dependence of ω_c well below the surface is similar to that of a polytrope. Therefore its contribution can be approximated by a frequency-dependent contribution to ω, the frequency variation arising from the rapid variation of ω_c in the outer layers. In the radiative interior, however, ω_c is substantially greater than the plane-parallel polytropic value ω_{cp}; but together with the buoyancy frequency it can be treated as a small perturbation to equation (7), equivalent to adding to the left-hand side the quantity $\omega^{-2}\Psi$, where

$$\Psi(w) \simeq \tfrac{1}{2}\int_{r_1}^{r_c} a^{-1}(\omega_c^2 - \omega_{cp}^2 - a^2 N^2/w^2)(1 - a^2/w^2)^{-\frac{1}{2}}\, d\ln r, \qquad (29)$$

where r_c is the radius of the base of the convection zone. The polytropic contribution ω_{cp}^2 which is rather smaller than ω_c^2 throughout most of the radiative interior, has been subtracted out because it was notionally accounted for as a contribution to α.

In principle the function Ψ can be deduced by a method similar to $\alpha(\omega)$. The latter is calculated first, using only those modes that are confined to the convection zone. With $\alpha(\omega)$ now determined, the known functional form of the correction $\omega^{-2}\Psi(w)$ can then be used to compute $\Psi(w)$.

It might even be possible to use the small quantity Ψ to determine ω_c and N. Noticing that $\omega_c \simeq N$ and that ω_{cp} is rather smaller than ω_c in the radiative interior (see Christensen-Dalsgaard, these proceedings), one could set $\omega_c^2 - \omega_{cp}^2 = N^2$ as a first approximation in equation (29). The resulting equation can be inverted, yielding

$$\omega_c^2 \simeq N^2 \simeq -\frac{4}{\pi}\frac{d}{d\ln r}\int_{a_c}^{a} (w^{-2} - a^{-2})^{-\frac{1}{2}}\frac{d\Psi}{dw}\, dw , \qquad (30)$$

where $a_c = a(r_c)$, which might be used as a starting value in an iterative procedure to solve the full equation (29).

Other corrections, such as the influence of the perturbations to the gravitational potential, must also be accommodated. Work on this in collaboration with W.A. Dziembowski is presently under way.

The utility of the asymptotic method depends on the property that the leading-order approximation (7) to the resonance condition (4) is readily invertible. The small corrections to that condition can be computed by a perturbation expansion; knowledge of the functional form of those corrections then permits them to be estimated directly from the frequencies of oscillation. Thus one is provided with a simple procedure whereby an approximation to the sound speed can rapidly be obtained. How much the results will be improved when the additional known corrections are incorporated is yet to be determined. If a substantial improvement is found, it will surely be tempting to compute higher-order terms in the asymptotic sequence. However, it is unlikely that all the information that is contained in the data could easily be extracted in this way. Therefore the asymptotic approach should be considered complementary to, and not an alternative to the techniques that employ the exact (numerically computed) modes of oscillation.

The asymptotic analysis upon which this diagnostic method depends arose as a direct consequence of an attempt to understand the dispersion law found by Duvall (1982). I am grateful to S. Frandsen for making me aware of Duvall's work prior to its publication, and I thank J. Christensen-Dalsgaard for his considerable contributions to all stages of the development of the method. I thank R.K. Ulrich for providing me with eigenfrequencies of three of his solar models, and J.Christensen-Dalsgaard for acting as a relay in Copenhagen because the UCLA and Cambridge computers failed to communicate directly with one another.

REFERENCES

Christensen-Dalsgaard, J. 1980. Mon.Not.R.astr.Soc. 190, 765-791
Christensen-Dalsgaard, J. 1982. Mon.Not.R.astr.Soc. 199, 735-761
Christensen-Dalsgaard, J. 1984. Theoretical problems in stellar
    stability and oscillations. (ed. M. Gabriel and A. Noels,
    Institut d'Astrophysique, Liège) 155-207
Christensen-Dalsgaard, J., Gough, D.O. and Morgan, J.G. 1979
    Astron. Astrophys. 73, 121-128; 79, 260
Christensen-Dalsgaard, J., Duvall Jr, T.L., Gough, D.O., Harvey, J.W.
    and Rhodes Jr, E.J. 1985. Nature 315, 378-382
Däppen, W. and Gough, D.O. 1984. Theoretical problems in stellar
    stability and oscillations. (ed. M. Gabriel and A. Noels,
    Institut d'Astrophysique, Liège) 264-268
Deubner, F-L. and Gough, D.O. 1984. Ann.Rev.Astr.Astrophys 22, 593-619.
Duvall Jr, T.L. 1982. Nature 300, 242-243
Gough, D.O. 1977. Proc.IAU Colloq.36 (ed. R.M. Bonnet and Ph.Delache,
    G. de Bussac, Clermont-Derrand) 3-36
Gough, D.O. 1984a. Mem.Soc.astr.Italiana 55, 13-35
Gough, D.O. 1984b. Phil.Trans.R.Soc.Lond. A313, 27-38
Gough, D.O. 1985. Solar physics and interplanetary travelling
    phenomena (ed. B. Chen and C. de Jager, Yunnan Obs., Kunming),
    in press
Keller, J.B. 1958. Ann.Phys. 4, 180-188
Keller, J.B. and Rubinow, S.I. 1960. Ann.Phys. 9, 24-75

Asymptotics and Quantum Chaos in Stellar Oscillations

J. Perdang
Institute of Astronomy, Madingley Rd, Cambridge CB3 0HA,
England
and
Institut d'Astrophysique, 5 Av de Cointe,
Cointe-Ougree B4200, Belgium*
* Permanent address

Abstract: Ray chaos in stars, in the asymptotic formalism of
geometric acoustics, is shown to arise under perturbations
breaking the spherical symmetry of the star.   If mild enough, ray
chaos   merely   induces   a   random   scatter   of   the   actual
eigenfrequencies of the perturbed star around the standard
asymptotic frequencies $\omega \propto (n + 1/2\ell + \varphi)$ of the radially
symmetric model. The statistical properties of this intrinsic
scatter are found to differ from the statistics of a scatter due
to observational noise.

I. Introduction

Depending on whether a classical mechanical system exhibits a
regular or a chaotic time-behaviour, the associated quantum
mechanical system shows basically distinct regimes.   As first
recognised by Percival and Pomphrey (Percival 1973,1974; Pomphrey
1974), in the quantal context, classical chaos generates a
modification in the distribution of the eigenvalues   and in the
structure of the eigenfunctions.   These modifications are now
termed quantum chaos in the literature.   In spite of efforts by
a variety of authors (Connes and Stormer 1975, Emch 1976,
Lindblad 1979, Kosloff and Rice 1981, Hose and Taylor 1982,
Shapiro and Goelman 1984), a precise definition of quantum chaos
is still lacking.   Following the general consensus in this field,
I shall understand by this concept the following features of the
asymptotic   eigenvalue   problem: (a)   the   eigenvalues   are
irregularly spaced; (b) they are strongly sensitive to parameter
changes in the Schrodinger equation; (c) the nodal pattern of the
wave function is erratic (Percival 1973, Berry 1977, Berry and

D. O. Gough (ed.), Seismology of the Sun and the Distant Stars, 141–171.

Tabor 1979, Stratt et al. 1979, McDonald and Kaufman 1979, Noid et al. 1980; cf also Zaslavsky 1981, Berry 1983).

The parallelism between the quantum mechanical wave equation and its limit of classical mechanics on the one hand, and the linear adiabatic stellar oscillation equation and its limit of ray theory (geometric acoustics) on the other hand, suggests that quantum chaos may show up in the problems of linear stellar oscillations. It is the purpose of this paper to classify the conditions under which this regime is expected to arise in stars. We shall show that in this regime the statistical properties of the quantum chaotic eigenfrequencies, rather than the individual eigenfrequencies, provide the most relevant astroseismological information.

The theoretical discussion of this paper is kept general enough to apply to all varieties of stellar oscillations. We shall have in mind, however, acoustic waves in most of the specific asymptotic interpretations of the wave equation.

II. The Stellar Wave Problem. Ray Equations

If we linearise the stellar hydrodynamics around a stationary state of the star (cf Ledoux and Walraven 1958) we generate a propagating wave equation for small disturbances. We shall write the latter schematically

$$\Omega(\underline{r}, -i\partial/\partial\underline{r})\ \Psi(\underline{r},t) = i\ \partial/\partial t\ \Psi(\underline{r},t)\ ,\quad \underline{r} \in V^3 \subset E^3\ , \tag{2.1}$$

where Ω is a partial differential operator in configuration space; $\underline{r}$ is the spatial position defined in a 3-dimensional bounded region V^3 of Euclidian space E^3 (the interior of the star). The wave function $\Psi(\underline{r},t)$ (which may be the density, pressure, ... disturbance) is subject to an initial and a boundary condition:

$$\Psi(\underline{r},t=0) = \Phi(\underline{r})\quad ,\quad \underline{r} \in V^3 \tag{2.2}$$

and

$$B(\underline{r}, -i\partial/\partial\underline{r})\ \Psi(\underline{r},t) = 0\ ,\quad \underline{r} \in \partial V^3 \tag{2.3}$$

where $\Phi(\underline{r})$ is a given function, and $B(\underline{r}, -i\partial/\partial\underline{r})$ is a boundary operator, defined over the surface of the star, ∂V^3. We transform this <u>propagating wave</u> problem into a <u>stationary wave</u> problem by separating time in the form

$$\Psi(\underline{r},t) = exp{-i\omega t}\ \ \psi(\underline{r})\ , \tag{2.4}$$

to obtain

$$\Omega(\underline{r}, -i\partial/\partial\underline{r})\ \psi(\underline{r}) = \omega\ \psi(\underline{r})\ , \tag{2.5}$$

with the boundary condition

$$B(\underline{r}, -i\partial/\partial\underline{r})\ \psi(\underline{r}) = 0\quad . \tag{2.6}$$

The set of values ω for which system (2.5-6) admits of a nontrivial regular solution defines the spectrum of frequency eigenvalues of the star.

Formally, we may regard Eq.(2.5) as a Schrodinger equation, in which $\Omega(\underline{r}, -i\partial/\partial\underline{r})$ plays the part of the Hamiltonian

operator, and ω is the counterpart of the energy.   By applying
the 'canonical quantisation condition' backwards, i.e. by the
substitution

$$-i \partial/\partial \underline{r} \rightarrow \underline{k} \qquad , \qquad (2.7)$$

where $\underline{k} = (k_x, k_y, k_z)$ is interpreted as the canonically
conjugate momentum to the position vector $\underline{r} = (x,y,z)$, we
associate a 'classical' Hamiltonian problem with the formal
'quantal' wave equation (2.5), of Hamiltonian

$$\Omega(\underline{r}, \underline{k}) = \omega \qquad . \qquad (2.8)$$

Physically, the momentum $\underline{k}$ is a wavevector and the Hamiltonian
is a frequency.   The standard Hamiltonian equations become

$$\dot{\underline{r}} = \partial/\partial \underline{k} \; \Omega(\underline{r},\underline{k}) \qquad , \qquad \dot{\underline{k}} = - \; \partial/\partial \underline{r} \; \Omega(\underline{r},\underline{k}) \qquad , \qquad (2.9)$$

which define the ray equations of the asymptotic approximation
of geometric acoustics, or, in Joyce's (1975) terminology, the
equations of motion of a classical phonon.   The relevant
accessible phase space (6-dimensional) can be written $\mathbf{V}^3 \times \mathbf{K}^3$
($\mathbf{K}^3$ being the 3-dimensional Euclidean region accessible to the
wavevector).   Eq. (2.8) is a first integral of the ray system
(2.9).
      Our formal derivation of the ray equations (2.8,9) differs
from the conventional procedures adopted by physicists (short
wavelength approximation; cf Eckart 1960), or by mathematicians
(propagation of weak discontinuities; cf Luneburg 1964).   It has
the advantage of exhibiting the formal equivalence between the
quantum mechanical and the stellar wave problem explicitly,
thereby implying (a) that both systems admit of same classes of
qualitative behaviour; and (b) that both systems can be
investigated by the same mathematical techniques.
      Since we distinguish two classes of behaviour in quantum
mechanics, namely quantum regularity and quantum chaos,
respectively associated with classically regular and chaotic
motions, we are likewise entitled by (a) to classify linear
stellar oscillations into 'quantum regular' and 'quantum chaotic'
ones.

III.   Differences with Genuine Quantum Mechanics
(1)   The stellar wave equation, when reduced to form (2.1) is in
reality made up of several (m) wave functions.   This in turn
implies that the associated Hamiltonian (2.8) is made up of m
branches; at the surface of the star, we switch from one branch
to another one (under conservation of frequency, Eq. 2.8).   We
illustrate this point by the standard 1-dimensional acoustic wave
equation

$$\left(\partial^2/\partial x^2 - 1/c^2 \; \partial^2/\partial t^2 \right) \Psi(x,t) = 0 \qquad . \qquad (3.1)$$

Assume for simplicity that the sound speed c is space
independent.   Then by factorising Eq. (3.1) we notice that this

wave equation is equivalent to the following two wave equations

$$(\partial/\partial x - 1/c \; \partial/\partial t) \; \Psi(x,t) = 0 \; , \tag{3.1'}$$

$$(\partial/\partial x + 1/c \; \partial/\partial t) \; \Psi(x,t) = 0 \; , \tag{3.1''}$$

in the sense that any solution of either (3.1') or (3.1'') is solution of (3.1), and conversely, any solution of (3.1) is also a solution of either (3.1') or (3.1''). Eqs. (3.1',1'') can then be written in form (2.1)

$$\Omega_{\pm} \, (x, -i\partial/\partial x) \, \Psi(x,t) = i \, \partial/\partial t \, \Psi(x,t) \; , \tag{3.2}$$

with $$\Omega_{\pm} \, (x, -i\partial/\partial x) = \pm c \, [-i \, \partial/\partial x] \tag{3.2'}$$

The operator Ω_+ thus generates a wave propagating in the positive x-direction (exp i k_+ x exp-iωt, $k_+ > 0$), while Ω_- produces a wave propagating in the opposite direction (exp i k_- x exp-iωt, $k_- < 0$). The existence of two branches of the space operator (Eq. 3.1') implies in turn that the classical Hamiltonian of the associated ray problem has two branches

$$\Omega_{\pm} \, (x, k) = \pm c k \quad . \tag{3.3}$$

Let further the wave propagation (3.1) be defined over a finite interval $V':-L < x < L$. At the boundary points $\partial V' = \{L, -L\}$, let $B_{\pm}(\pm L, -i\partial/\partial x)\Psi = 0$ be the respective boundary conditions. Consider then a wave propagating in the positive x-direction; its associated ray is obtained from the Hamiltonian Ω_+. At the boundary point x = +L, the infalling, positively propagating wave is transformed into an outgoing wave travelling in the opposite direction; the ray associated with the latter is found from the Hamiltonian Ω_-. Since the frequency is a first integral for the global wave propagation, we have from the two branches (3.3)

$$\Omega_+(L, k_+) = +c k_+ \; = \; \omega \; = \Omega_-(L, k_-) = -c k_- \quad , \quad x = L \; . \tag{3.4}$$

A similar relation holds at the boundary point x=-L. From Eq.(3.4) we obtain the reflection condition for the momentum : $k_+ = -k_-$. For the 3-dimensional version of the wave equation (3.1) it can be seen in a similar fashion that the wave vectors $\underline{k}_{in}$ and $\underline{k}_{out}$ of the infalling and outgoing wave must be connected by

$$\underline{n} \cdot (\underline{k}_{in} + \underline{k}_{out}) = 0 \tag{3.5}$$

n being the normal to the boundary. Just as in the 1-dimensional case, this connection relation is independent of the precise boundary condition (2.3). Standard stellar wave problems (propagation of acoustic waves, gravity waves, ...), essentially belong to the class of propagating 3-dimensional wave equations of type (3.1), so that the reflection condition (3.5) holds for the associated ray problems.

(2)   While the Schrodinger equation   is   Hermitian, the stellar wave equations are not necessarily so.   In particular, in the presence of rotation, the Coriolis force term introduces an anti-Hermitian contribution.   It does not seem, however, that this feature has any major incidence on the (asymptotic) wave problem.

(3)   Quantum chaos has been analysed so far in   two mathematically distinct categories of quantal problems only :

(a)   In Schrodinger equations with analytic classical Hamiltonians.   Typical instances are the Henon-Heiles and Contopoulos-Barbanis Hamiltonians (harmonic potential plus corrections of a third degree polynomial in the coordinates), the Percival-Pomphrey Hamiltonian (harmonic potential plus a correction of fourth degree), the Toda Hamiltonian (exponential potential), ... of two degrees of freedom which have been carefully studied by various authors over the past ten years (Pomphrey 1974, Percival and Pomphrey 1976, Noid et al. 1980, Weissman and Jortner 1981, Pullen and Edmonds 1981, Ramaswamy and Marcus 1981, Meyer et al. 1984).   These systems belong to the classically well understood class of generic quasi-integrable systems defined by Hamiltonians of the form

$$H(\underline{q},\underline{p};\varepsilon) = H_0(\underline{q},\underline{p}) + \varepsilon H_1(\underline{q},\underline{p}) \qquad (3.6)$$

(H_0, an integrable Hamiltonian; H_1, a nonintegrable correction; both functions H_0 and H_1 are analytic functions of their arguments).

Integrability means that the Hamiltonian system admits of F independent 'isolating integrals in involution' (Arnold 1978), J_1, J_2, ... J_F (F, number of degrees of freedom).   By Liouville's theorem, these integrals define smooth F-dimensional manifolds in the 2F-dimensional phase space; these manifolds, filling up the whole phase space, have the topological structure of tori T^F (direct products of F circles C^1 : $T^F = C^1 \times C^1 \times \ldots \times C^1$).   Any initial condition $(\underline{q}_0 \underline{p}_0)$ singles out a torus $T^F(\underline{q}_0 \underline{p}_0)$ carrying   the   orbit $\underline{q}(t)$ $\underline{p}(t)$ generated by this initial condition; such orbits are called regular solutions.   A generic initial condition $(\underline{q}_0 \ \underline{p}_0)$ produces an orbit that is dense in the orbit carrying torus $T^F(\underline{q}_0 \ \underline{p}_0)$; generic orbits, projected into configuration space, exhibit caustics (boundaries of the projection of   the orbit carrying torus).   Special initial conditions $(\underline{q}'_0 ,\underline{p}'_0)$ always exist that generate rational tori, i.e. tori carrying closed orbits $T^F(\underline{q}'_0 \underline{p}'_0)$; these closed orbits are 'nonisolated' on $T^F(\underline{q}'_0 \underline{p}'_0)$, i.e. there is a continuous infinity of them.   These closed orbits correspond to a periodic time-behaviour.   Although infinitely sparse (i.e. of Euclidean 2F-dimensional measure zero) among the totality of tori, the

rational tori densely fill the phase space in the case of generic
analytic Hamiltonians.

If the integrable Hamiltonian H_0 is perturbed by a
sufficiently small nonintegrable (analytic) correction (ε small
in Eq. 3.6), then the nonrational tori of the underlying
integrable Hamiltonian are just slightly distorted and they
continue to carry regular orbits. On the other hand, the
rational tori are broken up under the effect of the perturbation;
the total volume **D** of phase space occupied by these disrupted
tori has a small 2F-dimensional Euclidean measure provided that
ε is small enough (as ε = 0, this measure must vanish by the
observation made above). This is the essential content of the
Kolmogorov-Arnold-Moser (KAM) theorem (Arnold 1978).

The disruption of a rational torus typically produces
several next-generation tori surrounded in turn by
higher-generation tori; the region in between tori of successive
generations is filled by <u>chaotic orbits</u>. The distinguishing
property of chaotic orbits is that they are <u>exponentially
unstable</u>: two arbitrarily close initial conditions generate
exponentially diverging solutions in time. In contrast, two
arbitrarily close initial conditions on tori (regular orbits)
diverge at most like a polynomial of time. (For a more basic
algorithmic definition of chaotic orbits see Perdang 1985a). The
total region **K** in phase space, occupied by chaotic orbits, is a
subset of region **D**, so that the 2F-dimensional measure of **K**
itself is vanishingly small for ε small enough. If the perturbing
effect in Eq. (3.6) is small enough, the probability of catching
a chaotic solution is virtually zero.

For the potentials mentioned at the beginning of this subsection,
ε is a measure of the order of magnitude of the amplitudes of the
generalised coordinates, or equivalently, of the order of
magnitude of the <u>energy</u>. Therefore at sufficiently low energy,
practically all orbits are regular; chaotic orbits only arise in
the high energy range with significant probability.

Analysing these systems of two degrees of freedom numerically,
Kosloff and Rice (1981), Jaffe and Reinhardt (1982) and
Shirts and Reinhardt (1982) observe that in spite of the chaotic
character of the higher energy solutions, the latter keep an
apparently regular aspect most of the time; irregular behaviour
is only observed when the orbit crosses the neighbourhood of the
critical saddle points of the potentials, implying in turn that
<u>chaos is locally generated</u> in phase space for these
systems. These authors conclude therefore that such chaotic
orbits may be pictured as remaining confined to torus-like
manifolds in phase space over several and often over many periods
of oscillation, these manifolds being remnants of genuine
tori of an integrable Hamiltonian. However, on these manifolds
there are critical instability patches, such that if the orbit

crosses the latter, it is sent to another nearby torus-like manifold which exhibits again similar properties. Shirts and Reinhardt (1982) refer to these structures as <u>vague tori</u>; vague tori are defined analytically by F independent isolating pseudo-integrals of motion which are formally obtained as (nonconvergent) series in ε, using the Birkhoff normal form procedure.

The quantum mechanical energy eigenvalues of these systems are expected to be well approximated by applying a semiclassical quantisation method to the tori and vague tori. We review such a technique in Section VII.

For these analytic Hamiltonians, the Schrodinger equation is defined over the whole configuration space; the boundary conditions at infinity merely reduce to regularity conditions; they do not enter the classical Hamiltonian problem.

(b) A second class of quantal problems analysed for quantum chaos are the (2-dimensional) <u>billiards</u>. A billiard is defined by the Hamiltonian

$$H(\underline{q},\underline{p}) = K(\underline{p}) = \tfrac{1}{2}|\underline{p}|^2 \text{ if } \underline{q} \in V^F \quad \text{and } H(\underline{q},\underline{p}) = \infty \text{ if } \underline{q} \in \partial V^F . \tag{3.7}$$

Notice that if we define V^F analytically

$$V^F \; : \; S(\underline{q}) < S_0 \tag{3.7'}$$

($S(\underline{q})$ an analytic function, S_0 a constant), then the billiard Hamiltonian can be regarded as the limit $n \to \infty$ of the family of analytic Hamiltonians (n, integer)

$$H(\underline{q},\underline{p};n) = K(\underline{p}) + \left(S(\underline{q})/S_0 \right)^n . \tag{3.7''}$$

The nature of the orbits of these systems is conditioned by the structure of ∂V^F alone. To understand the character of the solutions of billiard systems, we start out with identifying V^F with the whole Euclidean space E^F, so that Eq. (3.7') can be ignored. Moreover we identify $q_i = -\infty$ with $q_i = +\infty$, i = 1,2, ..., F, so that any solution of this system lies on an F-dimensional torus labelled by the momentum $\underline{p}$, $T^F(\underline{p})$; hence the solutions of this system are to be regarded as regular. Introduce next the finite region V^F. Then any solution, in between two successive collisions with the boundary ∂V^F, still lies on a portion $R^F(\underline{p})$ of a torus $T^F(\underline{p})$, namely on the intersection of $T^F(\underline{p})$ with $V^F \times K^F$. Let $\underline{p}_0$ be the initial momentum and $\underline{p}_i$ the momentum just after the i th collision with the boundary, i = 1, 2, ... Then denoting by n the outer normal to the boundary, we have at collision i (cf Eqs. 3.7 or 3.7',7'')

$$\underline{n} \cdot (\underline{p}_{i-1} + \underline{p}_i) = 0 \quad , \; i = 1,2,3,\dots \quad \text{on} \quad \partial V^F . \tag{3.8}$$

Hence, the region in phase space explored by any actual solution of (3.7,7') is

$$R^F(\underline{p_0}) \cup R^F(\underline{p_1}) \cup \dots \cup R^F(\underline{p_{n-1}}) \cup R^F(\underline{p_n}) \cup \dots = M \qquad (3.9)$$

showing that different alternatives may arise.

(i) After a finite number N of collisions the solution returns to the original region $R^F(\underline{p_0})$, $R^F(\underline{p_N}) \equiv R^F(\underline{p_0})$. Then we have two possibilities: Either M has the topological structure of an F-dimensional torus, T^F, and the solution is regular. If all solutions are regular, the billiard is called integrable. Or M has a topology differing from that of a torus. In planar (F=2) billiards this possibility was first recognised by Zemlyakov and Katok (1975). For instance for an L-shaped billiard M can be shown to be a sphere with 5 holes; Richens and Berry (1981) who investigated this latter billiard numerically point out that the character of the time behaviour of such a system is intermediate between regularity and chaos. A full characterisation of the behaviour of such systems (for instance in terms of algorithmic complexity theory) is lacking.

(ii) The solution never returns to the original zone $R^F(\underline{p_0})$ (for any finite number n, $R^F(\underline{p_n}) \neq R^F(\underline{p_0})$); it ultimately explores a region of dimensionality higher than F. Such a solution is chaotic.

Besides these generalities a more specific rigorous result due to Lazutkin (1973) is known. This result is indicative that for generic billiards with smooth boundaries, both regular and chaotic solutions coexist, just as they do in generic analytic Hamiltonians of type (3.6):

If the billiard is two-dimensional (F=2) with the domain V^2 being convex, and if the radius of curvature of the boundary ∂V^2 as a function of arc length is of class C^{553} then a (discontinuous) family of caustics corresponding to regular orbits exists near ∂V^2 (cf Mather 1982 for a complementary result).

Lazutkin's theorem of existence of regular orbits close to the boundary, in billiards with smooth enough boundaries, was in fact anticipated, without a formal proof, in Keller and Rubinov (1960). Notice that the condition of smoothness of the boundary is vital, as transpires from the well documented example of the race-track billiard (Bunimovich 1974); for the latter it is known that almost all solutions are chaotic for any nonzero length L of the straight side of the boundary; more precisely, in the terminology of ergodic theory, this billiard defines a K-flow (cf Arnold and Avez 1968). On the other hand, if L vanishes, the stadium collapses to a circle (of radius R), all orbits becoming regular. For a readable analysis of this billiard see Berry 1981 who demonstrates numerically that for L small, the orbits although chaotic remain close to those of the circular billiard over a long time span; for instance for L/R ~ .01 chaos becomes conspicuous only after some 1000 bounces.

Keller and Rubinov (1960) have set forth arguments in favour of the existence of a second family of regular orbits in planar billiards with smooth boundaries, namely for orbits nearly coinciding with the shortest diameter of the billiard. The latter being stable by construction, they must be regular by virtue of our remarks on the connection between chaos and instability of the orbits. These two families of regular orbits are the counterparts of the well-known two classes of orbits existing in elliptical billiards, namely those which have as caustics ellipses confocal with the boundary, and those which have confocal hyperbolae as caustics (cf Steinhaus 1950; for a mathematical analysis of the elliptical billiard see Poritsky 1950).

Keller and Rubinow have also conjectured that both these regular families should survive in 3-dimensional billiards with smooth boundaries: The first family consists of orbits lying near the closed surface geodesic of shortest length; the second family corresponds to orbits in the neighbourhood of the shortest diameter of the billiard. Whether further regular families exist in the 3-dimensional case is unknown.

Lazutkin's (1973) result is indicative that in general in (planar) billiards with smooth convex boundaries two classes of solutions coexist, namely regular ones as well as chaotic ones; a chaotic solution is then in general. not ergodic, in the conventional sense that the energy surface in phase space is densely filled by such a solution, since part of the energy surface is occupied by regular orbits. In this respect the stadium, and the integrable billiards are atypical, since for the latter only one class of solutions are encountered (with probability 1). While a few generic planar billiards have been analysed numerically (Benettin and Strelcyn 1978, Robnik 1983), no results are available for 3-dimensional billiards.

The quantum mechanical level distribution in the stadium billiard has been studied numerically in McDonald and Kaufman (1979), Casati et al. (1980) and McDonald (1983). They find that the level spacing is consistent with a Wigner distribution

$$P_W(S) = S\,(\pi/2D^2)\,\exp\!\left[-(\pi/2D^2)\,S^2\right] \quad ; \qquad (3.10)$$

here S represents the spacing between two adjacent eigenvalues, D is the average level spacing, and P(S)dS is the probability to find a spacing in the range (S,S+dS). As was shown theoretically by Wigner (1967) distribution (3.10) is obtained as the generic lowest order approximation if the levels in the spectrum are interacting; in the absence of level interaction, the level spacings obey a Poisson distribution

$$P_p(S) = (1/D)\ exp(-S/D)\qquad\qquad\qquad (3.10')$$

In this theoretical discussion, the eigenvalue interval over which the distribution is sought is assumed to be small as compared to the order of magnitude of the eigenvalues. A detailed numerical discussion of the level spacings of a simple integrable billiard (rectangular 'irrational' billiard) was performed by Casati et al. (1985) who took account of all 10^5 lowest levels to compute a <u>global</u> probability distribution; these authors conclude that, while this global distribution still looks Poisson-like (Eq. 3.10'), the statistics ceases however to duplicate uncorrelated Poisson statistics (as is revealed by a study of the Δ_3-statistics of Mehta and Dyson); the latter only holds over small intervals of levels i.e. in the actual assumption of Wigner's original treatment.

The structure of the eigenfunctions of the stadium billiard has been analysed by McDonald and Kaufman (1979), McDonald (1983), Shapiro and Goelman (1984) and Mellor (1984). On the one hand, it is found that the overall nodal pattern of modes corresponding to high-order modes is irregular, crossings of nodal lines being avoided; the latter property reflects a general theorem by Uhlenbeck (1976): For almost all Hamiltonians of two degrees of freedom, the nodal lines of the eigenfunctions of the associated Schrodinger equation cannot intersect.

As recognised already by Pechukas (1972) the presence of self-intersections in the nodal pattern is a consequence of separability of the Schrodinger equation. Since separability is more restrictive than integrability of the underlying Hamiltonian - a classically integrable Hamiltonian can generate a separable Schrodinger equation only under special symmetry conditions - the presence of avoided crossings in the nodal pattern cannot rank as a sufficient indicator of quantum chaos. However, for classically integrable Hamiltonians, the nodal pattern remains very regular, while the nodal pattern of eigenfunctions associated with a chaotic range of a nonintegrable classical Hamiltonian is irregular. (The terms 'regular' and 'irregular' can be given a precise quantitative definition in the context of algorithmic complexity theory, cf Perdang 1985a). We may also say that the nodal pattern is unpredictable, in the sense that we do not know how to continue the nodal curve beyond a local wavelength (cf McDonald and Kaufman 1979).

On the other hand, a class of eigenfunctions are found which remain concentrated around <u>isolated</u>, <u>closed</u> classically <u>unstable orbits</u>. Such states, called 'scarred states', have been interpreted quasiclassically in terms of Gaussian wave packets propagating quasiclassically along those closed orbits (Heller 1984). A second class of eigenfunctions are found with a high concentration near the <u>nonisolated</u> classically <u>unstable closed orbits</u> of the stadium (orbits normal to the rectilinear section of the stadium); these states are referred to as 'superscars' in Heller (1984). A similar behaviour of the eigenfunctions of analytic Hamiltonians is found.

IV. Some Special Features of Stellar Eigenvalue Problems

The stellar oscillation eigenvalue problems differ in several respects from the two classes of quantal eigenvalue problems (a) and (b) tested for quantum chaos.

(1) The stellar problems typically involve features of both classes (a) and (b). On the one hand, the Hamiltonian $\Omega(r,k)$ is explicitly space dependent inside V^3, where it may be considered as analytic. On the other hand V^3 is a finite region, and boundary conditions are imposed at the surface ∂V^3, just as in the billiard problem. Therefore quantum chaos in the stellar oscillations can arise as a consequence of (i) the analytic structure of $\Omega(r,k)$ in V^3, (ii) the geometry of the surface (billiard effect) or (iii) possibly as a combination of the structure of $\Omega(r,k)$ and the nature of the surface.

(2) As can be seen from the instance of acoustic modes, it happens not infrequently that we have bounded rays corresponding to any positive value of the classical energy $\Omega = \omega$. This differs from the situation of the Hamiltonians of third degree in the coordinates, discussed above, where bounded classical solutions exist for energies less than a finite escape energy E_c. On the other hand, the energy itself (frequency) cannot, in general, be used as a measure of departure from a reference integrable Hamiltonian, as can be seen from simple models. The KAM theorem therefore remains certainly applicable to stellar ray problems regarded as described by an integrable Hamiltonian $\Omega_0(r,k)$ plus a slight nonintegrable correction $\varepsilon\Omega_1(r,k)$ (cf Eq. 3.6). If the strength ε is small, the asymptotic spectrum can be obtained through a quasiclassical quantisation procedure applied to tori and vague tori. Or in other words, quantum chaos is expected here to arise only for sufficiently large values of ε, irrespective of the frequency. For the majority of analytic Hamiltonians discussed above, the situation is quite different; there we have seen that an increase in the energy is equivalent to an increase in the strength of the perturbation; therefore, quantum chaos is eventually due to show up in the higher energy levels.

(3) The billiards investigated numerically in the literature all have very smooth boundaries, except possibly at a finite number of isolated points (cf polygonal billiards); more specifically the normal n to the boundary changes slowly. Consider now the situation of a boundary defined algebraically

$$\partial V^3 : \quad B_0(r) + \varepsilon B_1(r) = 0 \quad , \tag{4.1}$$

where $\varepsilon = 0$ defines an integrable billiard. In general, the addition of a perturbing term $\varepsilon B_1(r)$ of small strength does not guarantee that the ray pattern of the perturbed billiard remains comparable with that of the original integrable billiard. This is illustrated by the following example.

Take as the nonperturbed system a planar, circular billiard of radius R; introduce a boundary perturbation defined in polar coordinates by

$$r - R - \varepsilon R \sin N\theta = 0 \ . \qquad (4.2)$$

By choosing $\varepsilon \ll 1$, the correction to the boundary can be made arbitrarily small. However, if we select $N = O(1/\varepsilon)$, the direction of the normal to the boundary can be made to differ entirely from the direction of the normal of the original circular billiard. Therefore, the ray patterns of both billiards are completely different. By analogy with the system studied by Robnik (1983), this billiard is almost certainly ergodic for $N \gtrsim 1/\varepsilon$.

The wave problem associated with a billiard of type (4.2) raises a general issue. A wave of wavelength λ does not 'feel' the structural details of lengthscale $\ell < \lambda$; at a given position $\underline{r}$ it is affected by a spatial average of the properties of the medium it traverses; the average is calculated over an n-dimensional ball of radius λ (n = 1,2,3 depending on the nature of the problem), or by some similar procedure (cf Mandelbrot 1982). Therefore we are led to attach to a given classical billiard of boundary **B**, a 'quasiclassical billiard at wavelength λ ', of boundary $B(\lambda)$ defined by the averaging procedure indicated above. The orbits of the latter are then representative for calculating the eigenfrequencies by a quasiclassical technique. In the case of Eq.(4.2) an acceptable averaging procedure is given by

$$\mathbf{B}(\lambda): \ r(\theta) = R + \varepsilon R \ \frac{1}{2\Delta\theta} \int_{\theta-\Delta\theta}^{\theta+\Delta\theta} d\theta' \sin N\theta' \quad , \ \Delta\theta = \lambda/R \quad , \qquad (4.3)$$

$$= R \left[1 + \left(\frac{\varepsilon}{N}\right) \cdot \left(\frac{R}{\lambda}\right) \ \sin\left[N\frac{\lambda}{R}\right] \ \sin N\theta \right] \ . \qquad (4.3')$$

We see that the wave problem exhibits 3 regimes of wavelengths.
(i)   If $\lambda = 2\pi/|\underline{k}|$ is in the range

$$R \gtrsim \lambda \gg \ell , \qquad (4.4)$$

with $\ell = R/N$ the scale of the surface rugosity ($N \gg 1$), $\varepsilon . \ell/\lambda \ll 1$, the wave problem ignores the perturbation; the associated quasiclassical billiard is just given by the unperturbed billiard.
(ii)   If $\lambda \sim \ell$, we have a quasiclassical billiard explicitly depending on λ (Eq. 4.3') .
(iii)   If $\lambda \ll \ell$, the quasiclassical billiard is again independent of the wavelength and the averaged expression (4.3) reduces to the original perturbed expression (4.2). In the long-wavelength interval (i) (Eq. 4.4) the spectrum of frequency eigenvalues remains regular. Quantum chaos makes its appearance when $\lambda \sim \ell$ (mild quantum chaos); it is fully developed in the small-wavelength interval (iii). Notice that with the specific form of the perturbation (4.2) a large number of isolated

unstable closed rays survive, so that by Heller's (1984) argument 'scarred states' are expected to occur.

The situation we have sketched can be viewed as a crude model for the action of the surface on actual stellar waves. The surface may in fact exhibit a horizontal structure of lengthscale ℓ (cellular structure etc, cf in particular the Sun). The 3 regimes isolated above are then bound to manifest themselves in the frequency spectrum. Moreover, a realistic stellar surface structure is hardly expected to be regular as modelled by Eq (4.2), but rather the perturbing factor will be a complicated space function. Essentially, this means that the actual ray system in regime (iii) can be modelled by a formal ray system generated by a <u>random surface reflection</u> law on a nonperturbed surface.

A further speculation is in order. In the same fashion as the terrestrial surface is of a fractal character at scales $\lesssim 1$ km (cf Mandelbrot 1982), it is quite plausible that the stellar surface at lengthscales less than ℓ_f , and possibly over some range of scales (ℓ_F, ℓ_f) is a fractal set, of dimension $D > 2$. (Selfsimilarity is clearly suggested in the magnetic surface features of the sun; the level hierarchies of supergranulation , mesogranulation and granulation are likewise indicative of an approximate selfsimilarity). There is evidence - although unexplored in the specific stellar oscillation problems - that fractal behaviour modifies the <u>level density</u> of the frequency spectrum (cf Berry 1979 who extends the density formula in Baltes and Hilf 1977 to fractal configurations; in the stellar acoustic spectrum the effect of the vanishing surface sound velocity produces a change in the level density of a similar algebraic nature, Perdang 1982, so that singularities in the coefficients of the wave equation and fractal behaviour will be difficult to disentangle from a level density analysis).

In the presence of a fractal surface, a quasiclassical ray problem can again be defined as above, leading to three regimes: (i) $\lambda > \ell_f$, nonperturbated billiard; (ii) $\lambda \sim \ell_f$, transition to fractal regime and onset of mild chaos; (iii) $\lambda \ll \ell_f$, fully developed quantum chaos in the fractal range ($\mathbf{B}(\lambda)$ depends on the wavelength over the whole fractal range).

(4) The discussion under (3) extends to analytic Hamiltonians of the form (3.6) with H_0 integrable and H_1 a rapidly varying convoluted function of $\underline{q}$ (but a smoothly varying function of $\underline{p}$; in all stellar wave problems the total wave Hamiltonian is a polynomial of low degree in the components of $\underline{p}$). Following the procedure for billiards, we substitute to the actual Hamiltonian $H(\underline{q}, \underline{p}; \varepsilon)$ a space-averaged Hamiltonian. However, in typical stellar wave problems the <u>local</u> wavelength λ is strongly space dependent, and may even become complex. An averaging over a local wavelength is therefore not adequate; it seems more reasonable to select the <u>average</u> wavelength at a given energy level E (or frequency ω), $\bar{\lambda}(E)$, as the typical smoothing factor. This leads us to define the

'averaged quasiclassical Hamiltonian at average wavelength $\bar{\lambda}$ ' by

$$H(\underline{q},\underline{p};\varepsilon;\bar{\lambda}) = H_0(\underline{q},\underline{p}) + \varepsilon \overline{H_1(\underline{q},\underline{p})} \; , \tag{4.5}$$

$$H_1(\underline{q},\underline{p}) = \frac{1}{V^F(\bar{\lambda})} \int_{|\underline{q}'-\underline{q}|\leqslant\lambda} (d\underline{q}')^F H_1(\underline{q}',\underline{p}) \; , \tag{4.6}$$

where $V^F(\bar{\lambda})$ is the volume of the F-dimensional ball of radius $\bar{\lambda}$. An alternative definition, in closer analogy with the billiard example (4.3) would consist in averaging over angles only. Strictly speaking this procedure is circular: to define $\bar{\lambda}$ we need the solution of the ray problem in the first place. However, since in Eq.(4.6) we just need an order of magnitude of $\bar{\lambda}$, a crude estimate of the latter is sufficient. Thus, for stellar acoustic modes, we may take for $\bar{\lambda}$ the local wavelength $\lambda(r)$ at a given frequency ω averaged over the accessible region V in configuration space

$$\bar{\lambda} \sim \frac{1}{V} \int_V (d\underline{r})^3 \lambda(\underline{r}) \sim 6\pi \frac{\omega}{R^3-R_c^3} \int_{R_c}^R dr \, r^2 c(r) \quad ; \tag{4.7}$$

here R and R_c are the radius of the star and the radius of the deepest level of penetration of the rays and c(r) the sound velocity in the nonperturbed spherically symmetric wave problem.

As in the billiard case (3) this procedure shows again the existence of 3 regimes (i) $\ell \ll \bar{\lambda}$, ℓ being the characteristic scale of fluctuation of the perturbing Hamiltonian. The averaging procedure then wipes out all fluctuations ($\bar{H_1}$ = 0), so that the quasiclassical Hamiltonian reduces to the integrable component H_0 . At these energies (frequencies) the spectrum is quantum regular. (ii) $\ell \sim \bar{\lambda}$; the averaged quasiclassical Hamiltonian shows a progressively more complicated structure manifesting itself in a nonconvex shape of the energy surfaces in phase space as $\bar{\lambda}$ decreases; it is then almost sure that the associated orbits become chaotic, (cf a stability analysis of the type carried out by Brumer 1973); hence mild quantum chaos sets in. (iii) $\ell \gg \bar{\lambda}$; the averaging procedure has no effect on H_1 . Classical chaos implies the occurrence of strong quantum chaos. Contrarily to what happened for the smooth corrections H_1 discussed in section III, in the present problem orbital (ray) chaos is generated, in principle, at every point the orbit traverses in configuration space. Therefore the notion of vague tori fails to be applicable in this regime (while it may apply to the incipient quantum chaos in regime ii).

The special feature of stellar acoustic waves deserves a comment. Since in conventional stellar models the sound speed vanishes at the surface of the star, the local wavelength $\lambda(r)$ $\sim 2\pi c(r)/\omega$, vanishes at the surface; this implies in turn that such waves propagating close to the surface are experiencing the small scale structure of the surface, even though ω may remain small. Therefore regime (ii) will occur already in the low acoustic frequency spectrum if the perturbing Hamiltonian (i.e.

essentially the sound speed) has a complex structure in the stellar surface zones.

V. Integrable Reference Wave Equations

Sections III and IV indicate the generation of quantum chaos through slight perturbations of an integrable stellar wave equation. We here show that for virtually all realistic stars a representative integrable form of approximate wave equation does indeed exist. In fact any wave equation of spherical symmetry satisfies the integrability property. More specifically:

If a ray problem is defined in a spherical region, of centre O and if the ray Hamiltonian has spherical symmetry around O,

$$\Omega(\underline{r},\underline{k}) \equiv H(r,k) = \omega \tag{5.1}$$

(r distance from centre O; k, modulus of wavevector $\underline{k}$), then (a) any ray is confined to a 'meridian' plane and (b) the ray problem is regular, regardless of the precise physics included in the wave equation.

In the absence of surface reflections, the proposition is obvious. We sketch the alternative of reflections on the spherical surface.

Due to the symmetry of the Hamiltonian (5.1), the latter describes the motion of a particle in a central field (in between two encounters with the boundary). Therefore this problem is formally integrable. Any initial condition ($\underline{r}_o$, $\underline{k}_o$) generates a ray $\underline{r}(t)$, $\underline{k}(t)$, confined to a meridian plane E^2, passing through O, and parallel to the directions $\underline{r}_o$ and $\underline{k}_o$ in the configuration space. Since under reflection on the surface this plane E^2 is invariant, $\underline{r}(t)$, $\underline{k}(t)$ remains confined to E^2 forever. (In the nongeneric case of a radially directed initial wave vector $\underline{k}_o$, the plane degenerates into a straight line E^1 parallel to $\underline{k}_o$ and passing through O). More specifically, $\underline{r}(t)$ is constrained to stay in the disc D^2, intersection of the meridian plane E^2 with the spherical region. Hence property (a). The accessible phase space for this ray problem is then $D^2 \times E^2$; the original system of 3 degrees of freedom is reduced to a system of 2 degrees of freedom.

Select next polar coordinates in the meridian plane E^2. Write Eq. (5.1)

$$H(r,k) \equiv H\left[r, (k_r^2 + k_\theta^2)^{1/2}\right] = \omega \tag{5.2}$$

k_r and k_θ being the momenta conjugate to r and θ. The two integrals of motion of this formally integrable system are ω and k_θ. They define a 2-dimensional surface M^2 in the 4-dimensional phase space $D^2 \times E^2$; the formal topology of M^2 is the topology of a cylinder $C^1 \times D^1$ (direct product of a circle C^1 and an interval D^1, the latter being finite or possibly infinite). In fact, we can parametrise M^2 by an angle φ (varying over C^1) and by the momentum k_r (Eq. 5.2) conjugate to r (or more precisely by a uniform coordinate k_r' directly related to k_r) and varying over an interval such that (φ, k_r') uniquely specifies a point on M^2.

To take care of the effect of reflections, observe that ω and k_θ are invariants while k_r is turned into $-k_r$ at r = R, by virtue of condition (3.5). This shows that M^2 is an invariant surface under the reflection. Moreover, through the identification of the bases of the cylinder (k_r and $-k_r$) resulting from the reflection condition, the topology of the interval D^1 changes into the topology of the circle C^1. The reflection on the stellar surface thus transforms M^2 into a torus $T^2 = C^1 \times C^1$. All rays are therefore regular in the sense that they are carried by 2-dimensional tori.

This property directly applies to the ray equations associated with waves in the conventional setting of stellar models of spherical symmetry.

Ray patterns of acoustic waves in spherically symmetric stars have recently been computed with the purpose of applying quasiclassical quantisation techniques to obtain the asymptotic frequency spectrum. These ray patterns being regular, they necessarily exhibit (circular, as a consequence of the symmetry) caustics in the ray plane. Moreover, the calculations show that the inner zone of the caustic is avoided by the rays (Fluhr 1983, Gough 1984). The geometry of the acoustic ray patterns of a spherically symmetric star of arbitrary physics is therefore equivalent with the ray pattern of a spherical 'star' with constant sound speed, in the sense that there is a 1-1 correspondence between both patterns.

As a corollary of ray regularity, stellar eigenvalue problems of spherical symmetry are quantum regular. This regularity is revealed, (1) by the well known orthogonally self-intersecting regular nodal pattern of the stellar eigenfunctions; (2) it is also displayed by the structure of the distribution function of the nearest neighbour levels (cf Fig. 1). In fact, this distribution is locally reasonably consistent with a Poisson distribution (3.10'), and definitely not compatible with a Wigner distribution (3.10). The relatively large excursions of the experimental histogram (heavy lines) from the exponential law are in part due to the small number of sample points per bin; in fact if we increase the binsize by a factor 3 (dotted histogram) a smoother curve approximating more closely the exponential form is obtained. Besides we presume that the scatter is also due in part to a more subtle phenomenon stressed in Berry and Tabor (1977) in the context of billiards: rectangular billiards with rational ratios of side-lengths, a/b, typically exhibit larger fluctuations in their distribution of level spacings than those with a/b irrational; but the standard asymptotic frequency formula for p modes precisely shows an asymptotic degeneracy reminiscent of the degeneracy of a rational billiard of a/b = 2.

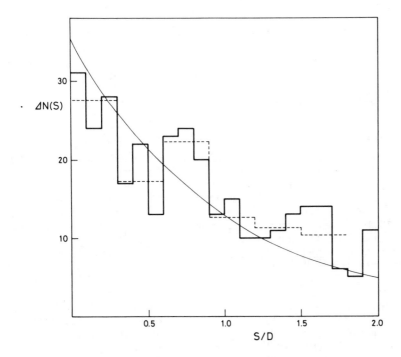

Fig. 1.- Level spacings of stellar acoustic modes in the
         range $27 < \omega < 33$ (units $GM/R^3 = 1$) containing
         353 frequencies; average spacing .017

VI.   Realistic Stellar Wave Equations

In real stellar wave problems spherical symmetry is obeyed in a
first approximation only.   Two classes of perturbations break
this symmetry: (a) <u>Large-scale</u> disturbances, such as tidal
effects in a binary system, rotation, ...; these effects are
built into the ray equation by adding a smooth (generically)
nonintegrable correction to the radial Hamiltonian.   (b)
<u>Small-scale</u> disturbances of lengthscale $\ell \ll$ R (local
circulations, convection organised in cells; in the particular
case of the sun, granulation, mesogranulation and
supergranulation); these effects are encoded in a spatially
rapidly fluctuating and in general highly irregular Hamiltonian
correction.

The full ray Hamiltonian for realistic stellar geometric acoustics is then of the form

$$\Omega(\underline{r},\underline{k}\,;\varepsilon_a,\varepsilon_b) = H_o(r,k) + \varepsilon_a H_a(\underline{r},\underline{k}) + \varepsilon_b H_b(\underline{r},\underline{k}) \,, \qquad (6.1)$$

where ε_a, ε_b are the strengths of the effects (a) and (b) respectively, and H_a and H_b are the corresponding Hamiltonian corrections, H_o being the radially symmetric Hamiltonian.

Following the procedure of section IV we introduce the quasiclassical Hamiltonian

$$\Omega(\underline{r},\underline{k}\,;\varepsilon_a,\varepsilon_b;\bar{\lambda}) = H_o(r,K) + \varepsilon_a H_a(\underline{r},\underline{k}) + \varepsilon_b \overline{H_b(\underline{r},\underline{k})} \,, \qquad (6.2)$$

the averaging affecting the small-scale component only. We then distinguish again the 3 regimes encountered above.

(i) $\bar{\lambda} \gg l$; we are left with a $\bar{\lambda}$-independent integrable Hamiltonian plus a smooth nonintegrable correction due to (a) above. As is seen from the instance of acoustic waves in conventional theoretical stellar models we are not allowed to apply the KAM theorem, since <u>at the surface the relevant Hamiltonians cease to be analytic</u> (sound velocity depending on depth h like $c \propto h^{1/2}$; in the sun we now have observational evidence in favour of this dependence; Christensen-Dalsgaard <u>et al</u>. 1985). Inside the star, the Hamiltonian can be regarded as analytic, so that possible departures from KAM behaviour are exclusively due to the effect of the surface on the ray pattern. Since the surface is acting as a billiard, the character of the ray pattern, in the limit of a small strength of the large-scale disturbances can be mimicked by billiards of surface identical with the perturbed stellar surface. Numerical work on 3-dimensional billiards is so far lacking. I have studied therefore in some detail the analytically simplest, (presumably) nonintegrable, representative of that class, namely a sphere deformed by a cubic term

$$(x/R)^2 + (y/R)^2 + (z/R)^2 + \varepsilon (x/R)^3 = 1 \,. \qquad (6.3)$$

(A mere deformation of a sphere into an ellipsoid does not affect the integrabililty and is therefore not instructive for our purposes). The cubic perturbation preserves the rotational symmetry around the Ox-axis, as well as the symmetry planes passing through the Ox-axis (meridian planes). This rotational symmetry induces an angular momentum integral; the latter in turn implies that any wave pattern is progressively rotated around the Ox-axis.

The numerical results show essentially 3 generic species of ray pattern.
(a) If the initial positions are chosen close to the surface of the star, and the initial velocity is roughly parallel to the surface, we obtain a very regular ray system that remains confined to the neighbourhood of the surface (Fig. 2).

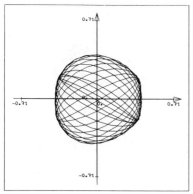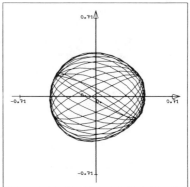

Fig. 2.- Regular surface rays (R = 1/2, ε = - 1/3);
      initial conditions x=-.32, y=-.19, z=o,
      p_x = p_y = o , p_z = .5

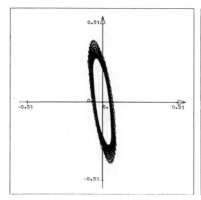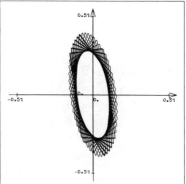

Fig. 3.- Regular surface rays (R = 1/2, ε = - 1/3);
      x=.o5, y=-.28, z=o,
      p_x = p_y = o , p_z = .5

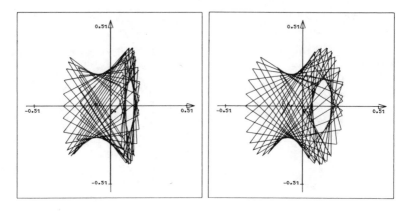

Fig. 4.- Regular shortest-width rays (R = 1/2, ε = -1/3);
x=.o7, y=.2, z=o,
$p_x = p_y = o$, $p_z = .5$

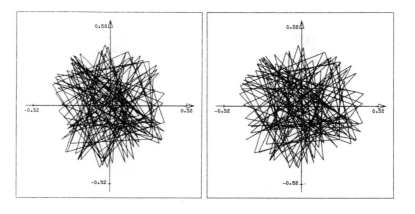

Fig. 4.- Chaotic ray pattern (R = 1/2, ε = -1/3);
x=.33, y=.17, z=o,
$p_x = .o7$, $p_y = -.8$, $p_z = .66$

As a consequence of the rotational symmetry, the intersection of a meridian plane with the surface defines a closed, stable limit-orbit of the billiard; the family of orbits illustrated in Fig. 2 can be regarded as small departures from these limit-orbits. A special instance of this class is shown in Fig. 3 which is generated by initial conditions deviating only slightly from the plane x = 0; since the intersection of the latter with the surface is the surface geodesic of shortest length, this orbit corresponds to the regular surface orbits conjectured by Keller and Rubinow (1960).

(b) If the initial positions are chosen approximately in the plane x = 0, with an initial velocity also approximately parallel to this plane, the ray pattern can be viewed as a 3-dimensionally perturbed ray pattern of a disc billiard (Fig. 4) (namely the planar billiard of surface $x^2 + y^2 = R^2$). The regularity of this pattern showing a hyperboloidal caustic surface is a consequence of the stability of the orbits in the x = 0 plane. This species of regular orbits can be interpreted either as a generalised version of Keller and Rubinow's conjecture of orbits close to the shortest inner diameter, or alternatively of the surface orbits close to the shortest geodesic.

(c) If the initial conditions are chosen without imposing any special constraint, we generate entirely irregular-looking ray patterns (Fig. 5). That these patterns are chaotic (in the sense of algorithmic complexity) is already suggested by the random character of the rays; it is furthermore seen to be consistent with the following probability property due to Czuber (1884): The average length of a collection of randomly distributed straight lines cutting a convex cavity of volume V and area A is given by

$$\langle L \rangle = 4 \, V/A \qquad\qquad (6.4)$$

(cf Santalo 1976 for a careful analysis of the 2-dimensional analogue of eq. 6.4).

We typically find for the irregular-looking ray patterns of species (c) average ray lengths between successive reflections around .64, while Czuber's expression yields .67 (to zeroth order in ε). In the case of regular orbits, the average ray length can take on any value between 0 and the maximum diameter ($\simeq 1$); for instance the orbit of Fig. 2 gives .12, while for Fig. 3 we have .58.

In contrast to the regular ray patterns (a) and (b), the species of chaotic rays seemingly explore most of the cavity, provided that we wait long enough. Over timespans of a few times the average interval between successive reflections the rays are seen to stay in a plane whose orientation changes slightly at each reflection, and this change is pretty irregular. The latter observation allows us to make a general estimate of the change of orientation, $\Delta\theta_n$, of the plane of motion after n reflections in the general case of a slightly perturbed surface showing no symmetry (and no additional first integral besides energy). Let $\Delta\theta_1$ be the amplitude of the change of orientation of the plane of the motion at each reflection; in order of magnitude $\Delta\theta_1 \sim \varepsilon$ as results from the reflection law. If, as the numerical experiment suggests, the direction of the change in orientation is unpredictable, the total amplitude after n reflections, $\Delta\theta_n$, can be regarded as produced by a random walk process, or $\Delta\theta_n^2 = n\,\Delta\theta_1^2$. Hence, the average time $\langle T_a \rangle$ required to cover all orientations ($\Delta\theta_n \sim \pi$) is given, to dominant order in ε_a, by (cf Eq. 6.4)

$$\langle T_a \rangle \sim \tfrac{4}{3} \left(\pi / \varepsilon_a \right)^2 R/v \qquad , \qquad (6.5)$$

v being the velocity of propagation. For acoustic modes in stars, this time – which should be shorter than the absorption time of these modes – is of the order of $\left(\pi / \varepsilon_a \right)^2 P_o$ (P_o, fundamental period). It shows that ray chaos in the sun due to large scale perturbations (lacking a continuous symmetry) should remain at present undetectable, unless ε_a were implausibly large ($> 1\%$). However, in contact binaries for instance, where ε_a may reach several times 10% this type of chaos can become detectable. Over times sufficiently shorter than $\langle T_a \rangle$, the concept of vague tori remains adequate.

(ii) $\bar{\lambda} \sim \ell$ and (iii) $\bar{\lambda} \ll \ell$: progressive dominance of small-scale effects $\varepsilon_b H_b$. It is unlikely that $\varepsilon_b |\partial/\partial q\ H_b| \sim |\partial/\partial q\ H_o|$ almost everywhere; we rather expect, at least in the case of acoustic modes, that small-scale surface effects are the most critical, the local wavelength being small close to the surface. The incipient role of small-scale perturbations is then again described by a billiard effect of the deformed surface. Provided that

$$\frac{\varepsilon_b\,R}{\ell} \gtrsim 1 \qquad , \qquad (6.6)$$

ray chaos sets in in regime (ii) through the surface reflections; the concept of vague tori remains meaningful for the onset of chaos. In regime (iii) a violent ray chaos reminiscent of the chaos in the stadium billiard is encountered (due here to the nonconvex surface). If, on the other hand,

$$\frac{\varepsilon_b\,R}{\ell} \ll 1 \qquad , \qquad (6.7)$$

ray chaos, if present, remains of a mild variety. The plane of the motion being deflected by $\Delta\theta_1 \sim \varepsilon_b R/\ell$ at each reflection, the time $\langle T_b \rangle$ required to let the plane take on all orientations is again estimated by relation (6.5) with $\varepsilon_b R/\ell$ substituted to ε_a; or

$$\langle T_a \rangle / \langle T_b \rangle \sim \left(\varepsilon_b / \varepsilon_b \right)^2 \left(R/\ell \right)^2 \quad . \qquad (6.8)$$

VII. Asymptotics
So far techniques to compute the asymptotic frequency

distribution are available for quantum regular frequency states only. It is the purpose of this section to show that for mild ray chaos however, statistical information on the asymptotic structure of the spectrum remains accessible. This justifies a posteriori our concern of tracing mild ray chaos in the stellar waves.

We start out by deriving the general quasiclassical frequency formula using Keller's (1958) versatile technique (cf also the rigorous mathematical treatment by Maslov 1972). Consider a pencil of rays on a torus T^3 in the phase space of an integrable ray system (2.8,9). Let the initial positions lie in a volume $dV(r_o)$ around r_o, and let the initial momenta be close to k_o. Denote by C the ray in configuration space, of initial conditions (r_o, k_o). A propagating wave, solution to Eqs. (2.1, 2) at time t and position r on C, lying on sheet j of the projection of T^3 onto configuration space (j = 1, 2, ... s, s being the number of sheets of the projection of T^3) is then sought in the form

$$\Psi(r,t) = \exp(-i\omega t) \; \psi_j(r) \;, \tag{7.1}$$

with
$$\psi_j(r) = P_j(r) \exp i \, S_j(r) \;. \tag{7.2}$$

The phase function $S(r)$ is assumed to be of the form

$$S(r) = N \; S^*(r) \;, \tag{7.3}$$

where N is an asymptotically large parameter, and $S^*(r)$ is a smoothly varying function with r; $S(r)$ is therefore a rapidly varying phase. The amplitude function $P(r)$ is by hypothesis a slowly varying real and positive function. Substitution of ansatz (7.1, 2) into the wave equation (2.1) shows that the phase function obeys the Hamilton–Jacobi equation to leading order in N, and therefore

$$\partial/\partial r \; S(r) = k \;. \tag{7.4}$$

The phase in Eq. (7.2) is thus fixed by the associated ray problem. To determine the amplitude function, interpret $|\Psi(r, t)|^2$ as a probability density of phonons. Then $[P(r)]^2$ $dV(r,t)$ represents the probability of finding a phonon at time t in the volume $dV(r,t)$ around position r. If $P(r_o)$ is the initial amplitude at r_o, and $dV(r,t)$ is the transformed initial volume $dV(r_o)$ occupied by the ray pencil at time t, then we have, by conservation of probability

$$[P(r)]^2 dV(r,t) = [P(r_o)]^2 dV(r_o) = c^{st} \;. \tag{7.5}$$

It follows that whenever the volume $dV(r,t)$ shrinks to zero, the amplitude function diverges, violating the assumption of slow variation of $P(r)$ in ansatz (7.2). Such a collapse of the volume of the ray pencil occurs at <u>caustics</u>, as well as at the <u>stellar surface</u>. But when the ray C crosses a caustic, or is reflected on the stellar surface, we switch, in the representation (7.2), from one sheet j (= in) to a contiguous sheet j (= out) of the projection of T^3. The conservation condition (7.5), even though inadequate on a caustic, or on the surface, enables us to link the amplitude of the infalling wave, P_{in}, with that of the outgoing wave, P_{out}.

Let $\underline{r}_{in}$ and $\underline{r}_{out}$ be two points on C close to the caustic, on the incoming and outgoing ray branch respectively, chosen such that $|dV_{in}| \equiv |dV(\underline{r}_{in})| = |dV(\underline{r}_{out})| \equiv |dV_{out}|$. Write

$$dV = dx_1\, dx_2\, dx_3 \quad , \quad dx_{k\,out} = \sigma_k\, \exp(i\,\pi\,\gamma_k)\, dx_{k\,in} \quad , \quad k = 1,2,3, \quad (7.6)$$

where dx_1 is parallel, and dx_2 and dx_3 normal to the direction of the ray; $\sigma_k > 0$ is a 'stretching factor' ($\sigma_1\,\sigma_2\,\sigma_3 = 1$), and γ_k a phase factor. In the neighbourhood of the caustic, these factors can be directly read off from the ray pattern. At an ordinary caustic $dx_{1\,out} = dx_{1\,in}$, $dx_{2\,out} = dx_{2\,in}$, $dx_{3\,out} = - dx_{3\,in} = e^{i\pi}\, dx_{3\,in}$, or $\sigma_1 = \sigma_2 = \sigma_3 = 1$, $\gamma_1 = \gamma_2 = 0$, $\gamma_3 = 1$; at a focus $dx_{1\,out} = dx_{1\,in}$, $dx_{2\,out} = - dx_{2\,in}$, $dx_{3\,out} = - dx_{3\,in}$ or $\gamma_1 = 0$, $\gamma_2 = \gamma_3 = 1$. From (7.5, 6) we then have

$$P_{out} = \exp(-i\,\pi/2\,\,\gamma)\,\,P_{in} \quad , \tag{7.7}$$

$$\gamma = \gamma_1 + \gamma_2 + \gamma_3 \tag{7.8}$$

($\gamma = 1$ at a caustic, $\gamma = 2$ at a focus).

At the surface two alternatives may arise: (a) condition (2.3) is a genuine boundary condition, in the sense that the coefficients in this differential expression are not all vanishing. Then we require that the sum of the incident and reflected wave, $\Psi_{in} + \Psi_{out}$, obeys the boundary condition at the point $\underline{r}$ of collision of C with the surface, or

$$P_{out} = - \frac{\lim\limits_{\underline{r}_{in} \to \underline{r}_b} B(\underline{r}_{in}, \underline{k}_{in})}{\lim\limits_{\underline{r}_{out} \to \underline{r}_b} B(\underline{r}_{out}, \underline{k}_{out})}\,\, P_{in} = \exp(-i\,\tfrac{\pi}{2}\,\gamma)\, P_{in}\,. \tag{7.9}$$

This relation defines the phase shift factor γ. (For instance the boundary condition $\partial/\partial n\, \Psi = 0$ yields $\gamma = 0$).
(b) Alternatively, if the surface condition is just a regularity condition, as is the case in many stellar wave problems, it merely serves to eliminate singular solutions. It can again be shown that the in- and outgoing amplitudes are then related by an expression of type (7.7), with γ usually taking on noninteger values fixed by the surface singularities (cf the discussion in Perdang 1985b). Since in Eq. (7.2) the amplitudes P_j are defined as real positive functions, the phase effects introduced in the connection formulae (7.7) are to be counted as contributions to the phase function.

We next transform the propagating wave (7.1) into a standing wave by rendering the wave function $\Psi(\underline{r},t)$ single-valued in configuration space. To this end C is chosen to be a closed ray. Traverse this ray from $\underline{r}_0$ in the propagation direction back to the initial position $\underline{r}_0$. Then the total phase changes by $\Delta\phi$, due to (a) the direct change in the phase function $\Delta S = \oint_c dS = \oint_c d\underline{r}\cdot\underline{k}$, and (b) the different contributions in the phase shifts, $-\frac{\pi}{2}\gamma$, at each crossing of a caustic or reflection at the surface. The wavefunction is then single-valued provided that $\Delta\phi$ is a multiple of 2π. Or

$$\oint_c d\underline{r}\cdot\underline{k} = 2\pi\left[K + \mu(C)/4\right] \quad , \; K, \; integer, \tag{7.10}$$

$$\mu(C) = \Sigma\,\gamma \tag{7.11}$$

The sum (7.11) lists the phase contributions due to the amplitude changes at each passage through a caustic or surface reflection; $\mu(C)$ is referred to in the literature as the Maslov index of orbit C; notice that this index is defined modulo 4 only.

The line integral (7.10) is invariant under a deformation of the contour C as long as we do not cross a singularity of the integrand. In our 3-dimensional wave problem we have 3 equivalence classes $[C_j]$, $j = 1,2,3$, of such loops (the projections into configuration space of 3 independent irreducible circuits on the orbit supporting torus T^3 in phase space). Therefore, we choose our initial conditions to let C coincide with representatives of each equivalence class. In each case we have

$$J_j \equiv \frac{1}{2\pi}\int_{[C_j]} d\underline{r}\cdot\underline{k} = K_j + \mu\left([C_j]\right)/4 , \; j=1,2,3, \tag{7.12}$$

where J_j, $j = 1,2,3$ are actions. The notation $[C_j]$ reminds us that to perform the integral we can select any loop of class $[C_j]$, not necessarily an actual ray path. The 'quasiclassical quantisation rules' (7.12) for our 3-dimensional stellar standing waves (assumed integrable) are seen to introduce in general 3 quantum numbers. In the particular situation of spherical symmetry we have seen that the ray problem is effectively 2-dimensional; for the latter, we have only 2 independent loops, and therefore two quantisation rules (7.12) involving two quantum numbers.

The requirement of integrability of the ray Hamiltonian means that the latter can be expressed in the actions alone:

$$\Omega(\underline{r},\underline{k}) \equiv H(\underline{J}) \tag{7.13}$$

Inserting the quantised actions (7.12) we obtain the asymptotic regular spectrum

$$\omega(\underline{K}) = H(\underline{J}(\underline{K})) \quad,\; \underline{J}(\underline{K})=\left(J_1(K_1),J_2(K_2),J_3(K_3)\right) \tag{7.14}$$

Incidentally this discussion shows that regular eigenfrequencies
are not attributes of individual rays; they appear as properties
of the ray carrying tori (i.e. families of rays).

The previous procedure remains meaningful for nonintegrable
wave problems, at least locally, as long as the existence of
orbit carrying vague tori is secured. The quantisation
conditions (7.12) are then obtained from the irreducible circuits
of the vague tori. In principle, the wave Hamiltonian $\Omega(\underline{r},
\underline{k})$ can be approximated locally by a truncated Birkhoff normal
form

$$\Omega(\underline{r},\underline{k}) \simeq H_B(\underline{J}) = \underline{a} \cdot \underline{J} + \underline{J} \cdot A \cdot \underline{J} + \cdots \qquad , \qquad (7.15)$$

where $\underline{a}$, A, ... are constants $(\underline{a} = (a_1, a_2, a_3);$ A a symmetric
matrix, ..). Hence the frequencies

$$\omega(\underline{K}) \simeq H_B(\underline{J}(\underline{K})) = a_0 + \underline{a}' \cdot \underline{K} + \underline{K} \cdot A' \cdot \underline{K} + \cdots , \qquad (7.16)$$

a_0, a', A', .. being directly related to $\underline{a}$, A, ... Eq.
(7.16) provides the best regular approximation to the actual
quantum chaotic spectrum.

Relations (7.14) and especially (7.16) exhibit the main
feature of a regular asymptotic spectrum. The eigenvalues are
smooth functions of the quantum numbers. Specifically, if we fix
two numbers, $(K_2, K_3) = K_2^{\circ}, K_3^{\circ})$, and let the remaining quantum
number K_1 range over an interval $(\overline{K} - \Delta K/2, \overline{K} + \Delta K/2)$, $\Delta K \ll \overline{K}$,
then we obtain a family of evenly spaced frequencies supported by
a straight line L_o in the $\omega - K_1$ plane. If moreover we select
different pairs of fixed values $(K_2, K_3) = (K_2^j, K_3^j)$ $j = 1, 2, \ldots,$
f, and let again K_1 range over the previous interval, we define f
straight lines L_j in the $\omega - K_1$ plane, which typically intersect
L_o. Frequencies of the latter family close to these intersection
points are quasi-resonant with frequencies of the families L_j.
The existence of such quasi-resonances flows from the Poisson
distribution (3.10') for the level spacing, which in turn can be
shown to be a consequence of representation (7.16) (Berry and
Tabor 1977). Moreover, the total number of frequencies in the
interval $(\omega - \Delta\omega/2, \omega + \Delta\omega/2)$, $\Delta N(\omega)$, increases rapidly with ω, $\Delta\omega$
being fixed. This follows also from representation (7.16) (for
an alternative examination of the density of stellar acoustic
frequencies see Perdang 1982). Consequently the average spacing
$D(\omega)$ of adjacent frequencies near ω decreases rapidly with ω.
Quasi-resonances are therefore a dominant feature of the
asymptotic regular spectrum.

Consider next the effect of a nonintegrable Hamiltonian
perturbation (cf 6.1) on a specific frequency ω_r of the regular
spectrum (of family L_o). Since we have a whole cluster, $\Delta N(\omega_r)$,
of frequencies in quasi-resonance with ω_r, we apply degenerate
perturbation theory. Denoting by $H_{ij} = \omega_i \delta_{ij}$ and H'_{ij} the
nonperturbed and perturbed matrix elements of the wave operators

associated with the Hamiltonians (6.1, a or b) and computed with respect to the eigenfunctions of the nonperturbed Hamiltonian, the frequencies ω of the cluster around ω_r are given by

$$dtm \left[(\omega_i - \omega) \delta_{ij} + \varepsilon H'_{ij} \right] = 0 \quad , \qquad (7.17)$$

the subscripts ranging over the states of the cluster. The relevant frequency range we have to consider ($\omega_r - \Delta\omega/2$, $\omega_r + \Delta\omega/2$) must have a width $\Delta\omega/\omega_r \sim \varepsilon$. Let $N(\omega)$ be the total number of states of frequencies less than ω (we shall asume that this number does exist, as is the case for stellar acoustic modes). Since $N(\omega)$ obeys a Weyl-type law (cf Perdang 1982) the number of states in the interval of interest is

$$\Delta N(\omega_r) \sim D\, N(\omega_r)\, \Delta\omega/\omega_r \sim D\, N(\omega_r)\, \varepsilon \quad (\gg 1) \; , \qquad (7.18)$$

D being the sensitivity of $N(\omega)$ to ω (in excess of 3 in the stellar acoustic problem, but essentially of order 1). We can now obtain some general statistical information on the character of the solutions of the secular equation (7.17). Since the states involved in the frequency interval are essentially independent, the matrix elements H'_{ij} (in the absence of symmetries and under inequality 7.18) can be viewed as random variables. Matrix H' then becomes a representative of a <u>random-matrix ensemble</u>. For the present exploratory purposes we adopt the simplest, namely the Gaussian orthogonal ensemble; all off-diagonal elements then have same dispersion h^2, all diagonal elements have dispersion $2h^2$, and all elements have zero average. If we then neglect the differences between the unperturbed frequencies in the interval of interest ($\omega_i \equiv \omega_r$), and set

$$s = (\omega - \omega_r)/\varepsilon \quad , \qquad (7.19)$$

then the values s are just the eigenvalues of the random matrix H'. The average eigenvalue of the latter is zero, while the standard deviation, $\langle s^2 \rangle^{1/2}$, is given by (for $DN(\omega_r)\varepsilon \gg 1$)

$$\langle s^2 \rangle^{1/2} = \varepsilon^{1/2} \left[D\, N(\omega_r) \right]^{1/2} h \qquad (7.20)$$

(cf. Brody <u>et al</u>. 1981). This discussion is not applicable if the perturbation keeps a strong symmetry; due to the presence of selection rules a large number of matrix elements then vanish, so that the Gaussian orthogonal ensemble picture becomes inadequate.

If we plot the true family L_o of perturbed frequencies ω versus K_1, then instead of following a straight line, frequency ω_r is now statistically shifted, with a dispersion given by (7.19,20). Any other unperturbed frequency of the original straight line suffers a similar shift; provided that the K_1-interval is not too extended ($\Delta K \leqslant K$), $N(\omega)$ and likewise h will not change appreciably over this interval, so that the dispersion (7.20) remains the same for all frequencies. The deformed L_o-family of frequencies presents itself in the ω-K_1 plane as a collection of frequencies irregularly scattered around the original unperturbed line, the overall dispersion being given by Eqs (7.19,20); alternatively the spacings S(1) between successive frequencies of neighbour K_1-values fluctuate around the constant nonperturbed spacing, with a dispersion $\sigma(1)^2$ = $2\langle s^2\rangle\varepsilon^2$. The spectral scatter signals quantum chaos.

I have indicated elsewhere, using a moire effect to display the magnified spacings between all frequencies, that the properties outlined above are found in the exact spectrum of the quantum mechanical Henon-Heiles Hamiltonian; in contrast, an approximation by a truncated Birkhoff normal form (Eq. 7.15), even if carried out to a high degree in the actions, fails to exhibit the small-scale fluctuations in the spacings (Perdang 1985c).

As a final point I show that although these quantum chaotic spectral fluctuations are noise-like, they can be distinguished from observational noise. Suppose that the observations are sufficiently uniform over the range of frequencies of the L_o family of interest, so that each frequency can be assigned an observational error σ_o (root mean square error), the same for all frequencies of the interval. Let S(n) be the spacing between n-th neighbours of the L_o family (n = $|K_1 - K_1'|$); since a measurement of a spacing requires 2 frequency measurements, the scatter (root mean square error) $\sigma_o(n)$ of n-th neighbour spacings must obey

$$\sigma_o(n)^2 = 2\,\sigma_o^2 \quad , \tag{7.21}$$

if exclusively of observational origin. A scatter depending on n, on the other hand, is indicative of quantum chaos. More precisely, in the specific statistical framework considered above S(n) is a sum of n statistically independent direct neighbour spacings S(1). Therefore, the intrinsic scatter of n-th neighbour spacings obeys

$$\sigma_i(n)^2 = n\,\sigma(1)^2 \quad , \tag{7.22}$$

with $\sigma(1)$ related to the theoretical parameters as indicated above. If noise and quantum chaos coexist

$$\sigma_{tot}(n)^2 = 2\,\sigma_o^2 + n\,\sigma(1)^2 . \tag{7.23}$$

For n large enough, the quantum chaotic effect - if present - can always be made to be dominant.

VIII. Conclusion

The theoretical arguments listed in this paper indicate that the possibility of quantum chaos in the asymptotic frequency spectrum of stars, especially in the acoustic spectrum is to be taken seriously.   The most obvious candidates are strongly deformed stars (contact binaries) in which chaos can manifest itself in the low spectrum   already.   On the other hand, and perhaps more important, every small-scale perturbing mechanism breaking the spherical symmetry is a potential source of quantum chaos.   Since such perturbations of the structure are likely to occur in the majority of stars, quantum chaos of the latter origin should be a pervading phenomenon of the sufficiently high acoustic spectrum.

It is reassuring to note that as long as quantum chaos remains of a mild type, it may be ignored for current astroseismological purposes.   The condition is that we do not attempt to match individual theoretical and observational frequencies but that we statistically fit the theoretical values with averaged observational data; in fact as shown in our discussion, a statistical average over a sufficiently large stretch of frequencies eliminates the chaotic contributions.   The irreducible scatter of the observational frequencies about a smooth distribution of frequencies (low order polynomial fit) then provides a measure of the quantum chaotic effect (7.20,23). This scatter in turn conceals astroseismological information – for instance on the statistics of the small-scale perturbations, if the latter are responsible for this chaotic effect – which can be extracted by a detailed numerical study of the perturbation determinant (7.17) calculated at different reference frequencies.

As a test of the statistical procedure (Eqs. 7.21,22,23) we have analysed the 9 highest order modes of the $\ell = 1$ family of the solar South Pole data (Grec et al. 1983). For the reasons discussed in section VI, it is not likely that quantum chaos should occur in this family.   In fact we find that ($\sigma_{tot}'(n)/S(1))^2$, n = 1,2,3,4, is irregularly fluctuating over the range 1.6 – 3.7 x 10^{-4} ; this result is compatible with Eq. (7.21), yielding $\sigma_{\rho}' \sim 1.6 \mu$ Hz; it is certainly not consistent with Eqs. (7.22,23).   As far as we can draw a conclusion from this admittedly poor   statistics, the procedure indicates the absence of quantum chaos.

It would be interesting to carry out a similar analysis for the families of modes of high degree ℓ , ideally $\ell > 500$.   For the latter, mechanism (b) is in fact an obvious candidate for generating quantum chaos.

The author wishes to thank Douglas Gough for his comments. He thanks the Institute of Astronomy, Cambridge, for the hospitality extended to him, and the Royal Society London, for the grant of a European Exchange Fellowship.

References

Arnold, V.I., 1978.   'Mathematical Methods of Classical Dynamics', Springer, N.Y.

Arnold, V.I., Avez, A., 1968. 'Ergodic Problems of Classical
    Mechanics', Benjamin, N.Y.
Baltes, H.F., Hilf, E.R., 1978. 'Spectra of Finite Systems',
    Bibliographisches Institut, Mannheim.
Benettin, G., Strelcyn, J.M., 1978. Phys. Rev. A 17, 773.
Berry, M.V., 1977. J. Phys. A 10, 2061.
Berry, M.V., 1979. in 'Structural Stability in Physics' p 51. eds.
    W. Guttinger, H. Eikemeier, Springer Berlin.
Berry, M.V., 1981. Europ. J. Phys. 2, 91.
Berry, M.V., 1983. in 'Chaotic Behaviour of Deterministic Systems'
    p 171. Les Houches, Session XXXVI, 29 June-31 July 1981. ed.:
    G. Jooss, H.G. Hellman, R. Stora.
Berry, M.V., Tabor, M. 1977. Proc. Roy. Soc. London A356, 375.
Berry, M.V., Tabor, M., 1979. J. Phys. A10, 371.
Brody, T.A., Flores, J., French,J.B., Mello, P.O., Pandey, A.,
    Wong, S.S.M., 1981. Rev. Mod. Phys. 53, 385.
Brumer, P., 1973. J. Comp. Phys. 14, 391.
Bunimovich, L.A., 1974. Funkt. Analiz. Jego Prilog. 8, 73.
Casati, G., Valz-Gris, F., Guarneri, I., 1980. Nuovo Cimento
    Lett. 28, 279.
Casati, G., Chirikov, B.V., Guarneri, I., 1985. Phys.Rev.Lett. 54,
    1350.
Connes, A., Stormer, E., 1975. Acta Math. 134, 289.
Czuber, E., 1884. Sitzungsb. Akad. Wissensch. Wien, Math.-
    Naturwiss. Kl. 2A90, 719.
Eckart, C., 1960. 'Hydrodynamics of Oceans and Atmospheres',
    Pergamon, Oxford.
Emch, G., 1976. Comm. Math. Phys. 49, 191.
Fluhr, T., 1983. Memoire de Licence, Universite de Liege
    (unpublished)
Gough, D.O., 1984. Philos. Trans. Roy. Soc. A313, 27.
Heller, E.J., 1984. Phys. Rev. Lett. 53, 1515.
Henon, M., 1983. in 'Chaotic Behaviour in Deterministic Systems'
    p 53, Les Houches, Session XXXVI, 29 June - 31 July 1981.
    ed.: G. Jooss, H.G. Helleman, R. Stora.
Hose, G., Taylor, H.S., 1982. J. Chem. Phys. 76, 5356.
Jaffe, C., Reinhardt, W.P., 1982. J. Chem. Phys. 77, 5191.
Joyce, W.B., 1975. J. Acoust. Soc. America 58, 643.
Keller, J., 1958. Ann. Phys. 4, 180.
Keller, J.B., Rubinow, S.I., 1960. Ann. Phys. 9, 24.
Kosloff, R., Rice, S.A., 1981. J. Chem. Phys. 74, 1340.
Lazutkin, V.F., 1973. Izv. Akad. Nauk. USSR 37, 186.
Ledoux, P., Walraven, T., 1958. Handb. der Physik 51, 353.
Lindblad, G., 1979. Comm. math. Phys. 65, 281.
Luneburg, R.K., 1964. 'The Mathematical Theory of Optics', Univ.
    of California Press, Berkeley and Los Angeles.
Mandelbrot, B.B., 1982. 'The Fractal Geometry of Nature', Freeman,
    San Francisco.
Maslov, V.P., 1972.   'Theorie des perturbations et methodes
    asymptotiques' Dunod, Paris.
Mather, J.N., 1982. Ergodic Theory and Dynamical Systems 2, 3.

McDonald, S.W., 1983. 'Wave Dynamics of Regular and Chaotic Rays'
    Ph.D. Thesis, Univ. of California, Physics Department and
    Lawrence Berkeley Laboratory. LBL-14837.
McDonald, S.W., Kaufman, A.N., 1979. Phys. Rev. Lett., 42, 1189.
Meyer, H.D., Haller, E., Koppel, H., Cederbaum, L.S., 1984. J.Phys.
    A17, L831.
Noid, D.W., Koszykowski, M.L., Tabor, M., Marcus, R.A., 1980.
    J. Chem. Phys. 72, 6169.
Pechukas, P., 1972. J. Chem. Phys. 57, 5577.
Percival, I.C., 1973. J. Phys. B6, L229.
Percival, I.C., 1974. J. Phys. A7, 794.
Percival, I.C., Pomphrey, N., 1976. Mol. Phys. 31, 97.
Perdang, J., 1982. Astrophys. Space Sci. 83, 311.
Perdang, J., 1985a. 'Existence of a Pseudo-period in Stellar
    Chaotic Oscillations' (in preparation).
Perdang, J., 1985b. 'Stellar Short-wave Asymptotics' (in
    preparation).
Perdang, J., 1985c. 'Quantum Chaos and Moire Patterns' (in
    preparation).
Pomphrey, N., 1974. J. Phys. B7, 1909.
Poritsky, H., 1950. Ann. Math. 51, 446.
Pullen, R.A., Edmonds, A.R., 1981. J. Phys. A14, L477.
Ramaswamy, R., Marcus, R.A., 1981. J. Chem. Phys. 74, 1379,
    1385.
Richens, P.J., Berry, M.V., 1982. Physica 2D, 495.
Robnik, M., 1983. J. Phys. A16, 3971.
Santalo, L.A., 1976. 'Integral Geometry and Geometric Probability'
    Addison Wesley, N.Y.
Shapiro, M., Goelman, G., 1984. Phys. Rev. Lett. 53, 1714.
Shirts, R.B., Reinhardt, W.P., 1982. J. Chem. Phys. 77, 5204.
Steinhaus, H., 1950. 'Mathematical Snapshots' p. 204. Oxford
    University Press.
Stratt, R.M., Handy, H.C., Miller, W.C., 1979. J. Chem. Phys. 71,
    3311.
Uhlenbeck, K., 1976. Am.J. Math. 98, 1059.
Weissman, Y., Jortner, J., 1981. Chem. Phys. Lett., 78, 224.
Wigner, E.P., 1967. SIAM Rev. 9, 1.
Zaslavsky, G.M., 1981. Phys. Rep. 80, 157.
Zemlyakov, A.N., Katok,.A.B., 1975. Matem. Zametki 18, (2), 291.

An improved asymptotic formula for solar gravity-mode periods

A. N. Ellis
Department of Applied Mathematics
and Theoretical Physics,
Silver Street, Cambridge CB3 9EW, UK

The problem of high–order low–degree gravity–mode oscillations of a star with an outer convection zone is not new.   Tassoul (1980) has obtained a second–order asymptotic formula for the periods of such oscillations for a theoretical model with the following property: the gradient of the squared buoyancy frequency, N^2, is continuous (and negative) at the base of the convective region.   Realistic solar models preclude such behaviour of N^2 however because of the closely adiabatic stratification of the lower layers of the convection zone.   Mixing–length theories lead to an N^2 with a discontinuity in gradient at the convective interface. Tassoul proceeds to derive the following *first-order* formula for such a case:

$$P = \frac{P_0}{\sqrt{\ell(\ell+1)}}(n+\frac{1}{2}\ell-\frac{5}{12})$$

where

$$P_0 = 2\pi^2/\int_0^{r_c} |N|\ d\ln r.$$

Here P is the period, ℓ is the degree of the mode, n is the order, r is the radial coordinate and the base of the convection zone is where $r=r_c$.   Tassoul derives this using an asymptotic method in the convective region.   Her result agrees with that of Zahn (1970) who has studied the problem of a star with a convective *core*.   However, Zahn uses an essentially numerical method in the convective region, so it is interesting that both methods agree to first order.

At the second order, Zahn's treatment of the convection zone must be used.   For low–degree modes ($\ell\leqslant4$) Tassoul's asymptotic method is valid only for very high orders ($n>1000$). This follows from the fact that $|N^2|$ is very small in the convection zone.   Using Zahn's method then, one obtains the following:

173

D. O. Gough (ed.), Seismology of the Sun and the Distant Stars, 173–175.
© *1986 by D. Reidel Publishing Company.*

$$P = \frac{P_0}{\sqrt{\ell(\ell+1)}}(n + \frac{1}{2}\ell - \frac{5}{12} - \frac{1}{\pi}\tan^{-1}\frac{\sin\pi/3}{K_\ell P^{2/3} - \cos\pi/3}) + (P_1 + P_2/\ell(\ell+1))\frac{P_0}{P}$$

where

$$K_\ell = \frac{(3/4\pi^2)^{1/3}\Gamma(2/3)}{\Gamma(1/3)}|\phi'(r_c)|^{1/3}\left[(\ln\xi\rho^{1/2}r^2)'|_{r=r_c+} + \frac{1}{10}(\ln\phi')'|_{r=r_c-}\right]^{-1}$$

and

$$\phi = \frac{\ell(\ell+1)N^2}{r^2}.$$

Here ρ is the density, ξ is the lagrangian displacement and primes denote differentiation with respect to r. The constants P_1 and P_2, like P_0, depend only on the stratification throughout the radiative interior. They are given in a forthcoming paper (Ellis, not yet published, hereafter called Paper I). They can also be deduced from Tassoul's equation (99). The above formula has been independently discovered by Provost and Berthomieu (in press).

It would at first sight appear that K_ℓ depends on P through the term involving ξ. But this term is to lowest order independent of P so that K_ℓ can be regarded as a constant. Furthermore, the actual value of this term can be derived from acoustic-mode estimates of the sound speed in the convection zone and by subsequent integration of the oscillation equations inward from the surface. (For the method see Christensen-Dalsgaard, these proceedings: for sound speed estimates see, for example, Christensen-Dalsgaard et al, 1985.) Once this term is estimated one is left with only two free parameters, namely the first and second derivatives of N^2 at $r=r_c$. Observations of gravity-mode periods, and hence of K_ℓ, will put a single constraint on these two parameters.

Numerical studies presented in Paper I show that the periods of theoretical models fit the asymptotic formula to a tolerance of 10^{-5} for $n/\sqrt{\ell(\ell+1)} > 10$.

An important final point should be made. Suppose the buoyancy frequency vanishes like ζ^β where ζ is the depth below the base of the convection zone. Then the above formula has been derived for the case when $\beta = 1$. In theoretical models, this value for β arises from a mixing-length treatment of convection. Roxburgh (public communication, these proceedings) has stated that predictions of mixing-length theory should not be relied upon in detail. In the spirit of this advice, and following Gough's suggestion (1984), we may regard β as an unknown, to be deduced from observation. How does the asymptotic formula depend on β? Preliminary analysis (Ellis, 1984) shows that the formula is affected at the first order, thus

$$P = \frac{P_0}{\sqrt{\ell(\ell+1)}}(n + \frac{1}{2}\ell - \frac{1}{4} - \frac{1}{2(\beta+2)})$$

The analysis is extended to second order in Paper I.

Conclusions

The improved formula introduces a new quantity K_ℓ that is in principle observable and so can be used to probe the base of the convection zone of the sun. Furthermore, a knowledge of the higher-order terms should lead to a better estimate of the global parameter P_0.

References

Christensen-Dalsgaard J., Duvall T. L., Jr., Gough D. O., Harvey J. W., Rhodes E. J., Jr., 1985, Nature, 315, 378

Ellis A. N., 1984, 'Theoretical Problems in Stellar Stability and Oscillations' (ed Gabriel, Noels), 25th Liège International Astrophysical Colloquium, 290

Ellis A. N., 1985, Mon. Not. Roy. ast. Soc., (submitted)

Gough D. O., 1984, 'Oscillations as a probe of the sun's interior' (ed Belvedere, Paternò), Mem. della Societa Astronomica Italiana

Provost J., Berthomieu G., 1985, Astr. Ap., (in press)

Tassoul M., 1980, Ap. J. Suppl., **43**, 469

Zahn J-P., 1970, Astr. Ap., **4**, 452

SOLAR G MODES: A METHOD TO FIND THE DEPTH OF THE CONVECTIVE ENVELOPE

M. Gabriel
Institut d'Astrophysique de l'Universite de Liège
Avenue de Cointe 5
B-4200 Liège

ABSTRACT. It is shown that the phase ε in the first-order asymptotic formula for g^+ modes of models with a radiative core and a convective envelope is not a constant, but is rather a function of σ^2 and ℓ whose limit is 5/12 when σ^2 goes to zero. Application to the sun shows that the law $\varepsilon(\sigma^2, \ell)$ is very sensitive to the depth of the convective envelope. Accurate measurements of periods for g modes will give information on the depth of the solar convective envelope. A fitting method to obtain $\varepsilon(\sigma^2, \ell)$ is given.

1. INTRODUCTION

The first-order asymptotic formula giving periods of high-order g^+ modes of degree ℓ for models with a radiative core and a convective envelope can be written (see for instance Tassoul, 1980):

$$P = \frac{\Delta P}{\sqrt{\ell(\ell+1)}} \, (n+\ell/2-\varepsilon) \quad . \tag{1}$$

It has been used by several authors (Scherrer and Delache, 1983; Scherrer, 1984; Fröhlich and Delache, 1984; Isaak et al., 1984; Berthomieu et al., 1984; Gabriel, 1984a,b; Cox and Kidman, 1984) to discuss solar g^+ modes. They all use $\varepsilon = 1/4$. As Gough (1984) pointed out, this value of ε is obtained supposing that the derivative of the Brunt-Väisälä frequency N is continuous at the bottom of the convective envelope. As shown by Zahn (1970) and Tassoul (1980), when the first derivative of N^2 is discontinuous and $dN^2/dr = 0$ at the base of the convective zone, $\varepsilon = 5/12$. Ellis (1984) has generalized this result to cases where

$$N^2 = N_{oi}^2 (r_e - r)^\beta, \quad \beta > 0 \tag{2}$$

r_e being the radius at the bottom of the convective envelope.

177

D. O. Gough (ed.), Seismology of the Sun and the Distant Stars, 177–186.
© 1986 by D. Reidel Publishing Company.

He has shown that then

$$\varepsilon = 1/4 + (2(\beta + 2))^{-1}. \tag{3}$$

However another hypothesis is made in the derivation of equation (1) which seems to have been overlooked. It is supposed that $|N^2/\sigma^2| \gg 1$ everywhere in the star except in the vicinity of the turning point at r_e: that is to say in the convective envelope too. This condition will be fulfilled in the envelope only if $\sigma^2 < 10^{-4}$ (σ^2 is the dimensionless eigenvalue) or if $P > 16,000$ min. This requirement is so drastic that it is not satisfied for any observed g modes.

A good approximation for g modes with frequencies larger than 10 μHz, as those studied by Frölich and Delache (1984b), is to set $N^2 = 0$ in the convective envelope. We show that then $\varepsilon \neq 5/12$ in the range of observed periods and that $\lim \varepsilon = 5/12$ when σ^2 goes to zero. We also show that the value of ε for observed g modes is very sensitive to the depth of the convective envelope, so that this could be found through accurate observational measurements of ε. Finally we give a four-parameter formula which permits one to find $\varepsilon(\sigma^2, \ell)$.

2. ASYMPTOTIC SOLUTION

2.1 Normal form

The normal form of stellar pulsation equations in the Cowling approximation can be obtained easily from equations (1) and (2) of Gabriel and Scuflaire (1979) and following the general method given by Gabriel (1984c) or the classical text books (see for instance Ledoux, 1969 or Smeyers, 1984). It can be written for g modes, assuming $\sigma^2 \ll \ell(\ell+1)/r^2 c^2$ (c is the sound velocity), thus:

$$\frac{d^2 y}{dr^2} + \left[\frac{\ell(\ell+1)}{r^2} \left(\frac{N^2}{\sigma^2} - 1 \right) - f(r) \right] y = 0 \tag{4}$$

with

$$y = \rho^{\frac{1}{2}} u = \rho^{\frac{1}{2}} r^2 \delta r \tag{5}$$

$$f(r) = \frac{1}{2} \rho^{-\frac{1}{2}} \frac{d}{dr} \left(\rho^{-\frac{1}{2}} \frac{d\rho}{dr} \right) - \frac{d}{dr} \left(\frac{1}{\Gamma_1} \frac{d\ell np}{dr} \right)$$

$$= \frac{1}{4} \left[\mathcal{A} + \frac{1}{\Gamma_1} \frac{d\ell np}{dr} \right]^2 + \frac{1}{2} \left[\frac{d\mathcal{A}}{dr} - \frac{d}{dr} \left(\frac{1}{\Gamma_1} \frac{d\ell np}{dr} \right) \right] \tag{6}$$

$$\mathcal{A} = \frac{d\ell n\rho}{dr} - \frac{1}{\Gamma_1} \frac{d\ell np}{dr}$$

To solve equation (4) we suppose that $N^2/\sigma^2 \gg 1$ in the radiative core, use Olver's (1974) methods to get asymptotic solutions. We also consider only the first-order approximation. This means that the solution of equation (4) is supposed to be identical to that of the comparison equation. Notice that it would be meaningless to go to much higher approximations, as terms in (σ/N) have been neglected in the derivation of equation (4).

2.2 Asymptotic solution starting from the centre

The comparison equation is:

$$\frac{d^2 z}{dx^2} + \left[1 - \frac{\ell(\ell+1)}{x^2} \right] z = 0, \tag{7}$$

where

$$x = \sqrt{\frac{\ell(\ell+1)}{\sigma^2}} \int_o^r \frac{N}{r} \, dr, \tag{8}$$

$$z = \left[\frac{\ell(\ell+1)}{r^2} \frac{N^2}{\sigma^2} \right]^{1/4} y. \tag{9}$$

Its solution is

$$z = x^{1/2} (A \, J_k(x) + B \, Y_k(x)), \tag{10}$$

$$k = \ell + 1/2,$$

where J_k and Y_k are the Bessel functions of first and second kinds. The regularity of the solution at the centre imposes the condition $B = 0$. For $x \gg 1$ one obtains

$$u = r^2 \delta r = A \left[\frac{\sigma^2 r^2}{\ell(\ell+1) N^2 \rho^2} \right]^{1/4} \sin(x - \pi\frac{\ell}{2}). \tag{11}$$

2.3 Solution in the radiative core starting from the bottom of the convective envelope

We suppose that near the bottom of the envelope

$$N^2 = N_{oi}^2 \cdot (r_e - r)^n = N_{oi}^2 \, r_1^2. \tag{12}$$

Then Olver's transformations applied to equation (4) lead to

$$\frac{d^2 z}{dx^2} + \left[\frac{\ell(\ell+1)}{\sigma^2} x^n + \psi(x) \right] z = 0, \tag{13}$$

$$\psi(x) = k^{-3} \frac{d^2 k^{-1}}{dr^2} - k^{-4}(f(r) + \frac{\ell(\ell+1)}{r^2}),$$

$$x = \left[\frac{n+2}{2} \int_o^{r_1} \frac{N}{r} dr_1 \right]^{2/(n+2)},$$

$$z = ky \qquad k^2 = \frac{dx}{dr_1} = \frac{N}{r} x^{-n/2}.$$

It is easily verified that $\psi(0) \neq 0$. If $n = 1$, equation (13) corresponds to Olver's case B. If $n \geq 2$, Olver's methods can be applied only if $\psi(x) = x^n h(x^{n+2})$, h being an arbitrary analytic function. This condition is not satisfied for equation (13) and we will discuss only the case $n = 1$.

The first approximation of the solution of equation (13) is obtained by setting $\psi(x) = 0$. In terms of $u = r^2 \delta r$ it becomes

$$u = \left[\frac{\sigma^2 r^2}{\ell(\ell+1)N^2 \rho^2} \right]^{1/4} x^{\frac{1}{2}} \left[C J_{1/3}(\chi) + D Y_{1/3}(\chi) \right], \tag{14}$$

with

$$\chi = \sqrt{\frac{\ell(\ell+1)}{\sigma^2}} \int_o^{r_1} \frac{N}{r} dr_1 \tag{15}$$

For large χ, equation (14) becomes

$$u = \left[\frac{\sigma^2 r^2}{\ell(\ell+1)N^2 \rho^2} \right]^{1/4} \frac{D}{tg \psi_o \pi} \sin(\chi - \frac{5\pi}{12} - \psi_o \pi) \tag{16}$$

where $tg \psi_o \pi = -C/D$.

For small χ, equation (14) gives

$$u = \left[\frac{4}{3} \frac{\ell(\ell+1)}{\sigma^2} \frac{N_{oi}^2}{r^2} \right]^{1/6} \rho^{\frac{1}{2}} \left[\frac{C}{\Gamma(4/3)} (\frac{\chi}{2})^{2/3} \right.$$

$$\left. \frac{D}{\sin\pi/3} (- \frac{1}{\Gamma(2/3)} + \cos\frac{\pi}{3} \frac{1}{\Gamma(4/3)} (\frac{\chi}{2})^{2/3}) \right] \tag{17}$$

2.4 Continuity of the core solution

The continuity of the solution requires that the values of u and u' given by equations (11) and (16) are equal at any point where both x and X are large. These conditions are fulfilled when

$$\sqrt{\frac{\ell(\ell+1)}{\sigma^2}} \int_o^{r_e} \frac{N}{r} dr = \pi \left[n + \frac{\ell}{2} - \frac{7}{12} + \psi_o \right] \tag{18}$$

where n is the number of nodes of u.

2.5 Continuity at the bottom of the envelope

With $N^2 = 0$, we cannot apply the asymptotic method in the envelope when ℓ is not large (for high enough ℓ, $\ell(\ell+1)$ can be taken as the large parameter) and eigenfunctions can be obtained only through numerical integrations. Let the eigenfunction at the bottom of the envelope be given by u_e and u_e'. Continuity imposes that they are equal to the values given by equation (17) for X = 0. When ρ is continuous at r_e, this leads to a value of $tg\psi_o\pi = - C/D$ given by

$$tg\psi_o\pi = tg\pi/6 - 1.584 \left[\frac{\ell(\ell+1)}{\sigma^2} N_{oi}^2 r_e \right]^{-1/3} \left[\frac{1}{2}\frac{d\ell n\rho}{d\ell nr} + \frac{d\ell nu}{d\ell nr} - \frac{1}{3} \right]_{r_e} \tag{19}$$

where

$$1.584 = \frac{3^{2/3}\Gamma(4/3)}{\Gamma(2/3)\sin\pi/3} = \frac{3^{5/3}(\Gamma(4/3))^2}{\pi} .$$

When the solution in the convective envelope is given by the first term of the asymptotic series (i.e. when $|N^2/\sigma^2| \gg 1$), equation (19) reduces to

$$tg\psi_o\pi = tg\pi/6 + (\frac{N_{oe}}{N_{oi}})^{2/3} \frac{1}{\sin\pi/3} . \tag{20}$$

If, near the bottom of the envelope,

$$N^2 = -N_{oe}^2 (r-r_e), \qquad r > r_e \tag{21}$$

Equation (20) gives $\psi_o = 1/3$ when $N_{oi} = N_{oe}$ and $\psi_o = 1/6$ when $N_{oe} = 0$.

If u_e is the solution of equation (4) (in the outer envelope) satisfying the surface boundary condition obtained with $N^2 = 0$ for $r \geqslant r_e$, it is independent of σ^2 and equation (19) shows that:

a) ψ_o is different both from 1/3 and 1/6,
b) $\lim\psi_o = 1/6$

as $\sigma^2 \to 0$.

As N^2 is very small in most of the convective envelope for realistic models, these results remain true when u_e is the solution of the full pulsation equations, though $((d\ell nu)/(d\ell nr))_{re}$ is then a function of σ^2.

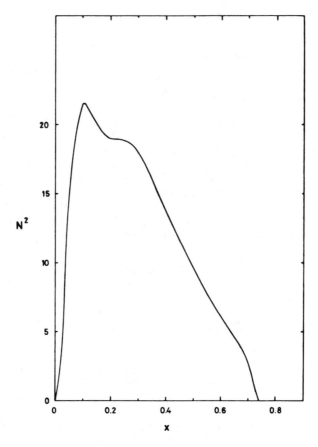

Figure 1.

It will be shown below that in the range of observed periods, even the
longest one studied by Fröhlich and Delache (1984b), $\psi_{o_2} > 1/6$. Equation
(19) shows that ψ_o is a function of the derivative of N^2 and of $\ell n y$ at
r_e.

          First, we notice that the dependence on N^2 is rather mild.
This situation is favourable, since N_{oi} must not be taken as

$$(\frac{dN^2}{dr})_{r_e}$$

but rather as an average over the domain where the condition $N^2/\sigma^2 \gg 1$
is not fulfilled. This makes the value of N^2 somewhat arbitrary. From
Figure 1, which shows N^2 in dimensionless units, we see that defining
N_{oi}^2 in the range $0.68 \leqslant x \leqslant x_e$, $N_{oi}^2 r_e \simeq 58$ while for $0.5 \leqslant x \leqslant x_e$ we
get $N_{oi}^2 r_e \simeq 30$. Nevertheless the uncertainty in $(N_{oi}^2 r_e)^{1/3}$ is only 25%,
and that in ψ_o only 6%.

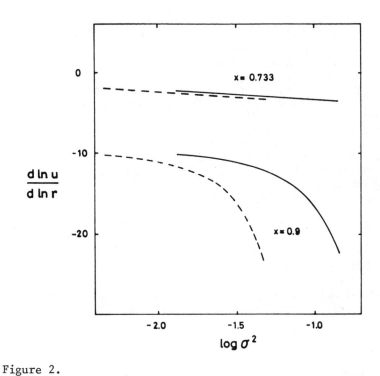

Figure 2.

TABLE 1. Values of the phase ψ_0 given by equation (19) for modes of
$30 \leqslant n \leqslant 100$ and $\ell = 1,2$ supposing that the bottom of the convective
envelope is at $x = 0.733$ and $x = 0.9$.

| | x=0.733 | | x=0.9 | |
n	$12\psi_0(\ell=1)$	$12\psi_0(\ell=2)$	$12\psi_0(\ell=1)$	$12\psi_0(\ell=2)$
30	3.78	3.80	5.25	5.25
35	3.63	3.65	5.04	5.04
40	3.51	3.54	4.88	4.88
45	3.41	3.44	4.76	4.76
50	3.33	3.36	4.65	4.00
55	3.26	3.29	4.56	4.56
60	3.19	3.22	4.49	4.49
65	3.13	3.16	4.41	4.41
70	3.09	3.10	4.35	4.35
75	3.03	3.02	4.29	4.29
80	2.98	2.98	4.24	4.24
85	2.94	2.94	4.19	4.19
90	2.91	2.93	4.14	4.14
95	2.86	2.89	4.09	4.09
100	2.82	2.86	4.05	4.05

Second, the last bracket in equation (19) varies rapidly with
the value of r_e. Figure 2 gives the value of $(d\ell nu)/(d\ell nr)$ in terms
of σ^2 for $\ell = 1$ and 2 at x = 0.733 and 0.9. At these two points,
$(d\ell n\rho)/(d\ell nr)$ is equal to -5.79 and -15.4 respectively.

Table 1 gives the phase ψ_0 computed from equation (19) for
$\ell = 1$ and 2, and $30 \lesssim n \lesssim 100$, supposing that the bottom of the convective envelope is at x = 0.733 and at x = 0.9. A value of $N_{0i}^2 r_e = 30$
was used. Evidently it should be possible to find the value of x_e when
periods are measured with an average accuracy of 10^{-3} unless N_{0i}^2 is very
poorly known. Then the two parameters would be entangled.

On the other hand, rather than using equation (1) with a
constant phase, observers should rather fit their data with an asymptotic formula which allows ε to vary. This introduces only two new
parameters, as the data in Table 1 can be fitted to the simple formula

$$\psi_o = 1/6 + \alpha n^{-k}.$$

We obtain k = 0.644 $\alpha = 1.325$ for x = 0.733,
        k = 0.383 $\alpha = 0.996$ for x = 0.9,
with an accuracy better than 1.5%.

We now outline a fitting procedure able to give ΔP, α and
k from high- and intermediate-order modes, and we show that it works.

3.   FITTING PROCEDURE

When observed periods are fitted to equation (1), ε will vary not only
because ψ_o is not constant but also because of the influence of higher-
order terms neglected in the discussion of the preceding chapter. So
we have to write quite generally

$$\varepsilon = 7/12 - \psi_o - \sum_{j=1}^{\infty} \psi_j \sigma^{2j},$$

where ψ_o is given by equation (22). For high and intermediate values
of n it seems appropriate to keep only the first term of the sum in
equation (23).

To check if the procedure works, we have made such a fitting
with the theoretical periods obtained from a standard solar model with
z = 0.018 and x_e = 0.733, with 651 points in the radiative core and
216 in the envelope. The computation method is that given in Gabriel
and Noels (1976).

Table 2 gives for a few modes of order n between 30 and 100
the dimensionless eigenvalues σ^2, the periods P, the computed phase
$\psi_c = 7/12 - \varepsilon$ obtained taking $\Delta P = 35.42$ min. and $\psi = \psi_o + \psi_1 \sigma^2$. Adopted
values for ψ_1 are 19 for $\ell = 1$ and 11.5 for $\ell = 2$. The good agreement
between the last two columns of these tables demonstrates that the
fitting procedure suggested above is able to provide values for α and k,
and finally the law $\varepsilon(n)$. From that law it will then be possible to
find the depth of the convective envelope.

Table 2a. The table gives the dimensionless eigenvalues σ^2, the periods, the computed phase $\psi_c = 7/12 - \epsilon$ and the phase $\psi = \psi_0 + \psi_1 \sigma^2$ for a few modes of $\ell = 1$.

n	σ^2	P(min.)	$12\psi_c$	12ψ
30	4.834604(-2)	759.13	4.72	4.70
35	3.568709	883.57	4.34	4.31
40	2.740731	1008.2	4.07	4.03
45	2.170312	1133.0	3.85	3.82
50	1.760867	1257.9	3.69	3.67
55	1.457093	1382.8	3.54	3.54
60	1.225580	1507.8	3.41	3.42
65	1.044984	1632.9	3.35	3.33
70	9.016123(-3)	1757.9	3.24	3.26
75	7.858205	1882.9	3.16	3.18
80	6.909552	2008.0	3.10	3.11
85	6.122917	2133.1	3.04	3.06
90	5.463224	2258.3	2.98	3.01
95	4.904569	2383.4	2.94	2.95
100	4.427628	2508.5	2.89	2.90

Table 2b. Same as Table 2a but for $\ell = 2$.

n	σ^2	P(min.)	$12\psi_c$	12ψ
30	1.398244(-1)	446.39	5.44	5.41
35	1.038040	518.09	4.95	4.84
40	8.006915(-2)	589.90	4.54	4.46
45	6.360981	661.83	4.23	4.17
50	5.173680	733.85	4.00	3.96
55	4.289958	805.90	3.79	3.78
60	3.614714	877.95	3.58	3.64
65	3.086410	950.13	3.48	3.51
70	2.665986	1022.3	3.37	3.41
75	2.326161	1094.4	3.21	3.29
80	2.047188	1166.6	3.12	3.19
85	1.815399	1238.9	3.09	3.15
90	1.620915	1311.1	3.02	3.12
95	1.456030	1383.3	2.97	3.06
100	1.315060	1455.6	2.94	3.01

4. CONCLUSION

We have shown that in the range of observed periods the phase ε appearing in the first-order asymptotic formula is not a constant. A fitting procedure is given to obtain it.

REFERENCES

Berthomieu, G., Provost, J. and Schatzman, E. (1984) Nature 308, 254.
Cox, A.N. and Kidman, R.B. (1984) in Theoretical Problems in Stellar
          Stability and Oscillations, 25th Liege Coll., 259.
Ellis, A.N. (1984) in Theoretical Problems in Stellar Stability and
          Oscillations, 25th Liege Coll., 290.
Frolich, C. and Delache, P. (1984a) Mem. Soc. Astron. Italiana 55, 99.
Frolich, G. and Delache, P. (1984b) in Solar Seismology from Space,
          JPL Publ. 83-84, 183.
Gabriel, M. and Noels, A. (1976) Astron. Astrophys. 53, 149.
Gabriel, M. and Scuflaire, R. (1979) Acta Astronomica 29, 135.
Gabriel, M. (1984a) Astron. Astrophys. 134, 387.
Gabriel, M. (1984b) in Theoretical Problems in Stellar Stability and
          Oscillations, 25th Liege Coll., 284.
Gabriel, M. (1984c) in Theoretical Problems in Stellar Stability and
          Oscillations, 25th Liege Coll., 64.
Gough, D. (1984) Mem. Soc. Astron. Italiana 55, 13.
Isaak, G.R., Van de Raay, H.B., Palle, P.L., Roca Cortes, T. and
          Delache, P. (1984) Mem. Soc. Astron. Italiana 55, 91.
Ledoux, P. (1969) in 'La structure interne des etoiles', 11th Saas-Fee
          course, 44.
Olver, F.W.J. (1974) Asymptotics and special functions, Academic Press.
Scherrer, P. and Delache, P. (1983) Nature 306, 651.
Scherrer, P. (1984) Mem. Soc. Astron. Italiana 55, 83.
Smeyers, P. (1984) in Theoretical Problems in Stellar Stability and
          Oscillations, 25th Liege Coll., 68.
Tassoul, M. (1980) Astrophys.J.Suppl. 43, 469.
Zahn, J.P. (1970) Astron. Astrophys. 4, 452.

THE INFLUENCE OF CONVECTION THEORY UNCERTAINTIES ON THE DEDUCTION OF THE SPEED OF SOUND IN THE SOLAR INTERIOR

Roger K. Ulrich

Abstract

Two models differing only in the surface layers are studied to determine how the derived frequencies are changed. These frequency changes can only be partially treated as being caused by a surface phase shift. Some portion of the uncompensated frequency change may be misinterpreted as a result of an internal difference between the models.

Introduction

The use of observations of the frequencies of solar oscillations to deduce the properties and dynamics in the solar interior is now reaching a point where detailed results can be derived. The information about localized properties of the solar interior is obtained by combining the results from as large a number of solar oscillation modes as possible. Those modes with periods near five minutes have been most thoroughly observed and their frequencies are reliably measured by a variety of observers.[1] Each mode of oscillation provides a different sampling of conditions in the solar interior and the range of information from the five minute oscillations is somewhat more restricted than we would like for use in deducing the properties of the interior. In particular, all the five minute mode frequencies are dependent both on the detailed properties of the solar surface as well as the properties of the interior. In fact the sensitivity to surface structure is greater than the sensitivity to interior structure so that the effects of the surface layers must be reduced or eliminated by combining data from many different five minute modes before the interior structure can be deduced.

The sensitivity structure of the modes is governed by the parameters which characterize the oscillations: ℓ, the degree of the spherical harmonic which describes the variation of the displacement amplitude over spherical surfaces and n, the order of the mode which gives the number of nodes in radial direction. To each value of n there is a corresponding ridge of oscillation power in the two dimensional plane which has as is horizontal axis the degree ℓ and as its vertical axis the frequency ν. In general terms the modes with smallest values of ℓ and highest values of ν sample the deepest parts of the solar interior. A particularly simple and revealing way to combine data from a

187

D. O. Gough (ed.), Seismology of the Sun and the Distant Stars, 187–198.

range in values of n was pointed out by Duvall[2] who showed that the
observed results from many of the ridges of power in the $\ell - \nu$ plane can be
combined by plotting $\pi \, (n + \alpha)/\omega$ against ω/L where $L^2 = \ell(\ell + 1)$.
After the proper choice of α the ridges nearly lie on top of each other.

The recent paper by Christensen-Dalsgaard, et al[3] has shown that the
observations of oscillation frequencies in the five minute band as summarized by
Duvall's relationship can be used to calculate the sound speed in the solar
interior from a direct inversion of an approximation to the dispersion relation.
This inversion method incorporates a representation of the effect of the surface
layers through a phase shift whose value is adjusted to provide the best fit to
the Duvall relationship. A variety of effects which are difficult to calculate
can influence the actual dispersion relationship of the solar atmosphere and the
success of the Christensen-Dalsgaard et al approach depends on the independence
of the deductions from the effects of the surface complications. Examples of
potential complications are the treatment of turbulent pressure, the calculation
of the detailed temperature profile in the super-adiabatic zone, the presence of
magnetic structures with field strength of 10 to 100 gauss and inhomogenieties
in the upper atmosphere. Each of these complications could alter the acoustic
properties of the medium and thus influence the oscillation frequencies. If the
complications are confined to a zone of extent less than the vertical wavelength
of the oscillation mode then the approximation of their effect through a phase
shift is probably adequate. On the other hand if the complications extend
through deeper layers then the net phase shift will depend in detail on the
structure of the oscillation and the depth dependence of the complication.
Since different modes have different structure near the surface, the effective
phase shift will depend on the specific mode under consideration. The inversion
then becomes vulnerable to misidentification of the location of the
complications i.e. the deduced internal properties of the sun could contain the
effects of phenomena which are actually present only on the surface. The
inclusion of modes in the inversion process which are restricted to the surface
layers can provide a calibration of the uncertain parameters.

This paper is discusses one set of surface uncertainties in order to
estimate the extent to which the solar interior properties deduced from the
Christensen-Dalsgaard et. al. procedure could be influenced by the surface
complications. One factor in the evaluation of the specific conclusions drawn
by Christensen-Dalsgaard et al is their use of an estimate of the error in the
observed frequencies for the high ℓ data which is too low. They used the
Harvey and Duvall[3] estimate of 10 μHz for the observational error for all
values of ℓ greater than 20. In fact that error estimate only applies to the
Duvall and Harvey data which extend only up to $\ell = 200$ and which is based on an
observing sequence of 5 days or about 110 hours. The frequency pixel size for
the lower ℓ data of about 2.5 μHz permits the isolation of individual modes of
oscillation. The higher ℓ data is based on an observing sequence of only two
days extending over about 36 hours. The frequency pixel size for this high ℓ
data is 7.7 μHz. Since the separation between successive modes is only on the
order of 5 μHz at $\ell = 300$ adjacent modes will blend into one another and make
it impossible to isolate individual eigenfrequencies. The inability to isolate
individual modes will undoubtably cause the error in this data to be larger than

10 μHz. Since the baseline of the high ℓ observations was not from limb to limb the number of modes included in a single spatial wavenumber bin encompases a range of about 6 in ℓ. Variations in the amplitude of the excitation of modes within each bin can then cause the measured frequency to vary by as much as ±15 μHz. Adding this variation in quadrature with the frequency bin width suggests that the error in the high ℓ data should be close to 20 μHz. The need to extrapolate the temperature profile to the surface as part of the inversion proceedure can magnify the random errors in the higher ℓ data. The importance of these and other errors is difficult to estimate because of the complex algorithm required between the input frequencies and the output distribution of sound speed. Compensation between various steps in the process could make the method appear less sensitive than it is in fact.

Compensation by the Surface Phase Shift

The basic dispersion relation which Christensen-Dalsgaard et al use is

$$k_v^2 = k^2 - k_v^2 = \frac{\omega^2}{c^2} - \frac{L^2}{r^2} \tag{1}$$

The vertical wave number k_v is a function of depth primarily because of the variation in sound speed c. An eigenfunction must have a $n\pi$ wavenumbers between the surface of the sun and the interior. After dividing by ω Christensen-Dalsgaard et al obtain a form of the dispersion relation as

$$F(w) = \int_{r_t}^{R} \left(\frac{r^2}{c^2} - \frac{1}{w^2} \right)^{1/2} \frac{dr}{r} = \frac{\pi(n+\alpha)}{\omega} \tag{2}$$

where $w = \omega/L$. The quantity α on the right hand side is the adjustable phase shift. The properties of the interior are deduced by inverting equation (2) to obtain an expression for the depth of each value of the sound speed in terms of an integral over F(w) which by the second part of equation (2) is an observable quantity. This approach will yield an interior distribution of c which is independent of surface effects only if F(w) is unaltered by any change which is confined to the surface regions alone. The phase shift α in $\pi(n+\alpha)$ can accommodate a one parameter family of variations in ω and the question we need to address is the extent to which the uncertain aspects of the surface regions cause changes that are incompatible with modifications in α.

To define the family of frequency changes which is compatible with altering the surface phase shift consider to models: Model 1 and Model 2. At a fixed value of $L = L_1$ the frequencies will be $\omega_{1,1}$ and $\omega_{1,2}$. In order for $F(\omega/L)$ to be independent of the change we must be able to pick another value of $L = L_2$ so that

$$\frac{\omega_{1,1}}{L_1} = \frac{\omega_{2,2}}{L_2} \tag{3}$$

At these new values of ω and L we must have

$$\frac{\pi(n+\alpha_1)}{\omega_{1,1}} = \frac{\pi(n+\alpha_2)}{\omega_{2,2}} . \tag{4}$$

The dependence of ω on L can be expressed in terms of an exponent

$$\beta = \frac{\partial \ln \omega}{\partial \ln L} . \tag{5}$$

The frequency $\omega_{2,2}$ can then be expressed in terms of

$$\delta(\ln L) = (L_2 - L_1)/L_1 \tag{6}$$

in two ways - first using equation (3) as $\omega_{2,2} = \delta(\ln L) \omega_{1,1} + \omega_{1,1}$ and second as $\omega_{2,2} = \omega_{1,2} + \beta \delta(\ln L)$. Equating these two expressions for $\omega_{2,2}$ then gives the required value of $\delta(\ln L)$. Equation (4) can then be used to relate $\delta(\ln L)$ to $\delta\alpha = \alpha_2 - \alpha_1$. Finally, the change in frequency $\delta\omega = \omega_{1,2} - \omega_{1,1}$ can be written as:

$$\frac{\delta\omega}{\omega} = \frac{1-\beta}{n+\alpha} \delta\alpha . \tag{7}$$

Equation (7) gives the pattern of frequency change which can be compensated by a change in α without altering the deduced internal distribution of sound speed. Patterns of frequency change due to the effects of surface uncertainties which deviate from equation (7) can lead to erroneous deduced internal structure.

The Convection Models

Of the various surface uncertainties only two mentioned thus far could potentially extend deep enough to be of concern to the inversion process - the convective theory and the possible presence of moderate strength magnetic fields. Since it seems somewhat unlikely that magnetic fields of high enough strength could extend over enough of the solar surface to alter the global frequencies, this discussion will concentrate on the issue of the detailed convection theory. The aspect of convection theory which is important in the context of the inverse determination of the sound speed is different from the usual goal of applications of convection theory. Normally, the adiabat of the isentropic zone of the convective region is the only quantity which is needed to compute stellar structure. This adiabat is one of the quantities which the inverse theory can presumeably derive if adequate data is available. The question at issue in this paper is instead the detailed distribution of the temperature and sound velocity within the superadiabatic zone just below the photosphere. The precise run of temperature and sound speed with depth is dependent on the specific formulation of theory which is used to derive the average properties of the solar atmosphere. Since the mixing length theory is

not regarded as adequate and no other firmly based theory is available to derive
this structure, we must regard the temperature and sound speed within the
adiabatic zone as unknown. The zone where this uncertainty is large extends
from the point of marginal convective instability near the photosphere to a
depth of a few pressure scale heights about 3000 km below the surface. At that
depth the difference between the true and adiabatic logorithmic temperature
gradients is only 0.002 while the value of the gradient is 0.16 and although the
convective flux continues to be uncertain, the residual effect on the thermal
structure is minimal. Unfortunately, the zone where the temperature is
uncertain is more than one vertical wavelenth thick and can not be treated as a
simple phase shift which is independent of ω.

In order to estimate the importance of the uncertainty in the temperature
stratification two models of the sun have been computed which have nearly
identical sound speed as a function of depth in the interior but which are based
on two different theories of convection at the surface. The first is based on
the standard mixing length theory as implemented by Henyey, Vardya and
Bodenheimer[4]. The second is based on the non-local theory described by
Ulrich[5]. The key feature of the second theory is the use of an iterative
procedure to globally adjust the thermal structure so that the convective flux
derived from a pair of integral equations is equal to the total flux minus the
radiative flux. Although there is some reason to believe that this treatment
should be a closer approximation to an exact description of the solar atmosphere
than the mixing length theory, it is not a fundamental theory because it retains
the usual semi-empirical mixing length parameter. The calculation of the
convective flux from an integral over the structure having an extent of at least
one mixing length instead of assuming that the local parameters are valid for a
mixing length is the principal improvement over the standard theory. Since the
superadiabatic gradient changes by more than an order of magnitude over a mixing
length, the elimination of the local assumption would seem likely to lead to an
improved theory. In fact the original hope by Ulrich that the interior adiabat
would be less sensitive to the mixing length when the non-local theory is used
was only marginally realized and the theory has not proven worth the effort to
implement and utilize extensively. The advent of Helioseismology as a probe of
solar atmosphere and interior structure now provides an opportunity to check
on the predictions of this non-local theory as well as other theories of
convection such as those of Spiegel, Unno and Gough[6]. For the present paper,
the theory of Ulrich has been used because a working code implementing that
approach exists at UCLA and required little effort to apply to the problem of
the deduction of the interior sound speed. Future work will consider the
question of checking on these theories through the high degree mode frequencies.
Here, the Ulrich theory is simply one possible way to calculate the structure of
the superadiabatic zone.

The principal change between the local and the non-local theories is the
displacement inward of the peak in the superadiabatic gradient by about a
quarter of a mixing length and the steepening of the temperature gradient.
These two effects occur because at the point where the local mixing length is
able to bring about an abrupt increase in the convective flux, the non-local
theory is unable to respond. Instead, the gradient steepens just interior so

that overshoot flux can make up the difference in the critical transition layer.
Figure 1 shows a comparison of the temperature distribution for the two models
and both effects are readily apparent. Both models join onto complete solar
interiors. Since the usual method of matching the observed solar radius and
luminosity is to adjust the average intitial helium abundance and the
atmospheric mixing length, the nearly identical interior temperature for the two
models might appear fortuitous. In fact, the treatment of the turbulent
pressure in the upper layers is also important and the ratio of the turbulent
pressure to the convective flux has been adjusted to bring about the match of
interior temperatures which is shown.

The Influence on the Inversion Process

The key quantity treated by the inversion process is the sound speed.
Figures 2 and 3 show the difference between the sound speed in the non-local
model minus the sound speed in the standard model. Figure 2 shows the
difference near the solar surface in a scale which is stretched in r and
compressed in $\delta c/c$. Athough the models are not identical in their interiors,
the size of the difference is smaller than the surface differences by a factor
of roughly 50. This contrast should be adequate to isolate the effects of the
surface from the effects of the interior - essentially the entire frequency
differences between the two models should be attributable to the surface rather
than the interior.

Figure 4 shows the frequency changes between the two convection theories
plotted against the parameter ω/k_h which is used in the inversion. Each
section of these figures shows a pair of lines - the heavier of the two is the
actual frequency difference while the lighter line is calculated from equation
(7) with $\delta\alpha = -0.003$. The figures show only those portions of the theoretical
results which are verifiable using data from Harvey and Duvall[7]. This
restricts the frequency range to 1800 μHz $< \nu < $ 4800μHz. The value of $\delta\alpha$
was chosen to provide a good fit for the higher n modes which seem to be better
behaved. Clearly not all the frequency changes can be accomodated by the
surface phase shift parameter.

The structure of the eigenmodes near the solar surface depends primarily on
the frequency and not on the value of ℓ. Consequently, the differences between
the models which are confined to the surface regions should alter all the
mode frequencies near each frequency value by the same amount. We can find such
surface induced effects by plotting the frequency shift against the frequency
instead of against the Duvall Law variable. Presumably, those frequency changes
which are independent of ℓ can be attributed to the surface alone. Figure 5
shows the frequency shifts plotted in this way. Again it is clear that a
portion but not all the frequency shift can be isolated in this way.

The results of the numerical experiment shown in figures 4 and 5 indicate
that at a level of 0.1 to 0.5 percent the details of the surface layers can
produce changes in the frequencies which are not compatible with the assymptotic
theory. Variations at this amplitude were noted by Christensen-Dalsgaard et al

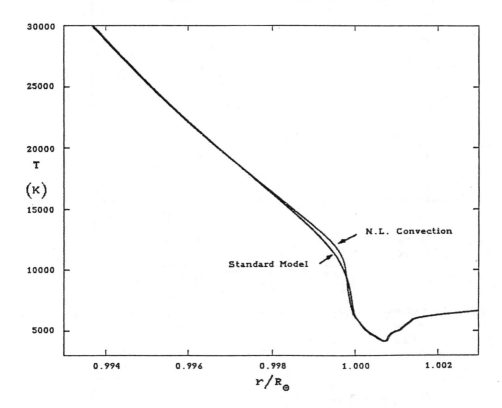

Figure 1.   Comparison of the temperature stratification near the solar surface computed with the local and non-local convection theories.

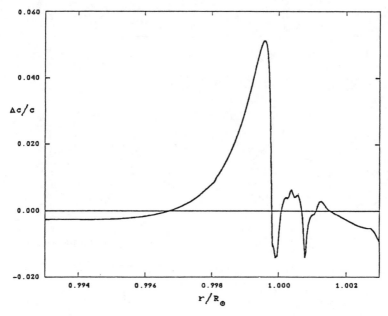

Figure 2. The difference between the sound speed in the non-local model minus
the sound speed in the standard local model. This figure shows the
structure near the solar surface.

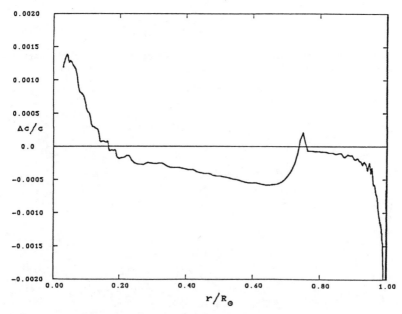

Figure 3. The difference between the sound speed in the non-local model minus
the sound speed in the standard local model. This figure shows the
structure throughout the model.

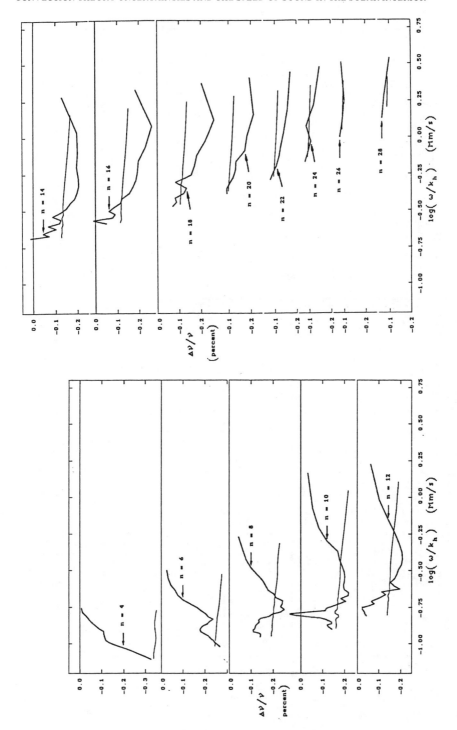

Figure 4.   The frequency change caused by going from the non-local model to the local model plotted against the variable ω/k_h

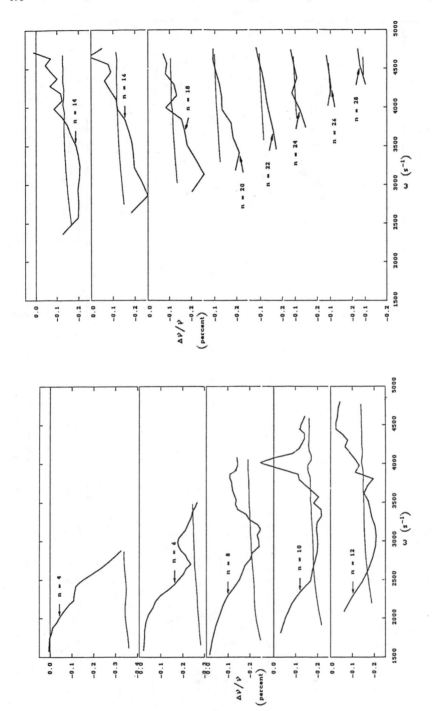

Figure 5.   The frequency change caused by going from the non-local model to the
local model plotted against the variable ω.

and attributed to errors in the asymptotic analysis. As long as the only modes used to analyze the solar structure involve both the surface and the interior, the deduction of interior structure alone will require the successful cancelation of the effects of the surface structure variations in order to avoid altering the deduced interior sound speed. Errors in the measured frequencies assume greater importance when the relatively large effects of the surface uncertainties must be canceled out. A more straightforward analysis would become possible if highly accurate frequencies were available for the highest ℓ modes which do not penetrate below the surface region.

Conclusions

Any conclusions drawn from the numerical experiment described above must be prefaced that they are based on a sample of possible variation rather than an upper bound. The variation in c shown in figures 2 and 3 is fairly large and may be representative of a significant variation in the theory. The temperature gradient is bounded from below by the adiabatic value and from above by the radiative value. Additionally, the envelope adiabat is constrained by the requirement that the model match the solar radius. One can then define a domain in the T vs P plane within which the actual atmosphere structure should remain. The two models presented above are close to opposite edges of this domain so that the difference between the standard model and the non-local model may be indicative of a large change rather than a small change. Within the context of this caveat the following conclusions can be drawn:

1. The detailed convection theory can introduce uncertainties in the deduced interior sound speed on the order of 0.2 to 0.4 percent.

2. The extrapolation from $\omega/k_h = 10^{-1}$ to $\omega/k_h = 10^{-2}$ at the solar surface is sensitive to the surface uncertainties. The underestimate of the error in the highest ℓ frequencies by Christensen-Dalsgaard et al makes this extrapolation a source of uncertainty.

The temperature profile deduced for the solar interior by Christensen-Dalsgaard et al is independent of the surface convection uncertainties at a level small enough to leave their overall shape intact. The extrapolation problem could cause the apparent depth of the pattern of sound speed to be in error. Improved data at the highest values of ℓ can reduce the extrapolation uncertainty.

References

1. Claverie, A., Isaak, G.R., McLeod, C.P., van der Raay, H.B., and Roca Cortes, T. Nature, 282, 591, 1979; Grec, G., Fossat, E. and Pomerantz, M., Nature, 288, 541, 1980; Duvall, T.L. Jr., and Harvey, J.W., Nature, 302, 24, 1983.

2. Duvall, T.L., Jr., Nature, 300, 242, 1982.

3.   Christensen-Dalsgaard, J., Duvall, T.L., Gough, D.O., Harvey, J.W., and
     Rhodes, E.J.,Jr., Nature, <u>315</u>, 378, 1985.

4.   Henyey, L., Vardya, M.S. and Bodenheimer, P. 1965, Astroph. J., <u>142</u>,
     841.

5.   Ulrich, R.K., Astr. and Sp. Sci., <u>7</u>, 71, 1970a; Ulrich, R.K., <u>7</u>,
     183, 1970b; Ulrich, R.K. 1976, Astroph. J. <u>207</u>, 564.

6.   E.A. Spiegel, Astroph. J., <u>138</u>, 216, 1963; W. Unno, Pub. Astron. Soc.
     Japan, <u>21</u>, 240; Gough, D.O., Astroph. J., <u>214</u>, 196, 1977.

7.   Harvey, J.W. and Duvall, T.L. Jr., 1984, Proceedings Conf. on Solar
     Seismology from Space, (eds. Ulrich et al), 165, (JPL Publication 84-84,
     Pasadena, 1984).

MEASURING THE SUN'S INTERNAL ROTATION USING SOLAR p-MODE OSCILLA-
TIONS

Timothy M. Brown
High Altitude Observatory/National Center for Atmospheric Research
P.O. Box 3000
Boulder, CO 80303
U.S.A.

ABSTRACT. 2-dimensional velocity images of the Sun were obtained for 5 days in June, 1984. This time series of images has been analyzed to obtain frequencies of solar p-mode oscillations with degrees between 8 and 50, with all azimuthal orders for each degree. I describe the analysis of these observations in some detail. The principal results of the analysis are measurements of the frequency splitting between modes with the same degree and radial order; these are related to the latitudinal variation of solar rotation, averaged over a depth range that depends on the degree. The observed spittings suggest that for 0.3 $R_\odot \leq r \leq 0.7\ R_\odot$, the solar latitudinal differential rotation is much smaller than at the surface (i.e., the rotation is roughly constant on spheres), and moreover than the rotation rate is close to the surface equatorial value. A preliminary analysis of the data shows no significant variation in the average mode amplitude with azimuthal order, for all degrees less than 50.

1. INTRODUCTION

The way in which the solar rotation varies with depth and latitude is important for our understanding of the structure and evolution of the Sun, and of other stars. For example, a clear picture of the Sun's internal rotation would provide a test of current theories of stellar convection and dynamo action (e.g. Gilman and Miller 1981), would allow one to assess theories of slow mixing in the solar interior (Schatzman 1969,1985, Michaud 1985), and may cast some light on the history of mixing episodes within the solar core (Dilke and Gough 1972). Moreover, recent advances in measuring and interpreting measurements of solar oscillations have made the solar internal rotation accessible to observation.

Duvall and Harvey (1984) have measured the rotationally-induced frequency splitting between oscillation modes with $m = \pm l$ for a range of n and l values, where n is the *radial order* of the oscillation mode, l is its *angular degree*, and m is its *azimuthal order*. These measurements allow one to infer the variation of solar rotation with depth at low latitudes (Duvall, et al. 1984). However, since these sectoral oscillation modes have large amplitudes only near the solar equator, they carry no information about the latitudinal variation of rotation. The observational technique used by Duvall and Harvey integrates over one spatial dimension on the Sun before the data are recorded; thus it is not possible to use this technique to measure frequencies simultaneously for all m values at a given l. In order to do such a measurement, it is necessary to obtain time series of 2-dimensional images of the Sun.

199

D. O. Gough (ed.), Seismology of the Sun and the Distant Stars, 199–214.
© *1986 by D. Reidel Publishing Company.*

The Fourier Tachometer (Beckers and Brown 1978, Evans 1981, Brown 1985a) (henceforth FT) is a new instrument capable of observing the solar velocity field with the precision required for detecting global solar oscillations, and with 2-dimensional spatial resolution. I have used the FT to obtain a series of observations that allow some statements to be made about the latitudinal variation of the solar rotation; a summary of these results is being published elsewhere (Brown 1985b). The purpose of this paper is to provide more detail on the findings themselves, on the methods used in obtaining them, and on various intermediate results.

2. NECESSARY THEORY

The principal effect of rotation on solar oscillations is one of advection; the sound waves responsible for the oscillation signal are carried along with the rotating Sun. As seen from an inertial frame, the observed frequencies of waves propagating in the direction of the solar rotation (prograde waves) are then increased, while those of retrograde waves are decreased. The magnitude of the frequency shift is proportional to m, since m is, in effect, the component of the wavevector aligned parallel to the rotational motion. If the solar rotation were constant on spheres, then (neglecting Coriolis forces) the variation of frequency within an oscillation *multiplet* (modes with the same n and l, but different m) would be given by

$$\nu(n,l,m) - \nu(n,l,0) = -m\,\overline{\Omega}\ , \tag{1}$$

where $\overline{\Omega}$ is an appropriate average of $\Omega(r)$ over the depth range sampled by oscillations with the given n and l. Compared to simple advection, Coriolis forces usually play a small role (see, for example, Brown et al. 1986), while the effect of other influences on the frequencies (centrifugal distortion of the Sun, magnetic effects) is expected to be smaller yet.

If the solar rotation is not the same everywhere, then the observed frequency shift for an oscillation mode is an average over the Sun of the local frequency shift, weighted according to the local energy density in the mode. Thus, modes with large l (which are confined to a shallow layer near the solar surface) are unaffected by the rotation in the solar core. Similarly, modes with $m = l$ are confined near the solar equator, and are sensitive only to the rotation at low latitudes; modes with $m = 0$ have their largest amplitudes near the poles, and respond principally to the rotation at high latitudes. These notions can be quantified using the results by Cuypers (1980). Assuming that the variation of Ω with $\cos\theta$ is negligible compared to that of P_l^m, and ignoring centrifugal and Coriolis forces, one may write

$$\nu(n,l,m) - \nu(n,l,0) = -m\,\frac{\displaystyle\int_{-1}^{1}\overline{\Omega}(\theta)\,[P_l^m(\cos\theta)]^2 d\cos\theta}{\displaystyle\int_{-1}^{1}[P_l^m(\cos\theta)]^2 d\cos\theta}\ . \tag{2}$$

Two interesting points arise immediately from the form of Eq. 2. Since the integrals in Eq. 2 are symmetric in m, the frequency perturbations for slow rotation are antisymmetric in m. Thus, in the presence of latitudinal differential rotation one should expect

$$\nu(n,l,m) - \nu(n,l,0) = -\Omega_0 m + \Omega_2 m^3 + \cdots \tag{3}$$

For the same reason, the part of $\overline{\Omega}(\theta)$ that is antisymmetric about the equator has no effect on the observed frequencies. In what follows, I will therefore assume that $\overline{\Omega}$ is equatorially symmetric.

The foregoing considerations lead directly to the following program for measuring the variation of solar rotation with latitude and depth:
(1) Measure $\nu(m)$ for a range of n and l values.

(2) Use the *linear* part of $\nu(m)$ to determine the part of $\Omega(r,\theta)$ that is *independent of latitude*.
(3) Use the *cubic* and higher terms in the expansion of $\nu(m)$ to determine the *latitudinal differential rotation*.
(4) Use the way in which these parameters *vary with l* to determine the *depth variation* of the rotation and the differential rotation.

3. FOURIER TACHOMETER CAPABILITIES AND PARAMETERS OF THE DATA SET

The FT is an interferometric instrument that measures the Doppler shift of a selected spectrum line with 2-dimensional spatial resolution and with high precision and stability. The details of its construction and operation will be published elsewhere (Brown, et al. 1986, in preparation). For the current purposes, only a few of these details are relevant. The detector used in the FT is a 100 x 100 pixel Reticon. With the image scale adjusted so that the entire Sun fits onto the detector, this yields a resolution of 19.5 arcsec per pixel. This implies that the maximum possible l accessible to the instrument at this image scale is about 150. At the time these observations were taken, the FT could measure apparent line-of-sight velocities with a precision of about 6 m/s for a single pixel and an integration time of 60s. Summing over the whole solar disk yielded values that were precise to typically 15 cm/s in a 60s integration. The stability of the measured velocities during this observing run was typically rather poor: zero point drifts of 2000 m/s during the day were common, and had to be removed from the data by least-squares fitting (see below). The stability and sensitivity of the FT have since been substantially improved.

The observations I will discuss here were taken between 8 June and 12 June, 1984. Observations were taken for typically 9.5 hr per day, with one data gap of 2 hr. duration on the first day, and no other gaps longer than 5 min. Images of the telescope objective lens were taken at the beginning and end of each day, providing a measure of the field variations in the instrumental Doppler shift. The observational cadence yielded one velocity image of the Sun every twenty seconds, but as the first step in the reduction, these were averaged in groups of three to give one image per minute. All of the observations measured the Doppler shift of the Fe I 5434 line, which is a moderately weak $g = 0$ line, formed about 300 km above $\tau = 0$ in the continuum (Lites 1972). This run was originally intended as an engineering run, but the results were sufficiently good that I was encouraged to use them for studying solar oscillations.

4. DATA ANALYSIS

Analysis of the velocity images to yield power spectra proceeded in three major steps. First, the raw velocity maps were reduced to properly oriented maps of *residual velocity*, containing only components of the velocity field with spatial scales smaller than about a solar radius. Next, these maps of residual velocity were decomposed into their spherical harmonic components via a spatial filtering process. Finally, timestrings of the amplitude of each spherical harmonic component were Fourier transformed, and the modulus squared of the result computed to yield the power spectra. In what follows, I will discuss each of these steps in more detail.

The first step in producing velocity residual maps was to remove the instrumental contribution to the observed Doppler shift. This contribution arises principally from errors in the interferometer construction, and appears as a roughly constant gradient of velocity across the field of view, with a total variation of about 3000 m/s from one side of the field to the other. This pattern can be measured quite accurately, using the image of the telescope objective as a source. This was done twice per day during the June 1984 run, once at the beginning of the day and once at the end. Differences between the morning and afternoon measurements consisted predominantly of a change in zero point of typically 1800 m/s, and a change in the gradient of about 200 m/s across the solar image. These variations were fairly repeatable from day to day, and undoubtedly resulted from the complete lack of thermal control of the interferometer, combined with the regular diurnal temperature variation inside the SPO Big Dome. Since, for this run, we obtained neither frequent calibration measurements nor temperature measurements of any sort, the instrumental correction was taken to be simply the value linearly interpolated in time between the morning and afternoon measurements.

An additional complication in removing the instrumental bias arises because the solar image rotates once per day with respect to the FT interferometer. This causes the bias pattern to rotate relative to the solar image, so it is necessary to use a numerical interpolation procedure to rotate the measured bias to the correct orientation before subtracting it from the raw velocity maps.

The FT detector rotates to follow the daily rotation of the solar image. However, during the June 1984 run, the zero point for this rotation was chosen incorrectly. Thus, after removing the instrumental bias, it was necessary to rotate the solar velocity images (again by numerical interpolation) so that the solar rotation axis was aligned with the y-axis of the data array. This rotation was through an angle of typically 15°, and the rotation angle itself could not be determined to better than about 1°. This error in the orientation of the image may have caused some problems in the ensuing analysis, but they are not expected to be large.

At this point, the images consisted of putatively correct maps of the total solar velocity field relative to the instrument, in a standard orientation. Next, the velocity fields due to the mean solar rotation and the limb red shift were removed. The former was taken to be the mean solar differential rotation described by Howard and Harvey (1970), while the latter was assumed to be described by $\delta v = 400 \, m/s \, \sin^4 \psi$ where positive velocities imply a red shift and ψ is the angle between the line of sight and the local vertical on the Sun.

The final corrections for known effects consist of removing the velocity field associated with the orbital and rotational motion of the Earth. These corrections are computed in a straightforward (if tedious) fashion from ephemeris data. Their maximum magnitude approaches 1000 m/s in the zero point offset, with up to 150 m/s variation in the apparent velocity across the solar disk.

Given a perfect instrument, the velocity maps should at this point consist of nothing but the line-of-sight component of the solar velocity field, measured relative to a mean differential rotation. Note that since no fitting operations had yet been done, time-varying motions of all spatial and temporal scales should still be recoverable from the data. Unfortunately, instabilities in the instrument and (perhaps) inadequacies in the analysis combined to give residual velocity images that were still dominated by large-scale, slowly changing structures with velocity amplitudes of typically 200 m/s. To remove these, it proved necessary to remove a least-squares fit biquadratic form in x and y (effectively sine latitude and sine meridian distance) from each velocity image. This procedure brought the velocity rms down to typically 70 m/s, but also filtered out all solar oscillations with $l = 0$–1, and greatly reduced the sensitivity for $l = 2$–3.

After producing the velocity residual maps, it was necessary to apply spatial filters to them in order to estimate the time-varying amplitudes of the various spherical harmonic components. This process was facilitated by the fact that the solar B angle passed through zero just before the observing run began, and had an average value through the run of less than 0.5°. Thus, for a reasonably large range of l, the solar rotation axis could be taken to lie in the plane of the sky. Spherical harmonic amplitudes were computed for all l and m values out to $l = 50$; this limit on l was chosen primarily to limit computation time.

The filtering method used first apodizes the velocity residual maps in the radial dimension: points further than $0.9 \, R_\odot$ from the image center are set to zero, and points between $0.8 \, R_\odot$

and $0.9R_{\odot}$ are multiplied by a cosine bell to provide a smooth transition between the interior and exterior of the image. Next, maps of velocity on an equally-spaced grid of longitude and y are constructed by interpolating along rows (lines of constant latitude) in the apodized image, and the interpolated values are extended with zeros to span 2π radians in longitude. Each row of this velocity map is then individually Fourier transformed, yielding a complex array containing amplitude as a function of y and the absolute value of the azimuthal order $\mid m \mid$. With the standard FFT routine used, the real part of each complex value represents the component of the velocity field symmetric about the Sun's central meridian; the imaginary part represents the antisymmetric component. It is important to note that, for a given m index, both parts are combinations of spherical harmonics with azimuthal orders of $\pm m$; it is not possible to distinguish the prograde from the retrograde modes except by considering their temporal behavior, i.e., until a temporal Fourier transform is performed.

After Fourier transforming in longitude, the final step is to Legendre transform the amplitude array along the latitudinal direction. This is done by computing the Legendre polynomials for all possible l values at each m, and forming a suitable dot product with the complex amplitudes at that m. In this case it is convenient to have the data values equally spaced in $\sin\theta$, so no interpolation is done in the latitudinal direction. The Legendre polynomials are computed as needed, using an efficient and stable recursion procedure described by Christensen-Dalsgaard (1985).

This filtering method (like any other linear method) does not completely isolate individual spherical harmonics from their neighbors, because the spherical harmonics themselves are not orthogonal on the observable solar hemisphere. Results using it on a number of test cases show that it is not much worse than the optimized response filters discussed by Christensen-Dalsgaard (1985). Nevertheless, the additional effort required to optimize these filters is probably worthwhile.

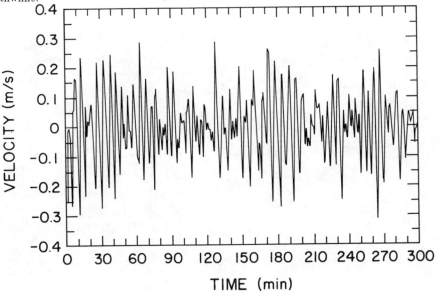

TIME (min)

FIG. 1. A time series of the imaginary part of the filter output for $l = 15$, $m = 2$, sampled once per minute for 5 hrs. The peak velocities appearing in series like this are about 0.3 m/s, and the rms noise (estimated from power spectra) is below 10 cm/s. The real and imaginary parts of the various filters are identical in their general form, though of course they do not resemble each other in detail.

In order that the spatial filters correspond accurately to spherical harmonics of the chosen degree and azimuthal order, it is important that the radius of the Sun assumed during the various interpolation and integration steps be correct. In fact, a 3% error was made at this point, causing spatial filters that should have responded to spherical harmonics of degree l (the *nominal degree*) to respond instead to those of degree $1.03 \cdot l$. This has two important effects. First, l and m are systematically misidentified, by amounts that are sometimes significant (see below). Second, it presumably causes the filters to have a greater sensitivity to nearby spherical harmonics than would otherwise be the case, particularly at large l. In what follows (including the Figures), I will discuss the results in terms of the nominal degree, uncorrected for the error in solar radius.

After spherical harmonic amplitudes were computed for all of the observed velocity maps, the data were reordered so as to produce timestrings of the complex amplitude of each spherical harmonic. Fig. 1 shows a 5-hour segment of the real part of one such timestring. Zeros were inserted to allow for the gaps between the individual days, and each time series was extended with zeros to a length of 8192 min (5.689 solar days). Small gaps in the time series (5 min or less) were filled using a bidirectional autoregressive prediction scheme similar to that described by Fahlman and Ulrych (1982). Low frequencies were filtered out of the timestrings by subtracting a 30-minute running triangular mean, and the first and last 60 min of each observing day were apodized with a cosine bell, in an attempt to reduce the influence of high-order sidelobes. Finally, each time series was Fourier transformed using a standard FFT routine, and the power spectrum was computed by taking the square of the modulus of the complex result.

Note that because the input time series in this procedure is complex, the resulting power spectrum generally is *not* symmetric about zero frequency. Recall that each computed spherical harmonic coefficient $A_l{}^m$ may be expressed as the sum of the amplitudes of oscillation modes for which m has the same magnitude, but different sign:

$$A_l{}^m = a_l{}^m + a_l{}^{-m} . \tag{4}$$

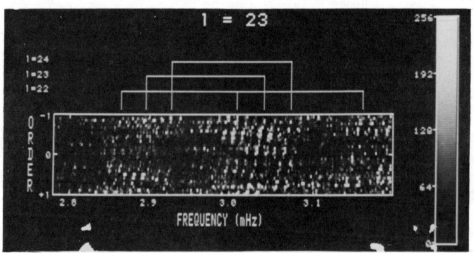

FIG. 2. Power (represented as intensity) as a function of temporal frequency for spherical harmonics with azimuthal order m, for a nominal degree of 23. Degrees $l = 22$ and $l = 24$ are also present, as well as frequency sidelobes arising from gaps in the observed time series. Tic marks indicate the frequencies of oscillations with $m = -l$ for $l = 22,23,24$. The maximum velocity amplitudes represented here are about 10 cm/s.

Since the oscillating quantities have the form $P_l^m \, exp\left[i\left(m\,\phi + \omega t\right)\right]$, the contribution of the various modes to the time series may be written

$$A_l^m(t) = a_l^m \, exp\left[i\left(\mid m \mid \phi + \omega t\right)\right] \;+\; a_l^{-m} \, exp\left[-i\left(\mid m \mid \phi - \omega t\right)\right] \; . \tag{5}$$

Inspection of Eq. 5 shows that prograde waves may be considered as having either negative values of m or negative values of ω. Since m is assumed positive in the spherical harmonic decomposition, the separation of prograde from retrograde waves occurs in the process of computing the temporal power spectrum, where prograde and retrograde waves have negative and positive frequencies, respectively.

For use in the later analysis, the power spectra (about 2500 of them, for $l \leq 50$) were assembled in groups with the same nominal l value, and with m running continuously from $-l$ to l. These $m-\nu$ diagrams give the clearest picture of the variation of frequency with azimuthal order at a given degree. A portion of one such diagram is shown in Fig. 2.

5. OBSERVATIONAL RESULTS

As expected from theory, the $m-\nu$ diagrams show that the oscillation power tends to be concentrated along narrow lines, or *ridges*, with each ridge arising from a single oscillation multiplet. The individual power spectrum peaks that constitute the ridges have widths that are typically slightly larger than the 2.6 μHz intrinsic frequency resolution of the timestrings. Moreover, on a plot such as Fig. 2, each ridge is slightly tilted with respect to the vertical. This tilt is a manifestation of the solar rotation, as described in Eq. 1. In general, the ridges tend to form clusters in frequency, corresponding to multiplets with the same n and different (but similar) l. This may not be readily apparent in Fig. 2, both because of the narrow frequency range shown there, and because at low degree, the width of the clusters in frequency is similar to their separation. The various ridges observed within each cluster result from the range of degrees to which the chosen spatial filter is sensitive, and from frequency sidelobes caused by the diurnal gaps in the observed time series.

The excess width of the observed peaks over the intrinsic resolution is rather small (less than 50%, on average), and may arise from a number of sources. The most important contributor is probably interference with high-order sidelobes of other modes, but several other mechanisms may also be significant: the combination of the variation of ν with m and contamination from modes with neighboring m values, finite lifetimes of the oscillation modes, and various sources of additive noise.

The ideal way to determine the dependence of ν on m would be to identify each of the peaks in each of the computed spectra, perform some (perhaps centroid) analysis to estimate the center frequency of each peak, and then use the peaks that could be confidently identified with oscillation modes to determine $\nu(m)$ for each n and l. Unfortunately, the number of clearly-defined peaks in the current set of spectra is so large (a few times 10^5) that pursuing this program by hand is completely infeasible. To reduce the effort of estimating frequencies, I adopted a cross-correlation technique. First, for each nominal l, I generated a spectrum averaged parallel to the ridges. Since the slope of the ridges was not known accurately to begin with, it was necessary to do the integration for a range of slopes, and then use that value of the slope which yielded the average spectrum with the largest amount of small-scale structure. I cross-correlated this average spectrum with each row of the $m-\nu$ diagram in turn; the lag at which this cross-correlation reached its maximum value was then an estimate of the frequency shift of that m value, relative to the average spectrum. The cross-correlations were performed over the frequency range from 2.58 mHz to 3.41 mHz, a range that spanned from 4 to 7 different n values, depending on the value of l.

This cross-correlation technique is simple and tends to provide a stable result, since the lag is determined from all of the ridges in a given $m-\nu$ diagram (including neighboring values of l and all their frequency sidelobes). Its virtues are not unalloyed, however. Averaging over n

and l entails a loss of depth resolution, since the depth to which an oscillation mode penetrates is determined by both these parameters. A more immediate problem occurs for degrees greater than about 40, where the separation between modes with neighboring l values becomes first equal to and then (at higher degrees) smaller than twice the separation between the diurnal frequency sidelobes. This causes sidelobes from adjacent modes to overlap and interfere in unpredictable ways, adding significantly to the noise. An automated version of the ideal analysis technique described above is currently in progress, and should alleviate this problem somewhat. Using longer timestrings will also help, since peaks that are intrinsically narrow are less apt to overlap. However, the only really satisfactory solution to this problem is likely to be the elimination of the sidelobes themselves, either by continuous observing or by appropriate analysis techniques (see, e.g., Noyes and Rhodes 1984, Brown and Christensen-Dalsgaard 1986).

The frequency shifts estimated from this procedure are illustrated in Fig. 3. ν is evidently an almost linear function of m, although a careful examination of the curve shows some evidence for systematic deviations from a straight line. In addition, there are (in this case one) m values for which there appear to be a very large errors. As mentioned above, the number of such points increases rapidly for degrees larger than 40. If the obviously bad frequency points are suppressed, the deviation of Fig. 3 from a straight line may be illustrated by subtracting a straight line corresponding to the surface equatorial rotation rate (synodic $\Omega = 430$ nHz, Snodgrass 1983) and smoothing the result with a 5-point running mean. The result is plotted in Fig. 4, along with the variation one would expect if the solar latitudinal differential rotation were the same as that observed at the surface, independent of radius.

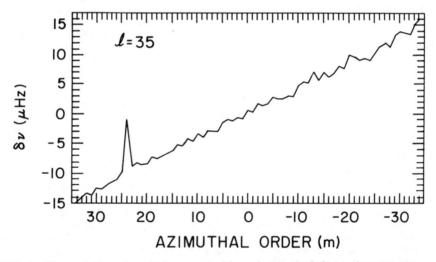

FIG. 3. The variation of mode frequency with m for nominal degree $l = 35$, measured using the cross-correlation technique described in the text. The mean slope is about -405 nHz for each increment in m, corresponding to the latitudinally-independent part of the synodic rotation rate. The siderial rate is obtained by subtracting 31 nHz from this value. Note the single anomalous value at $m = 24$. Bad values occur when spectra at a particular m value correlate poorly with the average spectrum, yielding a noisy cross-correlation. Problems of this sort become frequent for $l > 40$.

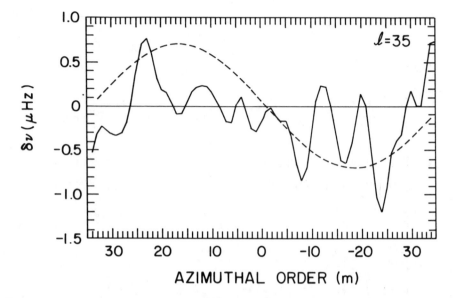

FIG. 4. The same as Fig. 3, except that the synodic solar rotation rate corresponding to equatorial magnetic features seen at the surface (-430 nHz per increment in m) has been subtracted, the bad data point has been edited, and the results smoothed with a 5-point running mean. The smooth curve is the variation one would expect if the observed surface differential rotation were independent of depth.

To extract quantitative information about the differential rotation from the frequency shifts, I performed polynomial fits to the measured values of $\nu(m)$. The procedure used was to fit a preliminary straight line to the data on the interval $-l \leq m \leq l$, with all points given equal weight. This provided a linear fit to the data, as well as an estimate of the standard error σ. I then fit a cubic over the same interval, giving each point a weight proportional to $\exp{-(\delta\nu/1.5\sigma)^2}$, where $\delta\nu$ is the difference between the observed frequency and the straight-line fit. Since this procedure is biased against points that deviate much from a straight line (e.g., points that might distinguish a straight line from a cubic), one should in principle iterate the process until a consistent result is obtained. However, for the current data set the contribution of the cubic component to the variance about a straight line was always much smaller than that of the random noise, so I did no iteration.

The quadratic and cubic coefficients (along with an estimate of the errors) are plotted in Fig. 5, and are tabulated in Table I. Both coefficients are scaled to show $1/l$ times the contribution of each term to the frequency difference $\nu(l,l) - \nu(l,0)$. Scaled in this way, the cubic coefficient corresponding to a particular value of the depth-independent differential rotation is almost independent of l. The errors indicated are twice the formal errors associated with the

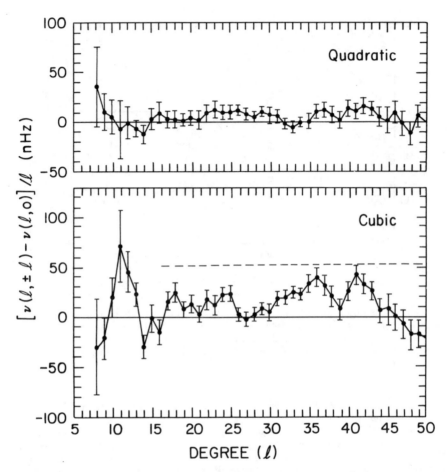

FIG. 5. The upper panel shows the results of the cubic fitting to $\nu(m)$ as a function of the nominal degree. The raw quadratic and cubic fit coefficients have been multiplied by l and l^2, respectively, so that the displayed values depend only weakly on degree. The dashed line in the lower panel is the cubic value that would result if the surface latitudinal differential rotation were independent of depth.

fitting process; the factor of two reflects an estimate of the systematic errors in the cross-correlation analysis. This estimate is based on the difference between the cubic fit results for the cross-correlation analysis and those for an interactive mode-identification procedure that I used on a small number of ridges. Both of these analysis methods gave similar formal errors, but they were typically in disagreement by about twice the formal errors. The errors tend to be large for degrees less than 10 because the cubic fits were made to relatively few data points. For degrees larger than 40, the errors grow again because of the interference effects already mentioned. Because of the large errors at degrees above 40, the negative trend in the cubic coefficient above $l = 45$ may be spurious. However, there seems no reason to doubt the cubic terms for $15 < l < 40$. The large feature centered at $l = 11$ may be real. The errors

TABLE 1. Cubic fit coefficients (scaled according to degree as shown) to the siderial frequencies $\nu(m)$ for degrees from 8 to 50. The errors refer to the 1-σ error between the fit and the frequency of an individul mode.

l	linear (nHz)	quadratic*l (nHz)	cubic*l**2 (nHz)	sigma (nHz)
8	449.2	36.4	-31.6	289.8
9	451.9	9.6	-21.5	157.2
10	426.8	5.0	16.8	170.6
11	412.5	-8.6	70.9	344.0
12	427.6	-1.0	44.7	214.1
13	442.2	-6.8	22.7	128.8
14	477.0	-13.3	-31.3	154.1
15	450.8	2.7	-1.3	203.5
16	460.3	8.7	-16.1	207.4
17	443.4	2.5	14.9	183.9
18	432.2	1.8	24.8	189.8
19	450.3	1.5	7.2	140.5
20	444.8	3.9	13.1	200.8
21	447.1	1.1	13.3	215.3
22	442.6	9.2	18.0	272.1
23	441.7	12.2	11.2	284.2
24	438.7	8.8	22.4	221.2
25	442.6	9.2	23.3	239.9
26	456.0	11.8	2.2	213.7
27	461.7	7.5	-3.0	226.9
28	457.8	4.2	2.1	249.4
29	447.9	10.0	8.6	237.4
30	451.9	6.6	4.0	388.6
31	445.8	5.8	18.7	325.8
32	446.8	-2.2	19.4	303.8
33	439.5	-6.2	24.9	303.2
34	446.5	0.0	22.6	307.6
35	435.9	0.2	33.5	462.4
36	432.9	10.8	40.1	519.4
37	435.2	12.5	31.2	534.0
38	442.9	7.0	19.8	608.2
39	446.4	1.4	7.2	651.5
40	442.5	13.9	26.2	637.6
41	422.9	10.8	42.8	617.2
42	433.7	16.4	32.5	693.0
43	439.4	13.3	26.3	583.8
44	447.9	4.7	5.5	879.7
45	448.0	0.8	8.5	1214.5
46	456.7	10.1	0.5	1087.4
47	464.3	-2.2	-6.9	1219.5
48	457.1	-11.8	-18.4	1247.6
49	464.2	6.9	-17.3	1236.5
50	472.4	0.0	-21.4	1188.2

associated with it are large, but there is no evidence that these errors are associated with a poor choice of the function being fitted, nor are there obvious problems with the frequency measurement or cross-correlation process near $l = 11$.

The quadratic term plotted in Fig. 5 is everywhere consistent with a zero result, though for degrees greater than 20 the values are almost all positive. This positive bias is probably a result of systematic errors in the analysis, though some solar process might be responsible. It is interesting that there is no evidence of the feature at $l = 11$ in the quadratic coefficient; if this feature were entirely a result of vagaries on the analysis, then one might expect some indication of these in the quadratic term as well as in the cubic.

It is interesting to perform a partial comparison between the current results and those by Duvall and Harvey (1984) by estimating the frequency splitting between sectoral modes with $m = \pm l$. This can be done by summing the linear and cubic terms in the polynomial fits just described, and plotting the resulting values of $[\nu(l,-l) - \nu(l,l)]/2l$ as a function of degree. This polynomial evaluation is necessary because the current observations do not have adequate frequency precision to allow a useful comparison if one considers only individual modes. Both results are plotted in Fig. 6. There are many qualitative similarities between the two data sets. Both show a slow decrease in frequency splitting as l decreases from 40 to about 15, and both show a large localized increase in the splitting near $l = 11$, followed by a sharp drop at lower degrees. The Duvall and Harvey data also show an increase in splitting for $l < 6$, rising to a value much larger than the surface value at $l = 1$. Unfortunately, the poor frequency resolution of the current set prevents a meaningful comparison for these low degrees.

There are also significant differences between the two data sets. The most obvious of these is that the splittings derived from the FT data are almost without exception larger than those from Duvall and Harvey, by an average of about 10 nHz. This is probably a bias resulting from the incorrect solar radius used in the spherical harmonic decomposition. Choosing the solar radius to be 3% too small causes one to underestimate m by the same amount. Since the splittings shown in Fig. 6 are derived by dividing a frequency difference by an m difference, underestimating m causes an overestimate in the splitting. The magnitude of this error should be about 13 nHz, suggesting that a correct analysis would yield results that are much more nearly in agreement with those of Duvall and Harvey.

Besides the difference in average value, the current splittings differ from those of Duvall and Harvey in that they show almost no decrease in splitting between degrees 15 and 25. This is a prominent feature in the Duvall-Harvey data, and the reason for its absence in the current set is unknown.

The splittings shown in Fig. 6 may also be compared with those reported by Hill (1985). The latter are based on observations of variations in brightness at the solar limb, and for periods near 300 s extend out to $l = 6$. Hill's rotational splittings are larger than those shown in Fig. 6 by a factor of about 4. Although the l ranges of the FT observations shown in Fig. 6 and those of Hill do not quite overlap, there is still no feasible way to reconcile the two sets, except as a misidentification of modes in one or both cases. For example, though the FT observations at $l = 6$ are too noisy to be plotted usefully on the scale of Fig. 6, they nevertheless differ from those of Hill at the same degree by more than 20 standard errors. Moreover, the overlap between the oscillation eigenfunctions for low degree is so large that no physically reasonable solar rotation profile could produce such disparate results for neighboring degrees. Since the mode identification on $m-\nu$ diagrams produced by the FT is so obvious and compelling, I believe that it is most likely that the oscillations described by Hill (1985) have been misidentified.

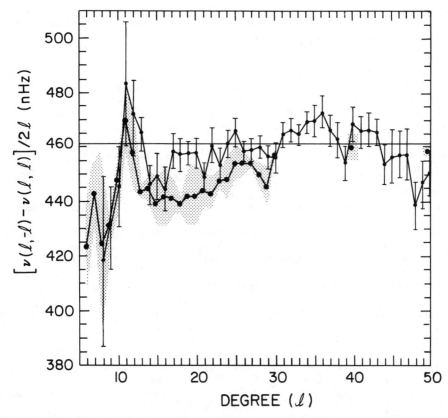

FIG. 6. Comparison between frequency splittings for sectoral modes $(m = \pm l)$ derived from the Fourier Tachometer observations (small circles with error bars) and observations by Duvall and Harvey (large dots with errors indicated by stippling). As described in the text, correction of solar radius value used in the analysis should lower the FT splittings by about 13 nHz, leading to better agreement between the two data sets. See the text for a detailed comparison.

6. INFERENCES ABOUT THE SUN

The dashed line in the lower panel of Fig. 5 indicates the magnitude of the cubic coefficient implied by depth-independent differential rotation with the latitudinal dependence observed at the surface. As one might suspect from the previous Figures, the magnitude of the cubic term is substantially smaller than that corresponding to the surface differential rotation. Barring unexpectedly large systematic errors in the data analysis, it seems clear that the depth-averaged latitudinal differential rotation must be considerably less vigorous than what we see at the solar surface. Moreover, there is some evidence in Fig. 5 that the magnitude of the cubic term decreases with decreasing degree. Since modes with low degree penetrate deeper into the Sun than those with high degree, this suggests that the differential rotation decreases with increasing depth. (Though without doing a proper inversion, one may not exclude the possibility that the latitudinal differential rotation is roughly constant within a layer with a rather sharp lower boundary. In this case, the decrease in the cubic term with decreasing l would occur because lower degree modes would see a larger contribution from the region below the boundary.) Fig. 6 shows that the low-latitude rotation rate at depth is very nearly the same as the surface equatorial rate, but much larger than the surface polar rate. Taken together, the two Figures imply that Ω at depth is more nearly constant with latitude than at the surface, and further that at depth, Ω at all latitudes becomes similar to the surface equatorial rate.

The meaning of "at depth" in the previous paragraph may be clarified by noting that all of the oscillation modes discussed here penetrate at least to the bottom of the convection zone: modes of degree 40 sample radii greater than about $0.6R_\odot$, while those with degree 10 reach to about $0.3R_\odot$. Thus, the above conclusions apply between radii of about $0.4R_\odot$ and $0.7R_\odot$. The outer few percent of the solar radius has a larger effect on the observed frequencies than these arguments might suggest, since the low sound speed near the surface causes the waves to spend a large fraction of their time there. Because of this heavy weighting of the surface layers, and because the latitudinal differential rotation is known to be large at the surface, it is possible that the observed cubic term in the $m - \nu$ relation may result entirely from a relatively thin surface layer, with the differential rotation small or absent through the bulk of the convection zone. This is not the only possible interpretation, however, and a definitive answer to this question must await a thorough analysis of better observations.

The frequency splittings and cubic terms for degrees near 11 are quite different from the neighboring l values. For $l = 11$, the equatorial rotation rate $[\nu(l,-l) - \nu(l,l))/2l]$ is larger than for nearby degrees, while the polar rotation rate $(-d\,\nu/dm$ evaluated at $m = 0)$ is smaller than its neighbors. The differential rotation implied by these values is larger even than the observed surface value. The straightforward interpretation of these facts is that there is a thin shell of material located near $r = 0.4R_\odot$ that rotates faster at the equator and slower at the poles than the material at larger and smaller radii. There are several arguments against this interpretation. It is unclear on physical grounds how such a narrow rotational feature could be produced or maintained within the Sun. Also, the overlap between the eigenfunctions for degrees near 11 is so large that it would be quite difficult to construct a rotational profile that affects the observed frequencies only for degrees between 10 and 13. This difficulty is aggravated by the smearing in l that is involved in the cross-correlation frequency analysis. Finally, the values of $\nu(m)$ for degree 11 have substantially more scatter about a cubic fit than do nearby degrees. The origin of this scatter is not known, but it suggests that the measurements near $l = 11$ may be subject to some data analysis problem that has not yet been identified. On the other hand, the feature at $l = 11$ is present in both the current data set and that of Duvall and Harvey. Thus, if it results from some problem with the observations or analysis, then the problem must be a fairly deep-seated one, and essentially independent of the observational methods employed.

Forces that distort the shape of the Sun or that cause certain sorts of asymmetry in the propagation speed of acoustic waves may cause a symmetric (e.g. quadratic) variation of ν with m. The observed quadratic term in the fit to $\nu(m)$ can be used to estimate upper limits on the importance of such forces. From Fig. 5, the maximum magnitude of the quadratic part of the frequency perturbation is about $10 \cdot l$ nHz. The most plausible solar cause for such a variation is a large internal magnetic field; this possibility has been discussed by Dicke (1982), and by Gough and Taylor (1984). The latter authors find that the change in shape of the solar equilibrium configuration is more important than the magnetic perturbation of the propagation speed, at least for a field configuration similar to that suggested by Dicke. Using their results with geometrical factors appropriate to $l = 9$ leads to a maximum field strength of $1-3 \times 10^7$ G, depending on whether the cutoff radius (at which the field strength effectively goes to zero) is

0.7 $R_\odot$ or 0.3 $R_\odot$. These limits are not very restrictive. However, it is important to note that a field configuration like that suggested by Dicke would result in a negative quadratic term, in contrast to that displayed in Fig. 5. Indeed, if one wishes to produce a positive quadratic term by distorting the Sun, it is necessary to make the Sun prolate. This tends to confirm the previous assertion that the nonzero quadratic term is a result of systematic errors in the analysis, and that no physical significance should be attributed to it.

7. MODE AMPLITUDES AS A FUNCTION OF m

Finally, it is of some interest to consider the relative amplitudes of the oscillation modes for $m \sim 0$ and for $m \sim l$. In the case of high-degree oscillations observed near the equator, these cases correspond to waves propagating North-South, and to waves propagating East-West, respectively. These amplitude comparisons require some care, because the spatial filtering process is not equally sensitive to all m values, and because of various asymmetries that may occur in the instrument.

The variation in spatial filter sensitivity with m arises because the oscillatory motions in the photosphere are essentially vertical. Modes with $m \sim 0$ have their largest amplitudes near the poles, where the motions are always nearly perpendicular to the line of sight. Contrariwise, modes with $m \sim \pm l$ have their largest amplitudes near the equator, so that the motions are often nearly parallel to the line of sight. Thus, the amplitudes of low-m modes tend to be systematically underestimated relative to those of high-m modes. To determine how to correct for this effect, I computed the observed velocity fields corresponding to spherical harmonics with a range of degrees and azimuthal orders, and applied the standard analysis technique to these artificial data. This computation showed that the ratio of the *power* response $R(m = 0) / R(m = \pm l)$ is about 0.47, almost independent of degree for $2 \leq l \leq 50$.

After correction for this geometric effect, the observed oscillation amplitudes are essentially independent of m for low degrees, with an increasing bias in favor of low m oscillations as l grows larger than about 20. At $l = 50$, the power seen in modes with $m \sim 0$ is typically about twice that for modes with $m \sim \pm l$. However, there is good evidence that this bias arises in the instrument. More recent observations taken on 6 June 1985 and 8 June 1985 show that rotating the Fourier Tachometer detector 90° relative to the solar image causes the bias to change its sense while maintaining about the same magnitude. The most likely explanation is that there is an asymmetric blurring in the detector itself (rather like astigmatism, but probably of electronic origin), which attenuates the higher spatial frequencies in one direction while passing the same spatial frequencies in the perpendicular direction.

Correcting for this instrumental effect leaves no measurable difference between the amplitudes of modes with low and high m, although the calibrations are sufficiently uncertain that one could not exclude a difference of perhaps 30% in the power. This result disagrees substantially with that reported by Kuhn (1984), who found that North-South propagating waves have smaller power than East-West propagating ones by about a factor of 2. Moreover, it is wholly inconsistent with recent results by Deubner (1986, this volume), which show a difference at low degree of about a factor of 10. The reason for these discrepancies is not clear; further observations will be required to resolve them.

I am grateful to J. Christensen-Dalsgaard, W. Däppen, T. Duvall, P. Gilman, D. Gough, and B. Mihalas for stimulating conversations, K. Streander, L. Gilliam, and the staff of the Sacramento Peak Observatory Big Dome for their assistance in obtaining the observations, and to P. Bandurian and D. Haynes for their assistance with the data analysis.

REFERENCES

Beckers, J.M. and Brown, T.M. 1978, in *Proceedings of JOSO Workshop on Future Solar Optical Observations*, Florence

Brown, T.M. 1985a, in *Proceedings of Snowmass Conference on Solar Seismology from Space* (ed. Ulrich, R.) (Jet Propulsion Laboratory, Pasadena) p.157.

Brown, T.M. 1985b, *Nature*, (in press).

Brown, T.M. and Christensen-Dalsgaard, J. 1986, *Astrophys. J.* (in press).

Brown, T.M., Mihalas, B.W. and Rhodes, E.J. Jr. 1986, in *Physics of the Sun* (ed. Sturrock, P.A.) (Reidel, Dordrecht).

Brown, T.M., Evans, J.W., Elmore, D.F. and Streander, G.W. 1986, (in preparation).

Christensen-Dalsgaard, J. 1985, in *Proceedings of Snowmass Conference on Solar Seismology from Space* (ed. Ulrich, R.) (Jet Propulsion Laboratory, Pasadena) p.219.

Cuypers, J. 1980, *Astron. Astrophys.* **89,** 207.

Deubner, F.-L. 1986, (this volume).

Dicke, R.H. 1982, *Solar Phys.*, **78,** 475.

Dilke, F.W.W. and Gough, D.O. 1972, *Nature* **240,** 262.

Duvall, T.L. Jr. and Harvey, J.W. 1984, *Nature* **310,** 19.

Duvall, T.L. Jr., Dziembowski, W.A., Goode, P.R., Gough, D.O., Harvey, J.W. and Leibacher, J.W. 1984, *Nature* **310,** 22.

Evans, J.W. 1980, in *Solar Instrumentation: What's Next?* (ed. Dunn, R.) (Sacramento Peak Observatory, Sunspot) p. 155.

Fahlman, G.G. and Ulrych, T.J. 1982, *Mon. Not. Roy. Astron. Soc.* **199,** 53.

Gilman, P.A. and Miller, J. 1981, *Astrophys. J. Suppl.* **46,** 211.

Gough, D.O. and Taylor, P.P. 1984, in *Oscillations as a Probe of the Sun's Interior* (eds. Belvedere, L. and Paterno, L.), Memoire della Societa Astronomica Italiana **55,** 215.

Hill, H.A. 1985, *Astrophys. J.* **290,** 765.

Howard, R. and Harvey, J. 1970, *Solar Physics* **12,** 23.

Kuhn, J.R. 1984, in *Oscillations as a Probe of the Sun's Interior* (eds. Belvedere, L. and Paterno, L.), Memoire della Societa Astronomica Italiana **55,** 69.

Lites, B.W. 1972, Ph.D. Thesis, University of Colorado.

Michaud, G. 1985, in *Solar Neutrinos and Neutrino Astronomy* (ed. Cherry, M., Lande, K. and Fowler, W.), AIP Conference Proceedings No. 126, New York, p. 75.

Noyes, R.W. and Rhodes, E.J. Jr. (eds.) 1984, *Probing the Depths of a Star: The Study of Solar Oscillations from Space* (Jet Propulsion Laboratory, Pasadena).

Schatzman, E. 1969, *Astron. Astrophys.* **3,** 331.

Schatzman, E. 1985, in *Solar Neutrinos and Neutrino Astronomy* (ed. Cherry, M., Lande, K. and Fowler, W.), AIP Conference Proceedings No. 126, New York, p. 69.

Snodgrass, H.B. 1983, *Astrophys. J.* **270,** 288.

ROTATIONAL SPLITTING OF ℓ=1 p MODES

H.B. van der Raay
Department of Physics
University of Birmingham, UK.

P.L. Palle and T. Roca Cortez
I.A.C., Universidad de la Laguna, Tenerife

ABSTRACT.   An analysis of the amplitude modulation of specific ℓ=1 p modes, has revealed a mean splitting of 0.74 μHz over the frequency range 2.4 to 4.0 mHz.   The effects of large scale surface phenomena rotating at the ~27 day surface rotation rate are also detected.   The lifetime of the ℓ=1 modes is seen to vary from >120 days at 2425.1 μHz to ~50 days at 3912.2 μHz.

INTRODUCTION

Consider the diagramatic representation of a m=4 mode illustrated in fig.1.   In a global measurement of the solar surface line of sight velocity, the observed velocity is that averaged over the exposed hemisphere. When considering sectoral harmonics (ℓ=m) a simpler representation is obtained by unwinding the solar surface and plotting the surface velocity as in the lower part of fig.1.   The region observed is, at a particular instant, shown by (a).   Here a significant positive velocity amplitude would be detected. As the phenomenon rotates relative to the observer, the situation depicted by (b) is obtained after a 1/4 rotation.   Here the effective velocity signal amplitude is zero since the instrument averages over its defining aperture. Hence when observing a particular p mode the velocity amplitude appears to be modulated; it is this amplitude modulation that gives rise to the frequency splitting of the modes by rotation.

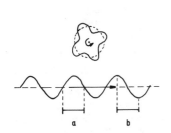

Figure 1 Diagramatic representration of m modes illustrating how amplitude modulation is observed.

215

D. O. Gough (ed.), Seismology of the Sun and the Distant Stars, 215–221.
© 1986 by D. Reidel Publishing Company.

In the general case consider a signal which is amplitude modulated:

$$v(t) = (B + \Sigma\, C_i \sin w_i t)\, \sin w_o t \qquad\qquad \ldots\ldots(1)$$

this may be expanded as

$$v(t) = B \sin w_o t + \Sigma\, \frac{C_i}{2} \sin (w_o - w_i)t + \Sigma\frac{C_i}{2} \sin (w_o + w_i)t \; \ldots\ldots(2)$$

This results in a central undisturbed frequency (w_o) with a series of symmetrically placed sidebands separated by frequencies w_i. In the special case of B=0, corresponding to an exact cancellation of the signal as observed by the instrument as illustrated in fig.1b, the central component (w_o) will be absent and only the $\pm w_i$ sideband frequencies will appear.

It should be noted that the amplitude of the central peak is enhanced by a factor of 2 relative to the sidebands and that the power is therefore enhanced four times.   Thus in an analysis of the power spectrum of solar doppler velocity signals the central w_o peak is greatly emphasised and would only be absent in the unique case that perfect cancellation of the observed spatially dispersed signals occurs.

As an alternative to determining the power spectrum of the observed velocity data and then interpreting the complex spectra revealed, the amplitude modulation of a particular frequency corresponding to a selected p mode may be found and by Fourier analysis the frequency components present in the modulation may be directly related to rotational effects.   In general the observed amplitude modulation of a frequency component may be caused by (a) rotation, (b) noise and (c) lifetime of the mode (phase coherence). The effects of (a) and (c) will be separated by considering differing subsets of the available data and the effect of (b) will limit the analysis to those velocity signals of larger amplitude and will inevitably create unwanted background peaks in the desired spectra.

METHOD

The present analysis will be limited to a data set of 166 day span containing 135 days of data of >10 hours per day.   These data were taken between 18th April and 30th September 1984 with a K769.9 nm resonant scattering spectrometer[1] situated at Izana on Tenerife. The relative intensities in the left and righthand limbs of the K absorption line are measured alternately at one second intervals and these data are then used to find the mean line of sight velocity averaged over 40 s intervals.   Taking into account the instrumental response function[2] the effects of the earth's rotation are removed, finally yielding a set of residuals containing the solar oscillation data.

These residuals are subjected to an interative sine wave fitting technique to determine the frequencies of the p modes.   Once the ℓ=1 mode frequencies have been found, a subset of 5 days is considered and the mean amplitude of a particular p mode frequency is found for day 1-5 inclusive.   Next the amplitude of that same frequency is found for days 2-6......n_i-n_{i+5}, until the complete data set is scanned. The intrinsic resolution of a 5 day data set is ~2 μHz and hence the precise determination of the mode frequency is not required.   The amplitude modulation of the ℓ=1 line at 2963.6 μHz determined in this way is illustrated in fig.2.

It should be noted that the amplitude is always positive, the phase of the modulation relative to the signal has not been used.   In order to utilise this phase information the precise frequency of the mode being investigated must be known since any error in the frequency determination will cause a progressive phase error with consequent incorrect sign reversals of the amplitude. This problem becomes particularly severe when long data stretches are considered; in the present case the 5 minute modes execute some $5{\times}10^4$ oscillations. However by omitting the phase information this problem does not arise.

It is seen in fig.2 that the observed signal does not go to zero and hence B in equation (1) is finite with a resultant w_o component in the frequency spectrum.   The reason for this is that (a) the amplitude found is the mean over 5 days and exact cancellation will only occur at one precise time, (b) a noise signal is always present and (c) the spectrometer does not strictly observe the integral solar disc. When using high resolution spectrometry in association with a rotating source, if the spectral resolution of the instrument is narrower than the rotationally broadened source, a spatially resolved region of the source will be selected[3]. This is the case with the present instrument and the solar surface. In addition, due to the observer's rotation on the earth's surface the region of the solar surface selected varies during the day.   Thus a partial scan of the solar surface is performed each day as a consequence of the spatial-spectral correlation and the intrinsic rotation of the sun and the earth.

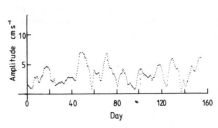

Figure 2 Amplitude modulation of the 2963.6 μHz. ℓ=1 line.

Before performing a Fourier analysis of the amplitude spectrum a moving mean over 15 days is subtracted from the data in fig.2. This leaves a total of 126 points which yield the spectrum shown in fig.3a. Several peaks are clearly visable indicating the presence of many frequency components which may be associated with the three effects discussed   earlier;   rotation,   noise,   lifetime.   However,   by systematically considering subsets of the data with differing starting times and durations a data string starting on day 68 and lasting for 66 days may be found which gives the spectrum shown in fig.3b.   Here there is one well defined peak at a frequency of 0.80 μHz and a second at 0.42 μHz. These are associated with rotational splitting and the effects of large scale surface phenomena respectively; the surface rotation rate corresponds to 0.42 μHz.

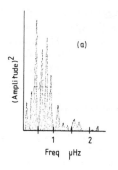

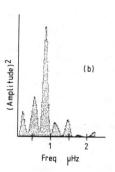

Figure 3   Frequency spectrum of the amplitude modulation of 2963.6 μHz
ℓ=1:   (a) full 126 days   (b) days 60–126 only.

A similar analysis of the ℓ=1 line at 3097.7 μHz shows a
prominent peak at 0.85 μHz for the first 90 days of data.   A second
peak is seen at 0.31 μHz as shown in fig.4.

At the lower frequency end of the ℓ=1 range it is found that an
analysis of the full 126 days available yields two well defined peaks
as illustrated in fig.5, whilst at 3368.6 μHz a stretch of 70 days
gives the spectrum shown in fig.6.   It should be noted that in all
cases the width of the individual peaks is consistant with the length
of data considered.

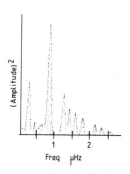

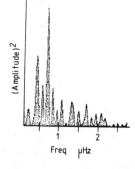

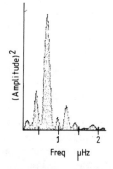

Figure 4   Frequency
spectrum of amplitude
modulation of 3097.7
μHz ℓ=1 line: days
1–90.

Figure 5 Frequency
spectrum of amplitude
modulation of 2693.6
μHz ℓ=1 line: days
1–126.

Figure 6   Frequency
spectrum of amplitude
modulation of 3368.6
μHz ℓ=1 line: days
1–70.

A summary of the results obtained for the ℓ=1 modes over the
frequency range 2400–4000 μHz is given in table 1.

The quoted coherence time is based on the number of days
contributing to the amplitude modulation frequency spectra and where
these include either the first or last day of the available data set,
should be considered as upper limits.   The results are shown in
graphical form in figs.7 and 8 where the ℓ=1 modes are plotted in
increasing frequency, #1≡2425.1 μHz and #12≡3912.2 μHz.

Table I   The rotational splitting (δf) large scale surface
feature splitting (δf′) and coherence times for
the ℓ=1 p modes.

Frequency μHz	Coherence days	δf μHz	δf^1 μHz
2425.1	126	0.51	0.40
2558.6	126	0.53	0.42
2693.3	126	0.70	0.44
2828.9	50	0.77	0.42
2962.7	66	0.80	0.42
3097.7	90	0.85	0.31
3232.1	60	0.84	0.52
3368.6	70	0.74	0.46
3504.2	60	0.74	0.42
3638.6	66	0.68	0.44
3773.0	60	0.72	0.46
3912.2	50	0.89	0.52

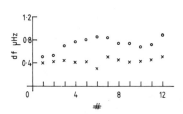

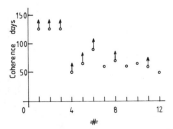

Figure 7   Splitting due to internal
rotation (O) and large scale surface
features (X), #1=2425.1 μHz.
#12=3912.2 μHz.

Figure 8   Phase coherence as a
function of line number
ó designates lower limit.

Having established the lifetimes of the various modes, and the
splitting due to rotation and large scale surface motions, a more
conventional representation of the data in the form of a power
spectrum of the data set should be considered.   The ℓ=1 line at
3368.6 μHz yields the detailed power spectrum for the full data set
shown in fig.9.   The multiplicity of lines is a clear indication of
the lack of coherence over this time span.   However, if a power
spectrum is obtained for those data corresponding to the coherence
period determined by an analysis of the amplitude modulation, the
results shown in fig.10 are obtained.   The frequency splitting due to
both large scale surface motions and those associated with rapid
internal rotation are clearly resolved.   The large central peak
corresponding to B≠0 is also seen. An anomolous peak spaced at 0.2 μHz
from the central peak can be linked to the basic ~0.2 μHz structure
present throughout the spectrum which is associated with the length of
the data set (1/T ~0.17 μHz). Finally, considering the full 126 days

associated with the 2558.6 μHz line the power spectrum indicated in
fig.11 is found.   The structure corresponds precisely to that
determined by the amplitude modulation analysis.

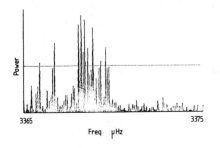

Figure 9 Power spectrum of
3368.6 μHz line for full data
set.

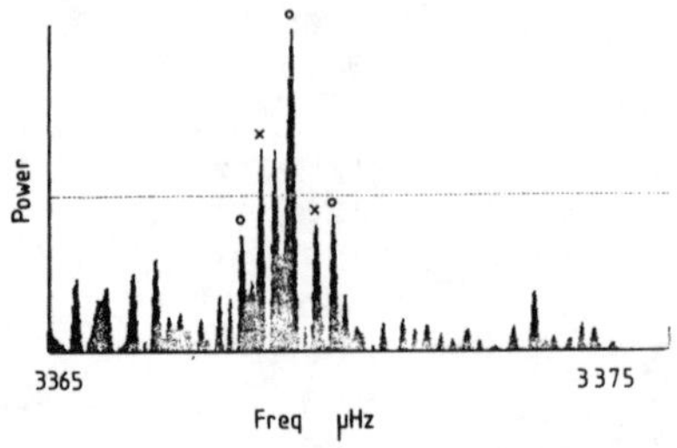

Figure 10 Power spectrum for 3368.6
μHz line days 2-65 showing internal
rotational splitting (O), and that
due to surface effects (X).

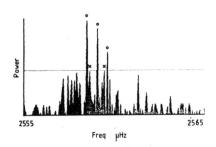

Figure 11   Power spectrum for 2558.6 μHz
line days 1-126 showing internal rotational
splitting (O) and that due to surface effects (X)

CONCLUSION

A study of the frequency spectrum of the amplitude modulation of $\ell=1$ p
modes has revealed the long coherence time of these modes in agreement
with that found in a different analysis of earlier data[4] The
rotational splitting for all 12 lines considered has been determined
and the mean value of 0.74 μHz is in good agreement with that found
for 28 days of data collected in 1980[5].   The previous and present
values also agree with that found by Duval and Harvey[6]. In addition,
with the longer data sets available, the effects of large scale
surface phenomena coupled with surface rotation have been detected.
These effects could not previously be resolved due to the shorter data
strings available.

The central unsplit peak, present in all the data, and also previously found in the 1980 data, may be ascribed to the way in which the integral solar surface is sampled by the resonant scattering spectometer.

ACKNOWLEDGEMENTS

The hospitality of Professor F. Sanchez at the observatopry at Izana is gratefully acknowledged.   The assistance of A. Jimenez, J. Perez and C. Regulo in the data collection and the continued efforts of the technical staff of the Birmingham solar group proved invaluable. This work was financially supported by a SERC grant.

REFERENCES

(1) J.R. Brookes, G.R. Isaak and H.B. van der Raay, Mon. Not. R. astr Soc. 185 (1978) 1

(2) H.B. van der Raay, P.L. Palle and T. Roca Cortez, Loc.sit.

(3) J.R. Brookes, G.R. Isaak and H.B. van der Raay, Mon. Not. R. astr. Soc. 185 (1978) 19

(4) H.B. van der Raay, Proceedings 25th Liege International Astrophysical Colloquium (1984) 215

(5) A. Claverie, G.R. Isaak, C.P. McLeod, H.B. van der Raay and T. Roca Cortes, Nature 293 (1981) 443

(6) T.L. Duval and J.W. Harvey, Nature 310 (1984) 19

LINEWIDTH AND ROTATIONAL SPLITTING OF LOW DEGREE MODES
IN THE 5 MINUTE REGION

G.R. Isaak
Department of Physics
University of Birmingham
P.O. Box 363
Birmingham B15 2TT, UK

ABSTRACT.   The linewidth of low degree modes in the 5 minute region
of the Sun's acoustic spectrum increases dramatically with frequency
over the range of 2 mHz to some 5 mHz, in approximate agreement with
theory.   Below some 2.5 mHz the linewidth is narrow enough to allow
rotational splitting of $\ell=1$ modes to be measured.   These measurements
seem to confirm earlier results.

1.   INTRODUCTION

Knowledge of the details of the acoustic spectrum of the Sun is likely
to follow the same historical path as the corresponding knowledge of
other spectroscopies e.g. that of the atom.   Measurements of
frequencies, line splittings and amplitudes were followed by those of
linewidths and line shapes.   It is likely that understanding of these
observables will also proceed in the same direction.   Discovery of
global solar oscillations in the 5 minute range (Claverie et al.
(1979)) was followed by that of a finer structure due to near
degeneracy of the $\ell=0,\ell=2$ and $\ell=1,\ell=3$ modes (Grec et al. (1980)) and
that of rotational, and possibly magnetic, splitting at a μHz level.
Claverie et al. (1981), Isaak (1982)).   Whereas the former two have
been absorbed into the theoretical framework of the Sun, there has
been a substantial refusal by the theoretical community to acknowledge
these findings concerning rotational splitting and possible
perturbations thereof.

This contribution is primarily a brief account of evidence which
suggests that the linewidth of modes near 3 mHz can now be measured
and that it increases by some two orders of magnitude over the 2 to 4
mHz region.   Oral presentations of these claims have been made by the
present author since the Dublin meeting in early 1982, whereas some of
his colleagues claimed that the coherence time was very long (e.g. van
der Raay (1982)) and substantially continue to do so until the present
(e.g. van der Raay et al. this volume).

D. O. Gough (ed.), Seismology of the Sun and the Distant Stars, 223–228.
© 1986 by D. Reidel Publishing Company.

Grec *et al.* (1983), from a 5 day observational span at the South Pole in 1980, also claim to have seen dramatic increases in the linewidth with frequency.   Woodard and Hudson (1983), using data of poor signal to noise ratio on luminosity fluctuations on the Solar Maximum Mission claim a coherence time of a few days and saw no variation with frequency.

A secondary aim is to use this information in order to select parts of the spectrum where the linewidth seems to be sufficiently narrow to ensure that removal of degeneracy by rotational splitting is directly measureable and to show a few individual modes.   The evidence seems to demonstrate that the measured splitting of individual $l=1$ modes is consistent with that claimed earlier from the evidence of the average of many $l=1$ modes (Claverie *et al.* (1981)). The reality of the triplet (or more complex) structure, however, is still open and requires even better quality data.

2.   OBSERVATIONAL METHODS

Optical resonance spectrometers, improved versions of the general type described earlier (Brookes *et al.* (1978a)), were viewing the Sun (i.e. Sun as a star) in the 7699Å Fraunhofer absorption line of neutral potassium through servo-controlled classical horizontal coelostats at Izana, Tenerife and Haleakala, Maui, Hawaiian Islands.

The Izana spectrometer was switched at 0.5 Hz between the steepest parts of the two wings of the solar line using a Pockels cell and used a single photon counting system operating near 10^6 counts/second.   Data corresponding to resonant scattering in the red (r) and blue (b) wings of the line were digitally recorded.   The normalised ratio $R(t)=(r-b)/(r+b)$ provides a sensitive and almost linear measure of the temporal variation of the global velocity of the solar disc (for caveat on the 'global', see Brookes *et al.* (1978b). Appropriate and small corrections for curvature of the spectral line near the operating points and differential atmospheric extinction across the solar disc are required before the daily calibration constant linking the observable $R(t)$ and velocity is derived from the diurnal variation of the Sun's velocity at the spectrometer sites.

The response of the Haleakala spectrometer was switched at 50 kHz with a photoelastic modulator.   An analogue system using a lock-in amplifier handled a photon flux some two orders of magnitude higher than that of the Izana spectrometer.   This higher switching rate and higher statistical accuracy provided a lower noise in the 5 minute region but the separate measurements of the observables (r-b) and (r+b) with distinct instrumentation reduced the stability at low frequencies.   Operation of the two spectrometers provided a data span of up to 22 hours in every 24, weather permitting, and measuremnts were made over an interval of 88 days in the summer of 1981.

3.   RESULTS AND CONCLUSIONS

After subtraction of the known velocity of the Sun relative to the observer, an apparent velocity due to the gravitational red shift and some other smaller terms, the velocity residuals formed a time series with a sampling every 42 seconds and some gaps which were filled with zeroes.   Figure 1 shows a typical day's result.

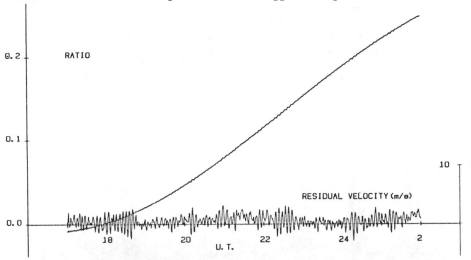

Figure 1.   The diurnal variation of the ratio R(t) and velocity residuals showing global solar oscillations on 2nd August 1981 measured from Haleakala, Hawaii.

A standard fast Fourier analysis was then used to obtain a spectrum from 0 to the Nyquist frequency of 11.9 mHz with a resolution of 0.09 μHz.   Figure 2 shows the spectrum between 2 and 4 mHz.   One can see that the lines are sharp near 2 mHz and considerably broader at higher frequencies.

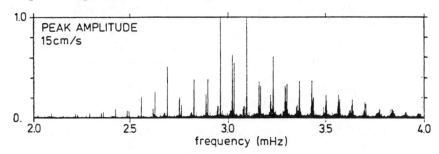

Figure 2.   Velocity power spectrum of 88-day data from Hawaii and Tenerife.

Figure 3 shows the power spectrum above 4 mHz.   A substantial broadening can be clearly seen.   The linewidth (and the amplitude) determine the precision of determining the centroids of the normal modes.   The precision ranges from less than 0.3 μHz at the lower to over 1 μHz at the higher frequencies.

The individual peaks have a width given by 1/T, where T is the observation time, whereas the envelope of that distribution is a measure of the linewidth.   Intuitively one would expect that the power spectrum will consist of some n components, if there are n sudden jumps in phase, and that it will approach a smooth curve which could be characterised by a linewidth as n increases.   Numerical simulations were carried out and confirm these ideas, as shown in Figure 4.

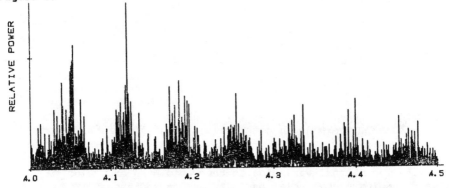

Figure 3.   Velocity power spectrum above 4 mHz shows very broad lines.

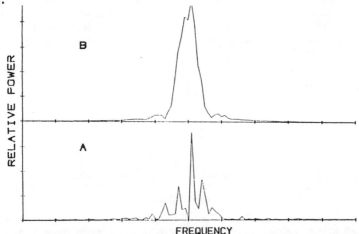

Figure 4.   Power spectrum of decaying wavetrain with re-excitation at random times and random phase (a) and a superposition of 50 such trials (b).

In figure 5 is shown the variation of linewidth with frequency of
ℓ=0 modes.

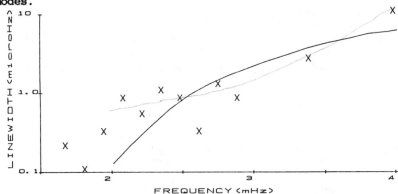

Figure 5.   The variation of the linewidth with frequency is shown,
together with contributions evaluated by Kidman and Cox (1984).

It can be seen that there is a dramatic increase of linewidth
with frequency.   The linewidths appear to be substantially smaller
than those due to damping calculated by Christensen-Dalsgaard and
Fransden (1983), but are consistent with the calculations of Kidman
and Cox (1984).   There is no clear evidence for additional broadening
mechanisms other than damping, although this is by no means excluded,
particularly at the lower frequencies.

The linewidths reach sufficiently low values to resolve
rotational splittings as shown in figure 6.

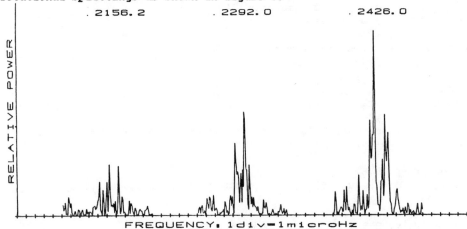

Figure 6.   Three ℓ=1 acoustic modes (n=14,15,16) of frequencies
2156.2, 2292.0 and 2426.0 μHz are shown.   The first two seem to show
a complex structure whereas the third appears to be a doublet of
separation of 1.5 μHz, consistent with the earlier claims concerning
rapid rotation of the solar core.

4. ACKNOWLEDGEMENTS

I wish to thank my colleagues in Birmingham, A. Claverie, C.P. McLeod
and H.B. van der Raay and in Tenerife, P.L. Pallé and T. Roca Cortes,
who helped collect the 1981 data set.   The work would not have been
possible without the hospitality of Professors J.T. Jefferies and
F. Sanchez at the Haleakala and Izana Observatories respectively.
This work was supported by an SERC grant.

5. REFERENCES

Brookes, J.R., Isaak, G.R. and van der Raay H.B.
Monthly Notices Roy. Aston. Soc. (1978a) **185,** 1–17

Brookes, J.R., Isaak, G.R. and van der Raay H.B.
Monthly Notices Roy. Astron. Soc. (1978b) **185,** 19–22

Christensen–Dalsgaard, J. and Fransden, S.
Solar Physics (1983) **82,** 165–204

Claverie, A., Isaak G.R., McLeod, C.P., van der Raay, H.B. and Roca
Cortes, T.
Nature (1979) **282,** 591–594

Claverie, A., Isaak G.R., McLeod, C.P., van der Raay, H.B. and Roca
Cortes, T.
Nature (1981) **293,** 443–445

Grec, G., Fossat, E. and Pomerantz, M.A.
Nature (1980) **288** 541–544

Grec, G., Fossat, E. and Pomerantz, M.A.
Solar Physics (1983) **82,** 55–66

Isaak, G.R.
Nature (1982) **296,** 130–131

Kidman, R.B. and Cox, A.N. (1984) 'Solar Seismology from Space' JPL
Publication 84–84.

van der Raay, H.B.
(1982) DISCO – Result of a Phase A Study European Space Agency
SCI(82)3

van der Raay, H.B., Palle P.L. and Roca Cortes, T.
(this volume)

Woodard, M. and Hudson, A.S.
Nature (1983) **305,** 589–593

SOLAR CORE ROTATION : A DOWN TO EARTH VIEW

Pierre CONNES
Service d'Aéronomie du CNRS
Verrières le Buisson, 91371, France

ABSTRACT. The 13-days periodicity solar oscillation discovered in potassium resonance-cell data has been shown to preserve coherence over several years ; the proposed explanation was a fast rotating solar core. We show here that this coherence is probably an observational artifact due to a very small non-linear interaction between the actual short-coherence solar phenomenon (arising from rotation of surface features) and the much larger annual and periodic radial velocity change of the observer.

1. THE 13-DAYS OSCILLATION PROBLEM

We are concerned here with the proposed evidence for a fast rotating solar core giving rise to a long coherence signal with 13 days periodicity. This has been discovered by the Birmingham group in the integrated disk radial velocity as measured by their K-line resonance cell [1,2] .In this paper we point out the possibility of an instrumental-observational effect connected with the Earth annual motion : the proposed frequency is precisely an harmonic of the terrestrial year. We are not concerned however with the fast-core evidence from [3] spatially resolved observations of Doppler velocities over the solar disk [3] in the so-called 5-min band of oscillations. Neither shall we attempt to discuss the Birmingham group evidence [4,5] from line splitting in the same band. Still, this last is more directly relevant to our problem, since it was collected from some of the very same integrated- disk observations used for the 13 d periodicity. Only three years of data have been used. The first (28 days in 1980, presented in [4]) showed singlet, triplet and quintuplet structure for the l = 0, 1 and 2 lines ; however it was later shown that the reduction procedure was not totally objective [6] ; furthermore no similar effect was found in the ACRIM global photometry data [7] . Then, the other years (1981 and 1983, with much greater number of days from two stations, presented in [5]) while giving some evidence of splitting nevertheless do not confirm the previous distinctive structures. It seems that as long as the situation has not been clarified, one should refrain from proposing any systematic

229

D. O. Gough (ed.), Seismology of the Sun and the Distant Stars, 229–235.
© *1986 by D. Reidel Publishing Company.*

explanation for the proposed phenomenon.

By contrast, the 13d periodicity is clearly seen so far as a permanent feature of our sunscape. Evidence has come in two stages. The first paper [1] presents results collected over 4 years (77, 78, 80, 81) ; amplitude, period and phase are given for each individual year. Since the maximum duration of the observing season was not quite two months, only a few consecutive periods of the oscillation could be observed ; the measured periods scattered from 13 to 16 d, with a best estimate of 13.1 ± 0.2 d, while the amplitudes were in the range of 5.2 to 9.8 m/s. The proposed explanation was a fast rotating solar core. Three successive papers [8,9,10] admitted the results, but disputed the interpretation, pointing out that 13 d is practically half the observed surface rotation period T. The essence of the explanation is the same in all three cases : any feature of the solar surface is carried along by the rotation, and remains visible during time T/2 ; the result is a modulation of the disk-integrated radial velocity, also with period T ; but the second harmonic, with period T/2, is enhanced. In the second stage [2] the Birmingham group has added one more year (1982) of particularly high quality data (103 quasi-continuous days), which confirms previous results ; still this does not invalidate the above explanations. Next, a new point is made of far greater importance (indeed the possibility had already been suggested, but not stressed, in [1]) : the oscillation appears to have been coherent from 1977 to 1982, and a far more accurate period is consequently computable. Two slightly different figures, 13.05 d and 13.035 d (later called here T_1 and T_2) , corresponding to two different data treatments are given. Lastly we are informed by van der Raay that "a preliminary analysis of the 1983 data (116 days) again confirms the period, phase and amplitude of the 13d period". Consequently, no surface features of any kind can be ultimately responsible, since none have adequate lifetimes : one has to postulate the action of a <u>solid</u> rotating body ; and the already proposed core is the obvious choice [2] .

2. DISCUSSION

The point to be made in the present paper is that some solid somewhere is indeed required, but it may well be the one under our feet. The starting point of our analysis had nothing to do with any speculation about solar physics. We have been concerned only with instrumental matters, and have proposed an "absolute astronomical accelerometer" [11,12,13] , which is specifically designed for the two problems of stellar seismology, and the search for extrasolar planetary systems [14] . There are two versions, one stellar and one solar. Studying the second led to an analysis of existing devices performance [15] , particularly of the resonance cell. This instrument is almost ideal for the detection of the solar gravitational red-shift, since it compares in the most direct possible manner identical Earth and Sun-bound atoms : it provides an absolute zero-velocity reference point. But the detection of oscillations is an altogether different problem. One is concerned solely with velocity changes, i.e. accelerations : a seismometer is nothing but an accelerometer. The resonance cell does <u>not</u> provide absolute velocity changes, and has been invariably calibrated from the known Sun-observer relative motion. The procedure is a reasonable one, and all the beautiful 5-min oscillation results by either the Birmingham or the Nice groups have been collected in that way. But we are now dealing with considerably lower frequencies : the whole calibration has to be performed again for each successive day ; hence many such operations are involved before the 13 d periodicity begins to appear ; by contrast many 5-min periods are seen <u>within</u> a single day. This recalibration would be required even if all instrumental drifts (e.g. on amplifier gain, Zeeman splitting etc...) were negligible, because the system operating point on the solar line profile shifts from day to day. The maximum change, about 8 m/s/day, somewhat perversely takes place in July, right in the middle of the preferred observing season, and the peak to peak velocity change is about 1 km/s ; of course it repeats year after year in periodic fashion. While this fact is unimportant as long as single years of data are treated, this is no longer true when successive years are coherently combined. Since any calibration error will repeat with a 1 year periodicity, we are led to expect that such a coherent treatment may show up the harmonics of the

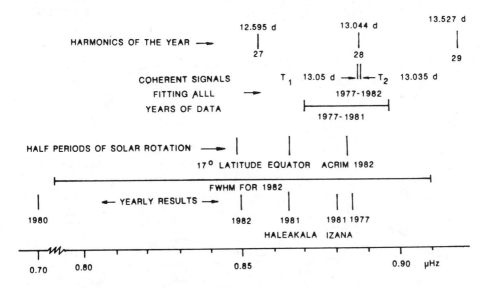

Fig. 1 – All relevant periodicities presented with a linear frequency scale. Yearly results : as given in [1,2]. FWHM of 1982 power spectrum (the narrowest one) : computed for 103 d of data. Half periods of solar rotation : as given by Allen Astrophys. Tables for 0° and 17° latitude, and by [10] for the ACRIM 1982 results. Coherent signals : as given by [1] for 1977-1981, and by [2] for 1977-1982. Harmonics of year : computed, with periods indicated.

(1/one year) = 0.0317 µHz frequency. We find from [1] that dates of
oscillation maxima over successive years tend to remain the same, and
that Harmonic 28 does fall very close to the period proposed in[2] :
one (anomalistic) year/28 = 13.045 d. Since the data are sampled with
a period which is not one exact year, but a whole number of days, we
may also consider 365/28 = 13.036 d which falls even closer to T_2 .
Fig 1 summarizes the situation. Since $F_1 = 1/T_1 = 0.8879$ µHZ and
 $F_2 = 1/T_2 = 0.8869$ µHZ ,   while   the   harmonic   separation   is
0.0317 µHZ the a priori probability for finding any harmonic within
the $F_1- F_2 = 0.001$ µHZ interval, is 1/32. Hence it is at least as
difficult to dismiss the coincidence as due to chance as it was to
treat the single-year results as independent of solar surface
rotation.

    Still, T_1 and T_2 do not correspond to error bars, merely to two
different computations from the same data. No error bars for the
best-fitting period have been given so far. May the solar origin of
the phenomenon be saved by any future reanalysis of the available
data, leading to an error bracket distinctly wider than the above
0.001 µHZ difference ? Hardly, as shown by Fig. 2 which presents the
difference between the measured phase and the computed phase for the
three most relevant periods. As expected from the data treatment [2] ,
13.035 d gives the best overall fit, while 13.05 d leads to a maximum

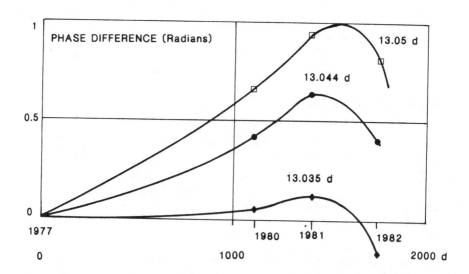

Fig. 2 - Difference between actually measured phase [2] , and phase
computed for a fully coherent signal with indicated period. The
middle curve corresponds to our Harmonic 28, and the others to the
periods T_1 and T_2 given in [2] . A one radian phase difference
corresponds to a date difference of two days.

1 rad error ; our harmonic 28 falls about half way. It is clear that using error bars larger merely by a factor of 2, leads to inadmissible phase shifts. In other words, the fit between proposed long-coherence solar effect and harmonic 28 is so good that, if we widen the error bracket enough to make a chance coincidence plausible, we at the same time kill our argument for any kind of long term coherence. The next question is : Why harmonic 28 rather than any other ? There is probably very little to choose between them ; both papers [1,2] stress that, from small-fraction-of-the-year observations, multiple solutions can be found ; and their frequency difference is of course the inverse of one year. This is about one quarter of the FWHM of the single-year power spectrum peak in the most favourable case, i.e. 1982. Hence the difference in the exactitude of the fits using Harmonics 27, 28 or 29 must be small, and our discussion would be unchanged if another harmonic had been selected.

Pinpointing the exact mechanism through which the results are perturbed by the one year periodicity is admittedly difficult, at least from the published documents. The case is complex : single-year observations do show a truly solar effect, for which indeed we have plausible explanations [8,9,10] ; hence Earth motion may at most contribute some amplitude and phase perturbations which simulate long term coherence. For instance, within year 1981, we have two independent and simultaneous observations giving almost the same frequency ; however the amplitudes are 6.6 and 9.8 m/s respectively. But the very same data show impressive amplitude agreement when the 5-min oscillation during a few hours are presented [16] ; the amplitude difference affects only the much longer period 13 d signal. Hence, while the gross features of the 13 d oscillation must be solar, a sizeable part may still be due to instrumental (or atmospheric), and/or reduction procedure effects. We have easily computed the basic non-linearity of the resonance cell signal by using the known solar line profile [17] and the Zeeman splitting of the spectrometer [18] ; the maximum distorsion within 10 h daily observing windows is about 23 m/s on May 15 and decreases to almost zero in August ; clearly such effects cannot be neglected. But this computation by itself is not enough to produce a complete explanation, because the reduction procedure [1] is quite elaborate, and has been but summarily described : while the data reduction for a single day (all we need to see the 5 min oscillation) involves finding 2 empirical constants, 3 more are required to unravel the 13-d periodicity. Trying to reproduce results without a detailed simulation would be risky at best. The prudent conclusion must be that it is only by a systematic, and completely described, reanalysis of the actual data that the respective solar and terrestrial contributions may be unraveled. We predict that when this is done, taking full account of all factors, no residual long coherence solar phenomenon will be found.

3. CONCLUSION

The major conclusion of this paper is instrumental. The just discussed difficulty arises because the Earth motion velocity changes are roughly 100 times larger than the solar 13-d amplitude ; and we may perhaps end up with an even worse situation for the 5-min lines, since the amplitudes of the weakest significant ones are only a few cm/s, i.e. another factor of 100 lower. (It is conceivable that part of the observed structure [5] arises in this way). Furthermore, when attempting to do stellar seismology, we shall have to deal with an Earth annual velocity change which is larger by a further factor of 60 than in the solar case. But there is nothing fundamental about the difficulty : it disappears if we have a fully linear device. This is the case for our proposed accelerometers [11,12,13] : then, there is no shift between a solar/stellar line profile and an instrumental one. The terrestrial velocity terms are cleanly removed at one stroke by operations involving frequencies only, which introduce no significant error. The absolute character of the measurement means that no calibration at all is required. It is noteworthy that ACRIM [7] is also an absolute device, for which no calibration from the data themselves is performed ; those data are interrupted by orbital gaps (just as ground based ones by nights), but are nevertheless being analysed as one continuous stretch. The data treatment is straighforward, and easily followed ; no empirical constants have to be substracted out. We believe it is no coincidence if the ACRIM spectra show neither the 5-min line splitting, nor a long coherence 13-d oscillation, nor indeed the 160 min one. Still, a full comparison is difficult : at least in the 5-min band, noise levels are larger with ACRIM. It seems that for solar type stars anyway, radial velocities will be an inherently more sensitive tool than intensity fluctuations [19]. Hence the need for absolute accelerometry.

We thank G.R. ISAAK for discussing solar/stellar oscillation problems, and H.B. van der RAAY for sending paper [5] and his conclusions about 1983 observations.

REFERENCES

1) CLAVERIE, A., ISAAK, G.R., Mc LEOD, C.P., VAN DER RAAY, H.B., Nat. 293, 443 (1981)
2) ISAAK, G.R., VAN DER RAAY, P.L., PALLE, P.C., ROCA CORTES, T, Mem. Soc. Astr. Italiana, 55, 353 (1983).
3) T. DUVALL, T.L., HARVEY, J.W., Nat. 310, 19, (1984)
4) CLAVERIE, A., ISAAK, G.R., Mc LEOD, C.P., VAN DER RAAY, H.B. PALLE, P.C., ROCA CORTES, T, Nat. 299, 704, (1982)
5) VAN DER RAAY, H.B., Proceed. of 25th Liège Int. Astrophys. Coll. (July 1984 ; in print)
6) GREC, G., FOSSAT, E., POMERANTZ, M., Sol. Phys. 82, 55 (1983)
7) WOODARD, M, HUDSON, H. Nat. 305, 589, (1983)
8) DURRANT, C.J., SCHROTER, E.H., Nat. 301, 589, (1983)

9) NYBORG ANDERSEN, B., MALTBY, P., Nat. 302, 808, (1983)
10) DUVALL, T.L., JONES, H.P., Nat. 304, 517 (1983)
11) CONNES J., CONNES, P., Presented at DISCO ESA Conference
    (Utrecht, 1982).
12) CONNES P., This meeting (abstract only)
13) CONNES, P., "Absolute Astronomical Accelerometry" Astrophys Spa.
    Sci., 110, 211, (1985); IAU Coll. 88 (Radial Velocities)
    Schenectady, Oct. 1984, in print.
14) CONNES, P., IAU Symp. 112 (Search for Extraterrestrial Life),
    Boston Univ., June 1984, in print.
15) CONNES, P., Proc. of Conf. on "Space Res. Prospects in Stellar
    Activity and Variability, 91, (1984) A. Mangeney Edit., Meudon
    Observatory
16) CLAVERIE, A., ISAAK, G.R., Mc LEOD C.P., VAN DER RAAY, H.B.,
    ·PALLE, P.L., ROCA CORTES, T., (Proc. of EPS Catania meeting
    1983).
17) SNIDER, J.L., Sol. Phys. 12, 352, (1970)
18) BROOKES, J.R., ISAAK, G.R., VAN DER RAAY, H.B., MNRAS, 185 (1978)
19) FOSSAT, E. Proc. Conf. on "Space Res. Prospects in Stellar
    Activity and Variability, 77, (1984) ; A. Mangeney edit., Meudon
    Observatory

The Internal Rotation of the Sun

Philip R. Goode
Department of Physics
New Jersey Institute of Technology
Newark, New Jersey, USA

ABSTRACT. The internal rotation rate of the Sun has been
calculated by Duvall, et al from the solar oscillation data
of Duvall and Harvey. The origin and implications of this
result are reviewed and compared with those of an earlier,
less reliable determination.

1. INTRODUCTION

Many years ago, Cowling and Newing (1948) suggested
that the oscillations of a star could be used used to infer
its internal rotation rate. Leighton (1960) observed
five-minute period oscillations of the surface of the Sun.
Ulrich (1970) speculated that these oscillations were a
surface manifestation of standing sound waves trapped
beneath the solar surface. Deubner (1975) showed,
observationally, that Ulrich's speculation was correct.
With this understanding of the five-minute period
oscillation, the field of solar seismology was opened.
With the observations of Duvall and Harvey (1983) the data
base became sufficient that Duvall, et al (1984) were able
to use it to infer the internal rotation rate of the Sun.
    We review here the work of Duvall, et al and its
implications. We include a detailed comparison of it with
the earlier work of Hill, Bos and Goode (1982) which
attempted to determine the internal rotation rate of the
Sun from a study of solar oscillations.

2. THE THEORY

The five-minute period oscillations are, in general,
non-radial and non-adiabatic. Four equations are necessary
to describe the oscillations. They are the equation of
continuity, Newton's second law, the first law of

D. O. Gough (ed.), Seismology of the Sun and the Distant Stars, 237–241.

thermodynamics and a subsidiary condition relating the
radiative flux to the mean radiative intensity. To find
the normal coordinate of the oscillation the four equations
are first linearized and then solved using a standard solar
model. The solution can be written in the form

$$\xi_{n\ell m}(r,\theta,\phi,t) = \{\xi_{n\ell}(r)\frac{\partial}{\partial\theta}, \; \eta_{n\ell}(r)\frac{1}{\sin\theta}\frac{\partial}{\partial\phi}\} \; Y_\ell^m(\theta,\phi)e^{j\omega t} \quad (1)$$

for a small amplitude oscillation, characterized by n,ℓ,m
and ω , the frequency of the oscillation. In an idealized
non-rotating non-magnetic star, each $(n\ell)$-oscillation is
$(2\ell+1)$-fold degenerate in frequency. Rotation removes this
degeneracy. For a slowly rotating star, the centrifugal
effect of rotation can be ignored and the degeneracy is
lifted in a way which is analogous to Zeeman splitting in
atoms. Different oscillations sample different regions of
the solar interior and, therefore, exhibit different
rotational splittings unless the Sun rotates rigidly.
Thus, with a sufficient variety of splitting data, we can
infer the interior rotation rate of the sun.
    To understand the nature of the five-minute period
oscillation, we consider the high frequency limit for
acoustic modes in which the four equations, when
linearized, reduce to

$$f'' + f\left[\frac{\omega^2 r^2}{c^2} - \ell(\ell+1) + O\left(\max(1; |\frac{d\ln\rho}{d\ln r}|)\right)\right] = 0. \quad (2)$$

The f-term represents the oscillation, ω is its frequency,
c is the local speed of sound and is the local density.
This equation defines a standard turning point problem for
the trapped modes. The inner turning point is closest to
the center for the lowest degree oscillations.
    The splitting can be calculated starting from the
equation of motion including rotation and magnetism

$$\frac{d\vec{v}}{dt} + 2\vec{\Omega}\times\vec{v} + \vec{\Omega}\times(\vec{\Omega}\times\vec{r}) = -\frac{1}{\rho}\vec{\nabla}P + \rho\vec{g} + \frac{1}{4\pi}(\vec{\nabla}\times\vec{B})\times\vec{B} \quad (3)$$

where P, g, Ω and B are the local pressure, gravity,
rotation rate and magnetic field, respectively and where

$$\vec{v} = \frac{d\vec{\xi}}{dt} \quad (4)$$

after perturbation. Under the assumption of a "small"
magnetic field and "slow" rotation, Equation [3] reduces to

$$\frac{d\vec{v}}{dt} + 2\vec{\Omega}\times\vec{v} = -\frac{1}{\rho}\vec{\nabla}P + \rho\vec{g} \quad (5)$$

From this, following Gough (1982) and assuming spherical

rotation, the relative frequency of an $(n\ell m)$-oscillation is given by

$$\sigma'_{n\ell m} = m \int_0^{R_0} K_{n\ell}(r)\Omega(r)dr,\qquad (6)$$

where $K_{n\ell}(r)$ is the splitting kernel. The splittting kernel is the sampling function for the oscillation. Solving this inverse problem yields the internal rotation rate of the sun. Of course, a reliable solution is predicated on a thorough sampling of the interior by the oscillations.

3.   RESULTS AND DISCUSSION

Duvall and Harvey (1983) identified the frequencies of many zonal (m=0) harmonics in the five minute period band. The data were obtained by averaging the image of the Sun to a line along its polar axis and detecting at 100 points along the line. A power spectrum could be obtained for each degree of oscillation by convoluting the signal with the appropriate degree Legendre polynomial and accounting for line of sight projection and limb darkening. With these identifications in hand, Duvall and Harvey (1984) collected new data with their detectors set-up along the equatorial line of the image of the solar disk. Their data were then sensitive to sectoral $(m = +\ell)$ harmonics. The mode identifications were made using identifications from the zonal mode data. The equatorial line detector set-up allows a clean separation into prograde and retrograde spectra. That is, for a particular $(n\ell)$-oscillation, there are two spectra--one for $m = +\ell$ and one for $m = -\ell$. Their frequency difference is due to rotational splitting. This means that rotational splitting can be determined even when spectral features are broad. With the rotational splitting, averaged over n, for most ℓ's between 1 and 100 in the five minute band, there is sufficient data to determine the internal rotation rate of the sun in its equatorial plane. The result of the inversion done by Duvall, et al (1984) is shown in Figure 1. The granularity of the solution reflects the granularity of the information that can be extracted from the data. The uncertainty increases as r/R_0 decreases because only the lowest degree oscillations penetrate to the deep interior.

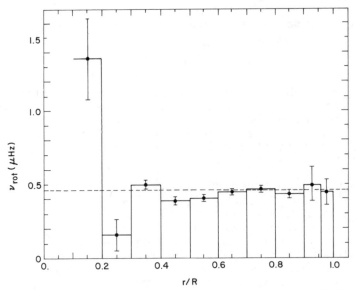

Figure 1. The internal rotation frequency of the Sun
(in$_\mu$H$_z$) as a function of fractional radius.

The calculated rotation rate is flat through the
convection zone and into the interior. The dip between
r/R$_o$ equals 0.2 and 0.3 is statistically significant, while
the rise inside 0.2 is only marginally significant. Using
the rotation frequency rate in Figure 1, the dynamical
quadrupole moment of the sun may be calculated under the
assumption that the interior rotation rate is adequately
described by its value in the equatorial plane. From this,
J , the unitless, dynamical quadrupole moment of the Sun is
(1.7+ 0.4)x10 . This result is consistent with that if
one assumed a rigidly rotating sun. The advance of the
perihelion of Mercury that would be induced by such a
quadrupole moment is ten times smaller than the error on
the observed advance of the perihelion from the radar
ranging data of Shapiro, et al (1976) and Anderson, et al
(1978). Therefore, this result for J does not upset the
consistency between the measured anomalous advance of the
perihelion of Mercury and that predicted by General
Relativity.
There have been two solar rotation laws deduced from
oscillation data -- that of Duvall, et al and that of Hill,
Bos and Goode (1982). The rotation law of Hill, Bos and
Goode (1982) implies a dynamical quadurpole moment which is
more than thirty times larger than that quoted here. The
conclusion to draw is that rotational splitting has not
been seen in at least one of the two data sets. Several
strong arguments in favor of the result of Duvall and
Harvey (1984) can be advanced:

1) The data of Duvall and Harvey (1984) allow a separation into prograde and retrograde spectra from which rotational splitting is trivially identified. No such separation occurs for the data of Hill and Bos (1983).

2) Compared to the data of Duvall and Harvey (1984), the data of Bos and Hill (1983) have about 100 times as many peaks per unit frequency regarded as being statistically significant--revealing an incredibly more complicated problem of interpretation using the Bos and Hill data.

3) The detector geometry of Bos and Hill allow a spectral separation for (ℓ+m)-even and (ℓ+m)-odd. Hill, Bos and Goode report (ℓ+m)-even spectra but the interlacing peaks were not found for (ℓ+m)-odd.

4) The Solar Maximum Mission data, as reported by Woodard (1984), puts a limit on rotational splitting--far below that of Hill, Bos and Goode but consistent with that of Duvall and Harvey (1984).

REFERENCES

Anderson, J.D., Kelsey, M.S.W., Lau, E.L., Standish, E.M., and Newhall, X.X. Acta Astronaut.5, 43(1978).

Bos, R.J. and Hill, H.A., Solar Physics 82, 89(1983).

Cowling, T.G. and Newing, R.A., Astrophys. J. 109, 149(1949).

Deubner, F.- L., Astr. Astrophys. 44, 371(1975).

Duvall, T.L. Jr. and Harvey, J.W., Nature 302, 24(1983).

Duvall, T.L. Jr. and Harvey, J.W., Nature 310, 19(1984).

Duvall, T.L. Jr., Dziembowski, W.A., Goode, P.R., Gough, D.O., Harvey, J.W. and Leibacher, J.W., Nature 310, 22(1984)

Hill, H.A., Bos, R.J. and Goode, P.R., Phys. Rev. Lett. 49, 1774(1982).

Leighton, R., Symposium on Cosmical Gas Dynamics, 4th, Varenna, ed. R. Thomas. IAU Symp. No. 12.

Shapiro, I.I., Counselman, C.C. and King, R.W., Phys. Rev. Lett. 36, 555(1976).

Ulrich, R., Astrophys. J. 162, 993(1970).

Woodard, M., Thesis, University of California, San Diego: Short Period Oscillations in Total Irradiance.

ON THE SOLAR ROTATION

M. Gabriel and F. Nemry
Institut d'Astrophysique de l'Universite de Liège
Avenue de Cointe 5
B-4200 Liège

ABSTRACT. It is pointed out that the solar rotation curve published by Duvall et al. has been obtained without considering the accuracy of the observations. When the limited accuracy, of the order of 10^{-2}, is taken into account, it is found that the data are compatible with a nearly solid rotation and that they do not contain enough information to give any indication on the rotation rate near the centre. Two rotation laws are considered; the rotation is supposed to be a function of the distance to the centre or to the rotation axis. Both laws lead to the same conclusion.

1. INTRODUCTION

Duvall and Harvey (1984) published the first extensive list of solar rotational splittings for 5 min. sectorial modes of degree $1 \lesssim \ell \lesssim 100$. These observations were analyzed by Duvall et al. (1984). Inverting the data they produced the first internal rotation curve of the sun obtained supposing that the angular frequency is a function of the distance to the centre only. We have done the same problem, choosing to represent the angular frequency ν_r by a polynomial of a variable x:

$$\nu_r = \sum_{j=0}^{J} \nu_j \ x^j. \tag{1}$$

Two choices were made for x.
     First, we considered the rotation to be a function of the distance to the centre of the sun. If we suppose that the observations have a high accuracy (10^{-8}), we reproduce qualitatively the results of Duvall et al. though with quantitative differences. However when we take into account the limited accuracy of the observations (chosen as 10^{-2}), the least-square fitting procedure shows that there is not enough information in the data to allow a fitting with a polynomial of order greater than 2. We then obtain a rotation curve fairly different from that of Duvall et al.
     Secondly, we considered rotation to be a function of the distance to the rotation axis. Given the accuracy of the data, the

243

D. O. Gough (ed.), Seismology of the Sun and the Distant Stars, 243–247.
© *1986 by D. Reidel Publishing Company.*

results in the equatorial plane are not significantly different from
the former case.

2. FITTING PROCEDURE

Equation (1) has been introduced in the formula for the rotational
splitting (see for instance Hansen et al. 1977), yielding

$$\Delta\nu(n, \ell, m) = m \sum_{j=0}^{J} \nu_j K_j(n, \ell, m) \qquad (2)$$

where $\Delta\nu(n, \ell, m)$ is the splitting for the mode of order n, degree ℓ and
azimuthal order m. K_j is the ratio of two integrals over the volume
of the star. When the rotation law is a function of the distance to
the centre, these two integrals can easily be solved analytically for
the angular variables, and only the radial integration has to be performed numerically. When the rotation law is a function of the distance
to the rotation axis, the angular integration can still be performed
analytically but it is more cumbersome. However, as only sectorial
modes have been observed, we considered only the case $m = \ell$. Then

$$Y_\ell^m \propto \sin^\ell\theta,$$

and we have only to perform angular integrals of powers of sin θ.
As observations give only an average splitting for each value of ℓ,
we also had to choose for each ℓ a value of n. We took it as the
average of n for the modes used to observe the splitting. Numerical
tests show that the choice of n has very little effect on the results
compared to the uncertainty in the observational values. Finally
equation (2) becomes

$$\overline{\Delta\nu(\ell)} = \sum_{j=0}^{J} \nu_j K_j(\ell). \qquad (3)$$

This equation was solved by least-squares fitting using the singular
value decomposition method (see Forsythe et al. 1977). It has the
advantage of pointing out when changes in the data less than their
accuracy can result in an entirely different set of coefficients ν_j
which satisfy the least squares criterion. These coefficients may be
considered as arbitrary and are set equal to zero.
        a) Results for a rotation law that is a function of the
distance to the centre.
        Figure 1 shows the results obtained supposing a relative
error of tbe observations of 10^{-8}. The degree of the polynomial J is
equal 9. The full line is for unaltered data. The dashed curve was
obtained by decreasing the splittings for $\ell = 1$ and 2 to 500 and 470
nHz respectively. They agree fairly well with the results obtained by
Duvall et al. Perhaps the differences might be explained by a different
choice of eigenfunctions and by the use of a different numerical method.

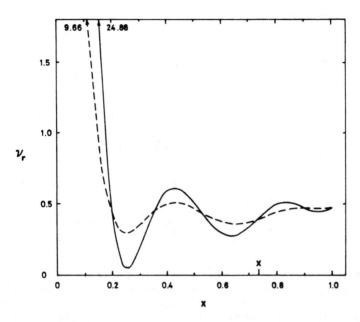

Figure 1. Rotation frequency as a function of x = r/R for rotation law that is a function of the distance to the centre. The full line is obtained for the observed splittings. The dashed curve is obtained taking splittings for ℓ = 1 and 2 equal to 500 and 470 nHz respectively. The data are supposed to have a very high accuracy.

Figure 1 shows that the programme works correctly, as it is able to reproduce the result obtained by Duvall et al.
        Figure 2 shows the results obtained taking a relative error of the observation of 10^{-2}. For most values of ℓ the observational error is even larger. In this case the programme tells us that all ν_j for j > 2 are arbitrary, and it sets them equal to zero. This means that given the accuracy of the data, it is meaningless to try fitting polynomials of degree J higher than 2. Figure 2 gives the rotation frequency for fitting obtained from various values of J (in all cases the programme cancels all ν_j for j > 2). All these curves are equally acceptable. From these results we may only conclude that if ν_r is a function of r alone, the solar rotation does not depart very much from solid rotation in most of the interior. Certainly it has no large increase below the bottom of the convective envelope (point x_e on Figure 2) as suggested by Hill et al. (1984). However, nothing can be said concerning the central core, which may rotate very fast or with nearly the surface value.

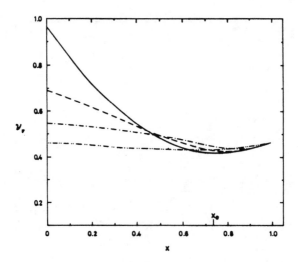

Figure 2.   Same rotation law as for Figure 1, but the observed splittings are supposed to have a relative accuracy of 10^{-2}.   The full curve
is for a second-degree polynomial, the dashed curve is for a fourth-
degree polynomial, the dot-dashed curve is for a polynomial of ninth
degree, the double dot-dashed curve is for a polynomial of fourth
degree but neglecting data for ℓ = 1, 2, 90 and 100.

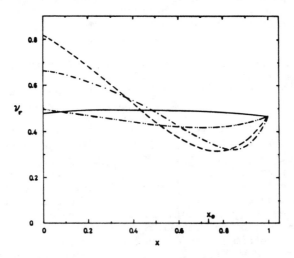

Figure 3.   Same as Figure 2 but for a rotation law that is a function of
the distance to the axis.

b) Results for a rotation law that is a function of the distance to the rotation axis.

Figure 3 shows results obtained supposing an accuracy of 10^{-2} of the data. Again the programme refuses to make fittings with polynomials of degree higher than 2. All the curves are equally acceptable. They show either that the solar rotation is near solid body or that it has a more or less pronounced minimum near the bottom of the convective envelope and then increases inwards.

3. DISCUSSION AND CONCLUSION

The two rotation laws we discussed are equally likely. In both cases the variance is (36 ± 2.1) nHz. Modes of $\ell = 1$ and 2 have a large contribution to this value. When the splittings for these modes are decreased, the variances drop to (13.9 ± 1.4) nHz. When data for $\ell = 1$, 2, 90 and 100 are neglected they drop to 9 and 10 nHz in cases (a) and (b) respectively.

Given the accuracy of the data, they do not contain enough information to allow a fitting with a polynomial of degree larger than 2. They are compatible with nearly uniform rotation, and they do not contain any indication of fast central rotation.

REFERENCES

Duvall, Jr, T.L. and Harvey, J.W. (1984) Nature 310, 19.
Duvall, Jr, T.L., Dziembowski, W.A., Goode, P.R., Gough, D.O.,
        Harvey, J.W. and Leibacher, J.W. (1984) Nature 310, 22.
Forsythe, G.E., Malcolm, M.A. and Moler, C.B. (1977) Computer Methods
        for Mathematical Computations. Prentice-Hall.
Hill, H.A., Rosenwald, R.D., Yakowitz, D.S. and Campbell, W. (1984)
        Proceedings of the fourth European meeting on Solar Physics,
        187.

THE INTERNAL ROTATION OF THE SUN

Ian W Roxburgh
Theoretical Astronomy Unit, School of Mathematical Sciences,
Queen Mary College, University of London.

ABSTRACT   Recent results on the internal rotation of the sun are difficult to explain.   These problems are reviewed and it is conjectured that a ^{3}He driven mixed shell near 0.3R$_\odot$ magnetically separates the inner core from the outer regions. Such a shell may rotate differentially and may separate a more rapidly spinning core from the rest of the sun.

THE SOLAR ROTATION

What is the internal rotation of the Sun?   Duvall et al (1984) deduce an internal profile for the angular velocity $\Omega(r)$, which is almost constant down to 0.4R$_\odot$ then increases slightly between 0.4 and 0.3R$_\odot$ then decreases to a minimum between 0.3 and 0.2R$_\odot$ rising in the centre to a value which may be twice the surface value.   In contrast, Hill (1985) finds a rapid increase by a factor 3 just below the convective zone and almost constant Ω from there to the centre. The results presented by Brown (1985) support the findings of Duvall et al, and further suggest that the latitudinal differential rotation decreases with depth but that the region around 0.35R$_\odot$ has enhanced differential rotation.

THE EFFECT OF MAGNETIC FIELDS

It is difficult to reconcile the Duvall et al (1984) rotational profile with even a tiny internal magnetic field unless this field is somehow separated into distinct zones, or if what we currently observe is a transient produced by a recent phase of mixing in the core, the internal magnetic field not having had enough time to equalise the angular velocity.   The decay time of the largest scale components of a global magnetic field is of order t$_\odot$ so in the age of the sun all but the few lowest order modes will have decayed away under Ohmic diffusion, but unless there is some form of enhanced diffusion there should be some relict of the original field trapped in the sun at the

249

D. O. Gough (ed.), Seismology of the Sun and the Distant Stars, 249–252.
© *1986 by D. Reidel Publishing Company.*

time of formation. Perhaps if the sun had an initial field it floated to the surface during a fully convective pre-main sequence phase, or was dragged with the outer convective zone as a radiative core was forming during approach to the main sequence.   However if the sun had a small ^{12}C driven convective core during its initial main sequence phase it would probably have generated a field; when convection ceased the field would diffuse outwards and decay, leaving the present sun with the relict of the largest scale components.

How a magnetic field affects the rotation has yet to be fully understood.   The condition that the angular velocity be constant on field lines is a steady state conditon, a weak field being wound up by differential rotation until it is large enough to react back on the motion; a field of only 10μG being large enough to produce uniform rotation in the life time of the sun. Perhaps some kind of instability limits the increase in field before it can react back on the motion, a problem that remains to be resolved. However even if some dissipative mechanism intervenes, the energy dissipated originates in the kinetic energy of rotation so that whatever the dissipation mechanism it reduces the kinetic energy of the rotation field (Mestel private communication); for a given angular momentum the lowest rotational energy state is that of uniform rotation so the mechanism must have the effect of equalising angular velocity.

ROTATIONAL INSTABILITIES

Since the present sun is losing angular momentum in the solar wind it is reasonable to suppose that the surface layers and convective zone have been slowed down. Indeed the evidence from other solar type stars suggests that the sun was originally rotating much more rapidly than at present and was slowed down during its early main sequence phase.   How is the outer convective zone coupled to the deep interior? Since the surface field is oscillatory it can only penetrate a very small distance beneath the zone, so coupling to the interior is probably through some intermediate zone between the two regions. The low surface abundance of Lithium is strong evidence that there is some mild turbulent mixing beneath the convective zone down to the layers where Lithium is burnt.

If there is no field in the interior then there is a range of instabilities that are driven by differential rotation produced by the spin down of the surface layers (Goldreich and Schubert 1967, Fricke 1968, Knobloch and Spruit 1983, Roxburgh 1984a).   Spruit et al (1983) used these instabilties to argue that the angular velocity distribution was close to marginal stability - which gave an almost constant $\Omega(r)$ down to $0.4R_\odot$ and then an inward increase - not exactly that deduced by Duvall et al but with the same overall behaviour except for the dip at $0.25R_\odot$. But such instabilties are not effective;   recently I showed (Roxburgh 1984b 1985) from consideration of the energetics that these instabilties would be unable to effectively mix the sun, the motion

motion being almost horizontal and leading to a latitude dependent
distribution of molecular weight that could sustain a much larger
angular velocity gradient.

However, this may not be true in the layers beneath the convective
zone and weak turbulence driven by a very small angular velocity
gradient may be effective in carrying out angular momentum in this
almost chemically homogenous region, and at the same time leading to a
decay in the surface abundance of Lithium (Dicke 1972, Roxburgh
1976). Since the energy to overcome the mild chemical composition
gradient comes initially from the kinetic energy of differential rotation,
such mixing will only be effective if the rotational energy is greater
than the difference in gravitational energy due to the composition
gradient. In standard solar models this is the case in the layers just
beneath the convective zone. (Dziembowski and Paterno, private
communication, have recently argued that gravitational diffusion leads
to a larger composition gradient, inhibiting such mixing.)

MIXING IN THE CENTRAL REGIONS

The drop in angular velocity at $0.3R_\odot$ deduced by Duvall et al is very
difficult to explain.   One intriguing possibility (Gough, private
communication) is that the central regions have recently turned over,
low angular momentum material moving from near the centre to $0.3R_\odot$
leading to a drop in angular velocity, and higher angular momentum
material moving to the centre, thus leading to a rapidly spinning small
core.   The only energy source capable of driving such a turnover is
the nuclear energy which revives the old conjecture of Dilke and
Gough (1972) that the ^{3}He instability produces such a non linear
instability.   Were this to have taken place it would produce radically
different solar models, since such a turnover would cause chemical
mixing thus changing the evolutionary history of the sun. The
objection to this conjecture is that a much more homogeneous solar
model has difficulty in explaining the oscillation data.

The results presented by Brown (1985) suggest that there is a zone of
latitudinal differential rotation in the same region as the radial
variation near $0.3R_\odot$. Such differential rotation suggests the presence
of latitudinal motion, turbulent or laminar, since the interaction of
such motion with rotation could produce a latitudinal variation of
angular velocity in much the same way as is thought to take place in
the outer convective zone.   One possibility is that the ^{3}He instability
drives mixing in a shell around $0.3R_\odot$ near the peak of the ^{3}He profile.

Were such a conjecture to be correct then the sun would be divided
into four zones: an inner core with approximately constant rotation; a
^{3}He driven mixed shell magnetically seperated from the core and
differentially rotating; an outer radiative region again approximately
uniformly rotating and magnetically separated from the mixed shell; a
differentially rotating outer convective zone magnetically separated

from the radiative region.   Within the inner core and the outer
radiative region angular momentum is transported by magnetic torques,
within the mixed regions transport is dominated by fluid motions and
at the boundaries between zones angular momentum is transported by
a mixture of overshooting and rotationally driven instabilities.
Whether or not such a model is at all viable has yet to be determined.

REFERENCES

Brown T, 1985, paper presented at Cambridge Conference on Oscillations

Christensen Dalsgaard J, Dilke F W W and Gough D O,
                1974 Mon. Not. Roy. Ast. Soc.,169, 429.

Dicke R H, 1972, Astrophys. J. 171, 331.

Dilke F W W and Gough D O, 1972, Nature 240, 262.

Duvall T L, Dziembowski W A, Goode P R, Gough D O, Harvey J W and
                Leibacher J W, 1984, Nature 310, 22.

Goldreich P and Schubert G, 1967, Astrophys J. 150, 571

Fricke K, 1968, Zeit. fur Astr. 68, 317.

Hill H A, Yakowitz D S, Rosenwald R D and Campbell W,
                1985 in *Hydromagnetics of the Sun*, ESA SP-220

Knobloch E and Spruit H, 1983, Astron and Astrophys. 125, 59.

Roxburgh I W, 1976 in *Basic Mechanisms of Solar Activity*
                ed. Bumba and Kleczeck, Reidel. p453.

Roxburgh I W, 1984a, Mem. Soc. Astr.Italy, 55,273.

Roxburgh I W 1984b, paper presented at 25th Liege Colloquium on
                Theoretical Problems in Stellar Stability and Oscillations.

Roxburgh I W, 1985, in *Solar Neutrinos and Neutrino Astronomy*,
                ed M L Cherry, W A Fowler and K Lande, p88, Am.Inst.Physics.

Spruit H C Knobloch E W and Roxburgh I W, 1983, Nature, 304, 320.

Magnetic torques and differential rotation

N. O. Weiss
Department of Applied Mathematics
and Theoretical Physics,
Silver Street, Cambridge CB3 9EW, UK

ABSTRACT. The rotation profile deduced from the measured splitting of
solar oscillations raises several problems. The slight decrease in
angular velocity within the convective zone can be explained and the
transition from differential to uniform rotation on spherical surfaces
must cause a further reduction in the radiative zone. The bump, associated with harmonics of degree 11, seems implausible and it is hard to
reconcile a rapidly rotating core with the presence of a significant
magnetic field. A possible explanation is that this core has been magnetically decoupled from outer regions since the sun evolved on to the
Hayashi track.

By measuring the rotational splitting of the frequencies of five-minute
oscillations on the sun, Duvall & Harvey (1984) and Brown (1985, 1986) have
provided the first reliable indications of the variation of angular
velocity, Ω, with radius, r (Duvall et al. 1984) and latitude λ, within
the sun. These results have important theoretical implications, and are
significant for our understanding not only of the sun but also of other
stars, whose magnetic activity depends on their rotation rates.

Within the convective zone, the new results are in accord with what
had been expected. The gross distribution of angular momentum is controlled by two effects (Weiss 1976). The first is redistribution of
angular momentum: for example, a steady axisymmetric meridional flow
leads to a state in which $r^2 \cos^2 \lambda \, \Omega$ is constant along streamlines. This
would imply that Ω doubles by the base of the convective zone, which is
ruled out by the observations (though this process may be significant
near the poles and just below the surface). Secondly, the Proudman-

D. O. Gough (ed.), Seismology of the Sun and the Distant Stars, 253–256.

Taylor constraint encourages convection in rolls aligned with the rotation axis, so that Ω can be uniform.   To a first approximation, this is
what is found.   The variations in Ω are relatively small and in agreement with numerical simulations, using the Boussinesq (Gilman 1979) and
anelastic   (Glatzmaier   1985)   approximations,   which   predict   that
$\partial\Omega/\partial r > 0$.   The only conflict is with hydromagnetic dynamo models,
which typically require $\partial\Omega/\partial r < 0$ in order that sunspot zones should
migrate towards the equator; this can be avoided by supposing that
dynamo action takes place in a shell just below the base of the convective zone.

The inferred variation of Ω in the radiative zone raises more
serious problems, which are discussed by Rosner & Weiss (1985).   Brown's
(1985,1986) results suggest that the latitudinal differential rotation
becomes quite small in the outer radiative zone.   In a stably stratified
region shear instabilities should be able to eliminate any significant
variation of Ω with λ (Zahn 1983).   The transition from a region where
$\Omega = \Omega(r\cos\lambda)$, at the base of the convective zone, to a region where
$\Omega = \Omega(r)$, by $r = 0.5\,R_o$, say, where R_o is the solar radius, produces a
distribution of angular velocity like that shown schematically in Fig.
1.   This variation, which is consistent with the observations, must lead
to meridional circulations.

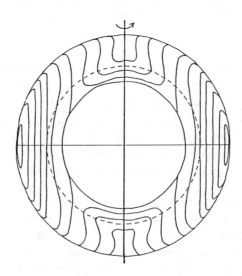

Fig. 1.   Sketch showing surfaces of constant Ω in the sun.   The base of
the convective zone is indicated by the dashed line.   (After Rosner &
Weiss 1985.)

The enhanced splitting for modes of degree ℓ = 11 has been found both at Kitt Peak (Duvall & Harvey 1984) and with the Fourier tachometer at Sacramento Peak (Brown 1985,1986). Thus it is a genuine effect. Nevertheless, we do not believe that any dynamical process could produce a local maximum in Ω at $r \approx 0.35\,R_o$ (Duvall et al. 1984). Some other (as yet unknown) mechanism must produce the splitting. This problem obviously calls for more attention.

Differential rotation in the radiative zone is reduced by magnetic torques and may generate hydrodynamic instabilities. Through most of this region Ω is almost uniform and gradients are too small to excite diffusive instabilities (Spruit, Knobloch & Roxburgh 1983; Spruit 1984). Magnetic fields are, however, more effective and it is easy to show that the timescale for restoring uniform rotation is the Alfven time $\tau_A = r/v_A$, where the Alfven speed $v_A = B_o / (\mu_o \rho)^{1/2}$. If the radial field B_o exceeds the interstellar value of 3×10^{-6} gauss then τ_A is less than 3×10^9 yr. Nobody knows the field strength in the radiative zone – 1 gauss is a reasonable guess – but it is clear that any plausible field would suffice to ensure that Ω is constant.

The evidence for a rapidly rotating core depends on measurements of the rotational splitting of modes with ℓ = 1 (Claverie et al. 1981; Duvall & Harvey 1984), which suggest that Ω increases rapidly to at least twice the surface value around $r = 0.2\,R_o$. This can only occur if there are no significant magnetic torques. Perhaps (as Mestel has suggested) the azimuthal field is limited by rapidly growing instabilities but the only plausible explanation we have found is that the core is magnetically decoupled from the rest of the radiative zone. This might be the case if the sun was never fully convective in the pre-main sequence Hayashi phase. Then the core might retain a primeval field which is isolated from the residual fields left behind as the convectively driven dynamo retreated outwards. At the interface, angular momentum would be transferred by the *ABCD* instability (Spruit et al. 1983), which is unaffected by gradients in molecular weight. Thus the core would rotate at about twice the surface rate as the sun was gradually spun down.

Any such explanation obviously requires more detailed investigation before it can be made convincing. It is also important to obtain better estimates of the value of Ω near the centre of the sun. If the measured splittings for ℓ = 1 are indeed correct and the core rotates twice as fast as the surface, theoreticians will have to think hard. Still, I

would be prepared to bet on a more boring outcome,with an angular velocity that is essentially uniform.

REFERENCES

Brown, T.M.: 1985, Nature (in press).

Brown, T.M.: 1986, these proceedings.

Claverie, A., Isaak, G.R., McLeod, C.P. & van der Raay, H.B.: 1981,
Nature 293, 443-445.

Duvall, T.L., Dziembowski, W.A., Goode, P.R., Gough, D.O., Harvey, J.L.
& Leibacher, J.W.: 1984, Nature 310, 22-25.

Duvall, T.L. & Harvey, J.W.: 1984, Nature 310, 19-22.

Gilman, P.A.: 1979, Astrophys. H. 231, 284-292.

Glatzmaier, G.A.: 1985, Astrophys. J. 291, 300-307.

Rosner, R. & Weiss, N.O.: 1985, Nature (in press).

Spruit, H.C.: 1984, in Hydromagnetics of the Sun, ed. P. Guyenne,
pp.21-27, ESA SP-220, Noordwijk.

Spruit, H.C., Knobloch, E. & Roxburgh, I.W.: 1983, Nature 304, 520-522.

Weiss, N.O.: 1976, in Basic Mechanisms of Solar Activity, ed. V. Bumba &
J. Kleczek, pp.229-242, Reidel, Dordrecht.

Zahn, J.-P.: 1983, in Astrophysical Processes in Upper Main Sequence
Stars, by A.N. Cox, S. Vauclair & J.-P. Zahn, pp.253-339, Swiss Soc.
Astron. Astrophys., Geneva.

INTERNAL ROTATION OF THE SUN AS A CONSEQUENCE OF INSTABILITIES IN THE
RADIATIVE INTERIOR

W.A. Dziembowski[1], L. Paternò[2,3], R. Ventura[3]

(1) N. Copernicus Astronomical Center, Warsaw, Poland
(2) Department of Physics, University of Catania, Italy
(3) CNR-National Group of Astronomy, Section of Catania, Italy

ABSTRACT. We consider the effect of the molecular weight gradient in
the Sun's radiative interior on the GSF instability, and deduce the
marginal gradient of angular velocity which can be tolerated. The same
gradient is also calculated for the ABCD instability which has the lowest threshold of the hydrodynamical instabilities. The curves we obtain for the angular velocity contradict those inferred from the observed splittings of p-mode oscillations. We suggest that angular momentum can be transferred from interior owing to some instability caused
by the presence of a global magnetic field in the Sun's core.

1. INTRODUCTION

In recent times the internal rotation of the Sun has been inferred
from the rotational splitting of p-mode oscillations (Duvall and
Harvey 1984; Brown 1986).
    The main features of the rotational curve indicate that rotation
below the convection zone down to $0.4-0.3$ $R_\odot$ is essentially rigid, and
it has an angular velocity close to the surface value with insignificant latitudinal variation. Moreover the analysis of data suggests the
presence of a sharp decrease of the equatorial angular velocity at 0.3
$R_\odot$ with a sudden increase deep down to the Sun's core (Duvall et al.
1984) and a thin layer located between $0.3-0.4$ $R_\odot$ where latitudinal
differential rotation raises to the surface value or more (Brown 1986).
    In view of these results we examine the influence of two axisymme-
tric instabilities, the Goldreich-Schubert-Fricke instability (GSF)
(Goldreich and Schubert 1967; Fricke 1968) and the axisymmetric baro-
clinic diffusive instability (ABCD) (Shibahashi 1980; Knobloch and
Spruit 1983), on the distribution of angular momentum in the Sun's interior. In particular we study the effect of the molecular weight stratification on the GSF instability, and derive the variation of the angular velocity of rotation with depth in the case of marginal stability for both GSF and ABCD situations.
    The same argument has recently been discussed by Spruit et al.
(1983) and Roxburgh (1984).

D. O. Gough (ed.), Seismology of the Sun and the Distant Stars, 257–263.

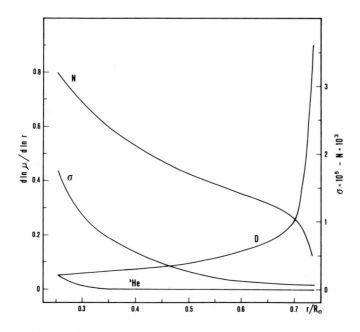

Figure 1. The gradient of mean molecular weight (left vert. axis) is plotted as a function of the fractional radius $r/R_\odot$. The two curves labeled
^{3}He and D indicate the effects of ^{3}He diffusion and barodiffusion respectively. In the same Figure we plot the behaviour of the Prandtl number σ and Brunt-Väisälä frequency N in the Sun's radiative interior
(right vert. axis).

2. THE GRADIENT OF MOLECULAR WEIGHT

We deduce the present mean molecular weight stratification in the
solar radiative interior caused by two main effects: i) ^{3}He formation
in the outer part of the solar core; ii) diffusion of elements heavier
than Hydrogen from the base of the convection zone deep down to the solar core.

We use a standard model of the Sun with initial abundances X = 0.74
and Z = 0.02. We essentially compute the rate of ^{1}H destruction due to
the reaction ^{1}H (p, e $^+$ν) ^{2}H, since the next reaction ^{2}H (p, γ) ^{3}He runs
very fast with respect the previous one:

$$\frac{\partial X}{\partial t} = - \rho \frac{X^2}{A} N_A <01> \tag{1}$$

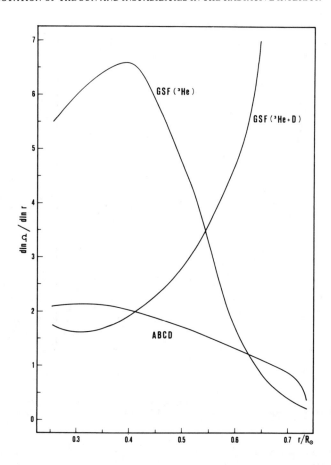

Figure 2. The marginal gradient of the angular velocity in the Sun's radiative interior is plotted as a function of the fractional radius $r/R_\odot$ for ABCD and GSF instabilities. For this latter we consider the effects of ^{3}He barrier alone and ^{3}He + barodiffusion barrier of molecular weight gradients.

where ρ is the density, A the atomic mass, and $N_A <\sigma l>$ the appropriate reaction rate as suggested by Harris et al. (1983).

   At the equilibrium, the mean ^{3}He stratification should reflect the mean H stratification, thus giving $dY_3/d\ln r = - dX/d\ln r = -dN/d\ln r$, where Y_3 is the Helium content by mass and $N = 1/\mu = 2X + 3Y/4 + Z/2$, with μ the mean molecular weight. Since $dN/d\ln r = - Nd\ln\mu/d\ln r$, we obtain:

$$\frac{\partial}{\partial t} \left(\frac{d\ln\mu}{d\ln r} \right)_{^3He} = -\frac{1}{N} \frac{d}{d\ln r} \frac{\partial X}{\partial t} \qquad (2)$$

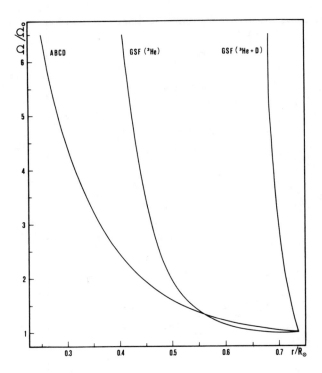

Figure 3. The same as in Figure 2 but for the normalized angular velocity.

Using expression (1) and integrating over the life of the Sun we have an approximate expression for $d\ln\mu/d\ln r$ as a function of $r/R_\odot$ for the present Sun. In Figure 1 the curve derived from expression (2) is labeled with ^{3}He.

As far as the diffusion of heavy elements is concerned, we follow Aller and Chapman (1960) and Alcock and Illarionov (1980) to compute the diffusion velocities V_{iH} of the elements with respect the background of Hydrogen. We consider only to elements: ^{4}He and ^{28}Si, this latter as an average representative of all heavier elements.

Using the continuity equation, after some manipulations we obtain:

$$\frac{\partial E_i}{\partial t} = \frac{1}{\rho r^2}\ \frac{d}{dr}\ (\rho\ E_i\ V_{iH}\ r^2) \tag{3}$$

where E_i stands for the abundance by mass of ^{4}He or metals. Since $X + Y + Z = 1$, we thus have:

$$\frac{\partial N}{\partial t} = -\frac{5}{4} \frac{\partial Y}{\partial t} - \frac{3}{2} \frac{\partial Z}{\partial t} = \frac{1}{\rho \, r^2} \frac{dJ}{dr} \tag{4}$$

where $J = - (5Y \, V_{YH}/4 + 3ZV_{ZH}/2)\rho \, r^2$ is the mass flux and V_{ZH} and V_{ZH} are the diffusion velocities of ^{4}He and metals respectively. Analogously to the expression (2), it follows readily:

$$\frac{\partial}{\partial t} \left(\frac{d\ln \mu}{d\ln r} \right)_D = - \frac{1}{N} \frac{d}{d\ln r} \frac{1}{\rho r^2} \frac{dJ}{dr} \tag{5}$$

We integrate equation (5) over the life of the Sun and obtain the variation of the molecular weight gradient as a function of $r/R_\odot$, as we show in Figure 1 by the curve labeled D.

3. MARGINAL STABILITY

It is well known that molecular weight barrier can prevent the GSF instability and allow for steep gradients of angular velocity (Knoblock and Spruit 1983). Following Tassoul (1978) it is possible to derive the following expression for marginal stability:

$$\frac{d\ln \Omega}{d\ln r} = 2\sqrt{2} \left(\frac{g}{2 \, \Omega^2 \, r} \left| \frac{d\ln \mu}{d\ln r} \right| \right)^{1/2} \tag{6}$$

where Ω is the angular velocity and g the gravity. In Figure 2 we show the behaviour of the marginal Ω gradient for the effects of ^{3}He barrier alone and ^{3}He + diffusion barrier.

The expression for the ABCD marginal stability is taken by Spruit et al.(1983) and Roxburgh(1984). We have:

$$\frac{d\ln \Omega}{d\ln r} = 2\sqrt{2} \, \sigma^{1/2} \frac{N}{\Omega} \tag{7}$$

where σ is the Prandtl number and N the Brunt-Väisälä frequency. The Prandtl number has been computed by means of the following expression:

$$\sigma = C_p \, \eta/\chi$$

where C_p is the specific heat at constant pressure, η the dynamical molecular viscosity and χ the radiative heat conductivity as given by Lang (1974):

$$\eta = 2 \, 10^{-15} \, T^{5/2} A^{1/2}/Z^4 \, \ln \Lambda$$
$$\Lambda = 1.3 \, 10^4 \, T^{3/2} / N_e^{1/2}$$
$$\chi = 16 \, \sigma_s \, T^3/3 \, \kappa\rho$$

where Z is the ionic charge, N_e the electron density, σ_s the Stefan-Boltzmann constant, and κ the opacity.

In Figure 1 we show the behaviour of N and the Prandtl number σ as functions of $r/R_\odot$ for our solar model, and in Figure 2 the behaviour of the gradient of Ω in the case of ABCD marginal stability.

From expressions (6) and (7) it is possible to derive by numerical integration the profiles of the marginal angular velocity as a function

of depth in the Sun's radiative interior for the above discussed instabilities. We then obtain for GSF and ABCD instabilities respectively:

$$\Omega_{GSF} = \Omega_o + 2\sqrt{2} \int_r^{r_o} (- \frac{g}{2r} \ \frac{d\ln\mu}{d\ln r})^{1/2} \ \frac{dr}{r} \tag{8}$$

$$\Omega_{ABCD} = \Omega_o + 2\sqrt{2} \int_r^{r_o} \sigma^{1/2} \ N \ \frac{dr}{r} \tag{9}$$

In Figure 3 we show the behaviour of Ω/Ω_o when only ^{3}He barrier is active, when ^{3}He and diffusion are considered for GSF instability, and for
ABCD instability. We assumed for Ω_o at the base of the convection zone
the value of $2.83 \ 10^{-6}$ rad s^{-1} as Spruit et al (1983). In this regard we
point out that recent models of differential rotation of the convection
zone including radiative transport and overshooting at the base of the
convection zone (Pidatella et al. 1985) indicate that Ω decreases
slightly inwards and reaches the value of about $2.78 \ 10^{-6}$ rad s^{-1} at the
base of the convection zone.

4. CONCLUSION

Our results confirm the previous conclusions of other authors that
the ABCD instability is the most severe dynamical instability, since the
GSF instability can be suppressed even by a modest μ-gradient as in the
case of ^{3}He diffusion alone. When the effect of barodiffusion is also
considered the radiative interior can suffer extremely large Ω-gradients.
    However the ABCD instability itself tolerates significant Ω-gradi-
ents below 0.5 R$_\odot$ and allows for rotation rates significantly larger
than those at the base of the convection zone.
    These results contradict the conclusions which can be drawn from
the analysis of the mode splittings. They suggest that angular momentum is effectively transferred from the interior leading to an almost
rigid rotation of the radiative envelope of the Sun.
    Since the ABCD instability has the lowest threshold of hydrody-
namical instabilities, we are force to think that a global magnetic
field in the Sun's core can offer a wide source of instabilities thus
explaining the inferred rigid rotation. This is consistent with results
concerning the excitation of internal gravity modes in the Sun obtained
recently by Dziembowski et al. (1985,1986).

REFERENCES

Alcock, C., Illarionov, A.: 1980, Astrophys.J. 235, 534
Aller, L.H., Chapman, S.: 1960, Astrophys.J. 132, 461
Brown, T.M.: 1986, These Proceedings
Duvall, T.L.jr., Harvey, J.W.: 1984, Nature 310, 19
Duvall, T.L.jr., Dziembowski, W.A., Goode, P.R., Gough, D.O., Harvey,
    J.W.: 1984, Nature 310, 22
Dziembowski, W.A., Paternò, L. Ventura, R.: 1985,   Astron. Astrophys.,
    in press
Dziembowski, W.A., Paternò, L., Ventura, R.: 1986,   These Proceedings

Fricke, K.J.: 1968, Z. Astrophys 68, 317

Goldreich, P., Schubert, G.: 1967, Astrophys. J. 150, 571

Harris, M.J., Fowler, W.A., Coughlan, G.R., Zimmerman, B.A.: 1983, Ann.
 Rev. Astron. Astrophys. 21, 165

Knobloch, E., Spruit, H.C.: 1983, Astron. Astrophys. 125, 59

Lang, K.R.: 1974, Astrophysical Formulae, Springer-Verlag, New York

Pidatella, R.M., Stix, M., Belvedere, G., Paternò, L.: 1985, Astron.
 Astrophys., in press

Roxburgh, I.W.: 1984, in Oscillations as a Probe of the Sun's Interior,
 eds. G. Belvedere and L. Paternò, Mem. Soc. Astron. Italiana 55,
 273

Shibahashi, H.: 1980, Publ. Astron. Soc. Japan 32, 341

Spruit, H.C., Knobloch, E., Roxburgh, I.W.: 1983, Nature 304, 520

Tassoul, J.L.: 1978, Theory of Rotating Stars, Princeton Univ. Press,
 Princeton

FINITE AMPLITUDE LIMIT OF THE ^{3}HE INSTABILITY

I.W.Roxburgh
Theoretical Astronomy Unit
School of Mathematical Sciences,
Queen Mary College
London E1 4NS

ABSTRACT. The Models of solar type stars are unstable to non-radial oscillations driven by burning ^{3}He. These oscillations modify the distribution of ^{3}He leading to finite amplitude oscillations with typical temperature perturbations $\delta T/T \sim 0.1$.

THE ^{3}HE DRIVEN INSTABILITY

Several years ago Dilke and Gough (1972) pointed out that the sun could be unstable to growing amptitude oscillations driven by the gradient of ^{3}He produced by the burning of hydrogen. Subsequent investigation by several authors (c.f. Christensen Dalsgaard et al 1974) have verified this result by detailed analysis of low order g modes, finding that a solar model is unstable after about 3.10^8 years.

The exciting mechanism can be understood by studying the behaviour of a 'blob' of fluid of scale ℓ, executing almost adiabatic oscillations with the natural bouyancy frequency. If radiative losses and energy generation are neglected such a blob executes simple harmonic motion about the equilibrium level where its density is the same as its surroundings. If radiative losses are include the motion is damped; when it is below the equilibrium level it is less dense and therefore hotter than its surroundings, on losing heat to the surroundings the density difference is reduced which reduces the restoring force and the blob returns to its equilibrium level with a smaller velocity upwards than it had when travelling downwards. Energy generation works in the opposite sense; in the phase when the temperature is hotter than the surroundings the energy generation is enhanced, this increases the temperature excess which leads to an enhanced restoring force so the blob returns to the equilibrium level with a larger upward velocity than it had on the downward phase. In the general case therefore the oscillation is excited if the energy gain from nuclear reactions exceeds the losses from radiation. If the blob has dimensions ℓ, this requires

D. O. Gough (ed.), Seismology of the Sun and the Distant Stars, 265–269.

$$\rho \ell^3 \delta \epsilon \quad > \quad \frac{K}{\rho} \delta T \, \ell^2 \tag{1}$$

where the radiative energy flux is $F = -K\nabla T$, $K = 4acT^3/3\kappa\rho$.

If the energy generation comes from the proton-proton chain and if for the purpose of illustration we confine our attention to the $^3He + {}^3He$ branch, the energy generation rate can be expressed as

$$\epsilon = \epsilon_1 \rho X_1{}^2 T^{\eta_1} + \epsilon_3 \rho X_3{}^2 T^{\eta_3} = \epsilon_{11} + \epsilon_{33}. \tag{2}$$

If we now use the relation $\mu = 4/(3 + 5X_1)$ and define $\nabla = (\partial \ln T/\partial \ln P)$ the condition for growing amplitude oscillations is

$$\left[\frac{2\epsilon_{11}}{\epsilon} - \frac{5X_1}{3+5X_1} \right] \frac{1}{H_1} + \frac{\epsilon_{33}}{\epsilon} \frac{2}{H_3} > \left[\frac{KT}{\rho\epsilon\ell^2} - \frac{\epsilon_{11}}{\epsilon} \eta_1 - \frac{\epsilon_{33}}{\epsilon} \eta_3 + 1 \right] \frac{(\nabla_{ad} - \nabla)}{H_p} \tag{3}$$

where H_p is the pressure scale height and H_1, H_3 are respectively the scale heights of X_1 and X_3.

The abundances of X_1 and X_3 are determined by the evolution equations

$$\frac{N}{A_1} \frac{\partial X_1}{\partial t} = -3 R_{11} X_1{}^2 + 2 R_{33} X_3{}^2 \tag{4}$$

$$\frac{N}{A_3} \frac{\partial X_3}{\partial t} = R_{11} X_1{}^2 - 2 R_{33} X_3{}^2 \tag{5}$$

and in the region where energy generation is significant X_3 is in local equilibrium so that

$$X_3 = X_1 \left[\frac{R_{11}}{2R_{33}} \right]^{1/2}, \qquad \frac{N}{A_1} \frac{\partial X_1}{\partial t} = -2 R_{11} X_1{}^2 \tag{6}$$

where to a good enough approximation we may take

$$R_{11} = a_1 \rho T^{\eta_1}, \quad R_{33} = a_3 \rho T^{\eta_3}; \quad \eta_1 \simeq 4, \quad \eta_3 \simeq 16 \tag{7}$$

$$\left[2 - \frac{5X_1}{3+5X_1} \right] \frac{H_p}{H_1} > \left[\frac{KT}{\rho\epsilon\ell^2} - 9 \right] (\nabla_{ad} - \nabla) - 6\nabla \tag{8}$$

In a homogenous solar model the right hand side is positive, H_1 is infinite and the sun is stable, as the sun evolves H_1 decreases and instability sets in. The evolution of X_1 is governed by the equation

$$\frac{N}{A_1} \frac{\partial X_1}{\partial t} = -2 R_{11} X_1{}^2, \qquad R_{11} \propto \rho T^{\eta} \tag{9}$$

It is a satisfactory approximation to the variation of X_1 with time to take

$$\frac{1}{X_1(t)} - \frac{1}{X_1(0)} = 1.5 \left[\frac{T}{T_c}\right]^n \left[\frac{\rho}{\rho_c}\right] \left[\frac{t}{t_0}\right] \tag{10}$$

which gives a central value of 0.35 when $t = t_0$ for $X_1 = 0.73$ at $t = 0$. It therefore follows that near the centre $\rho \simeq \rho_c$, $T \simeq T_c$ we have

$$\frac{1}{H_1} = 1.5 X_1 \left[\frac{n}{H_T} + \frac{1}{H_\rho}\right] \left[\frac{t}{t_0}\right] \tag{11}$$

Eliminating H_ρ and H_T, and neglecting the small contribution from the variation of the molecular weight gives

$$\left[\frac{t}{t_0}\right] > \frac{1}{1.5 X_1} \left[\left[\frac{KT}{\rho\epsilon\ell^2} - 9\right](\nabla_{ad} - \nabla) - 6\nabla\right]\left[\frac{1}{(n_1 - 1)\nabla + 1}\right]\left[\frac{6 + 5X_1}{3 + 5X_1}\right] \tag{12}$$

Taking $\nabla = 0.25$, $\nabla_{ad} = 0.4$, $X = 0.7$, $n = 4$ gives

$$\left[\frac{t}{t_0}\right] > 0.053 \left[\frac{KT}{\rho\epsilon\ell^2} - 19\right] \tag{13}$$

Detailed numerical calculations give instability when $(t/t_0) = 0.065$.

FINITE AMPLITUDE LIMIT

The existence of the instability has been confirmed by several investigations but its consequences have not been fully explored – indeed it is usually ignored. One possibility is that the sun settles down to finite amplitude g mode oscillations, and it is this finite amplitude limit I shall now consider.

Again confining our attention to just the ^{3}He $+ ^3$He branch of the p-p chain, the evolution equation for ^{3}He gives an equilibrium distribution

$$X_{3E}^2 = \left[\frac{R_{11}}{2R_{33}}\right] X_1^2 \propto X_1^2 T^{n_1 - n_3} \tag{13}$$

and it is this profile that leads to the onset of instability. However, the equilibrium value is changed by the oscillation and it is this feed-back mechanism that limits the growth. If the temperature variation during an oscillation is given by $\delta T = \delta T_0 \exp i\omega t$, then the equilibrium distribution of the ^{3}He is obtained by integrating equation (5) over a period, hence

$$X_{3E}^* = \left[\frac{\bar{R}_{11}}{2\bar{R}_{33}}\right]^{1/2} X_1, \qquad \bar{R}_{ij} = \frac{\omega}{2\pi} \int_0^{2\pi/\omega} R_{ij}\, dt \tag{14}$$

If the oscillation is almost adiabatic then $\delta\rho/\rho = 1.5\ \delta T/T$ and so

$$\overline{R_{ij}} = R^0{}_{ij}\left[1 + \frac{(n+1.5)(n+0.5)}{4}\left[\frac{\delta T_0}{T}\right]^2\right] \tag{15}$$

The equilibrium distribution of the ^{3}He is then

$$X_{3E}{}^* = X_1\left[\frac{R^0{}_{11}}{2R^0{}_{33}}\right]^{1/2}\left[1 + \frac{1}{32}[(2n_3+3)(2n_3+1)-(2n_1+3)(2n_1+1)]\left[\frac{\delta T_0}{T}\right]^2\right] \tag{16}$$

With $n_1 \simeq 4$, $n_3 \simeq 16$ this gives

$$X_{3E}{}^* = X_1\left[\frac{R^0{}_{11}}{2R^0{}_{33}}\right]^{1/2}\left[1 + 33\left[\frac{\delta T_0}{T}\right]^2\right]^{-1} \tag{17}$$

and the energy generation rate ϵ is increased to

$$\epsilon^* = E_{11}\bar{R}_{11}X_1{}^2 + E_{33}\bar{R}_{33}X_3{}^2 = \epsilon_0\left[1 + 6\left[\frac{\delta T_0}{T}\right]^2\right] \tag{18}$$

The finite amptitude oscillation has two important effects, it changes the rate of energy generation, and it produces a flatter ^{3}He profile in the central region. It is this change in ^{3}He profile that produces the feed back that limits the growth of the amplitude.

To determine the finite amplitude we write

$$X_3{}^* = X_3\ f(r)\ ,\quad \frac{1}{f(r)} = 1 + 33\left[\frac{\delta T_0}{T}\right]^2,\quad \frac{1}{H_3{}^*} = \frac{1}{H_3} + \frac{1}{H_f} \tag{19}$$

where H_f is the scale height of $f(r)$, then on replacing H_3 by $H_3{}^*$ in the previous analysis we obtain the condition

$$\left[\frac{2\epsilon_{11}}{\epsilon} - \frac{5X_1}{3+5X_1}\right]\frac{1}{H_1} + \frac{2\epsilon_{33}}{\epsilon}\frac{1}{H_3} + \frac{2\epsilon_{33}}{\epsilon}\frac{1}{H_f} > \left[\frac{KT}{\rho\epsilon\ell^2} - \frac{\epsilon_{11}n_1}{\epsilon} - \frac{\epsilon_{33}n_3}{\epsilon} + 1\right]\left[\frac{\nabla_{ad}-\nabla}{H_p}\right] \tag{20}$$

Following the previous analysis the instability condition at time t becomes

$$\left[\frac{t}{t_0}\right] - 0.37\ \frac{H_p}{H_f} > 0.053\left[\frac{KT}{\rho\epsilon\ell^2} - 19\right] \tag{21}$$

If we now approximate the amplitude δT_0 by
$$\frac{\delta T_0}{T} = \alpha\ \sin\left[\frac{2\pi r}{R}\right] \tag{22}$$

then in the central regions

$$\frac{H_p}{H_f} \; = \; 33\alpha^2 \; \frac{6\pi P_c}{G\rho_c^2 R^2} \; = \; 41\alpha^2 \tag{23}$$

where the numerical values are those of a polytrope of index 3. Since detailed calculations with $\alpha = 0$ give instability when $t/t_\odot = 0.065$, for larger t the amplitude α is given by requiring marginal instability hence

$$\alpha^2 \; = \; 0.066 \left[\frac{t}{t_\odot} - 0.065 \right] \tag{24}$$

Of course, the above calculation is only indicative of the way the oscillation may evolve, a better approximation would be to use the actual eigenfunction for $\delta T/T$ rather than the simple approximation used here, to then compute the ³He equilibrium profile and to test for marginal stability. However, even this is likely to prove inadequate since the expected amplitudes are large we need to consider the non-linear interaction with other modes and the enhanced radiative transport. The observable surface effects also need careful investigation since the g modes are evanescent in the convective zone and whilst the first unstable mode has a relatively large surface amplitude this is not true of higher order modes which may be important in the non-linear limit. If overshooting beneath the convective zone is important the effect of a sharp change in the temperature gradient on the reflection of g modes may be significant.

One prediction of this analysis is that the energy generation rate is enhanced, this will change the profile of chemical composition and thus affect the evoilutionary sequence of models leading to the model of the present sun; this in turn will change the predicted run of sound speed in the solar interior. It is interesting to note that the inversion of recent oscillation data gives a somewhat higher speed ($\simeq$ 2%) than the prediction of standard models in the region where $r/R_\odot \simeq$ 0.5 (c.f. Christensen Dalsgaard et al 1985). The detailed theoretical prediction of a model that has undergone - or maybe still undergoing finite amplitude oscillations is not yet available.

REFERENCES

Christensen Dalsaard J, Dilke F W W and Gough D O,
                1974 Mon. Not. Roy. Ast. Soc., 169, 429.
Christensen Dalsgaard J, Duvall T L, Gough D O, Harvey J W and
                Rhodes E J, 1985, Nature, 315, 378.
Dilke F W W and Gough D O, 1972, Nature 240, 262.

IS THE EXISTENCE OF A STRONG MAGNETIC FIELD IN THE SUN'S CORE PLAUSIBLE?

W.A. Dziembowski[1], L. Paternò[2,3], R. Ventura[3]
(1) N. Copernicus Astronomical Center, Warsaw, Poland
(2) Department of Physics, University of Catania, Italy
(3) CNR-National Group of Astronomy, Section of Catania, Italy

ABSTRACT. We estimate the energies of the identified solar gravity modes from the observed radial velocity amplitudes, and deduce that the
energies of some modes are of the some order of the total energy of the
convective motions in the solar envelope. This fact rules out the possibility of direct or undirect excitation of these modes by convection.
We suggest that the most plausible excitation mechanism rely upon a magnetic torque caused by the presence of a global magnetic field of the
order of a megagauss in the Sun's core.

1. INTRODUCTION

We use the results of Delache and Scherrer (1983), who reported recently the detection and the identification of 14 gravity modes in the low
frequency region of the solar oscillation spectrum, for the determination of the individual energies of these modes.

The method for computing the energy of the modes has already been
described in detail by Dziembowski et al. (1985).

The oscillation energy of a mode, characterized by a frequency ν,
and an angular dependence of perturbation $Y_\ell^m (\theta,\phi)$ is given by:

$$E_{\nu,\ell,m} = \tfrac{1}{2} \int |\vec{\mathcal{V}}_{\nu,\ell,m}|^2 \rho \, d^3 \, \vec{x} \tag{1}$$

where the local velocity field $\vec{\mathcal{V}}$ can be expressed in terms of radial
eigenfunctions $y_{\nu,\ell}(r)$ and $z_{\nu,\ell}(r)$ independent of m, since we can neglect the effect of rotation for the present purpose. The eigenfunctions
are calculated by means of numerical integration of the equations for
adiabatic oscillations using as equilibrium model a standard model of
the Sun with X = 0.73, Z = 0.02, and a depth of the convection zone of
$1.95 \ 10^5$km.

Our final formula for energy is:

$$E_{\nu,\ell,m} = M_\odot A^2_{\nu,\ell,m} B_{\ell,m} I_{\nu,\ell} S^{-2}_{\nu,\ell} \tag{2}$$

D. O. Gough (ed.), Seismology of the Sun and the Distant Stars, 271–274.

where A. is the radial velocity amplitude, B is an expression involving only the coefficients of spherical hormonics that is of order unity, S is the appropriate spatial filter function for Delache and Scherrer measurements (Christensen-Dalsgaard and Gough 1982), and I is a dimensionless integral over the radial eigenfunctions, which therefore only depends on the equilibrium model.

The results of our computations give energies ranging from a minimum of $2.22 \ 10^{37}$erg for the mode with $\ell=2$ and $\nu= 50.72$ µHz to a maximum of $1.69 \ 10^{40}$erg for the mode with $\ell=1$ and $\nu= 96.94$ µHz.

Though the identification of modes is certainly not free of uncertainties, we cannot avoid the fact that their individual energies are of the same order as the total energy of turbulence stored in the convection zone. This fact raises doubts on the applicability of a mechanism based on the excitation by turbulence (Goldreich and Keeley 1977), since this process results into a tendency to energy equipartition between gravity modes and resonant convective eddies. Other mechanisms based on nuclear burning effect (Christensen-Dalsgaard et al. 1974; Boury et al. 1975; Shibahashi et al. 1975) or radiative heat losses in the Sun's core (Keeley 1980) seem also to be unefficient.

Therefore we propose a fresh approach to this problem, based on the existence of a strong magnetic field in the Sun's core.

2. MAGNETIC EXCITATION AND DRIVING

We assume the presence of a global magnetic field in the Sun's core, which produces a mainly horizontal turbulence, owing to some kind of instability. We furthermore assume that turbulence balances the effects of magnetic field and rotation in order to mantain on average steady state in the core. The azimuthal component of the forces is given, in axial symmetry, by:

$$r \sin \Theta \frac{\partial \Omega}{\partial t} = - < (\vec{v} \cdot \vec{\nabla}) \vec{v} >_\phi + \frac{1}{4\pi\rho} \ (\vec{\nabla} \times \vec{B}) \times \vec{B} >_\phi \qquad (3)$$

where Ω is the angular velocity of rotation and $\vec{B}$ the magnetic field.

If the Reynolds stresses arise primarily from small scale horizontal turbulence characterized by a typical velocity ν_T and scale λ, and the magnetic torque T is mainly produced by mean magnetic field, it is easily deduced that steady state equation (3) implies as an order of magnitude the following relationship (Dziembowski et al. 1985):

$$\nu_T^2 = \lambda \ T/r \qquad (4)$$

where $T \simeq B^2/4\pi\rho$ and r is the distance from the Sun's center.

In our approach, the rate at which a given mode grows or decays is therefore given by the competing effects of driving by magnetic torque γ_T and damping by turbulent viscosity γ_{vis}, that in the linear approximation gives:

$$\frac{\partial A}{\partial t} = (\gamma_T + \gamma_{vis})A \qquad (5)$$

These two terms are approximately given by the following expressions (Dziembowski 1984):

$$\gamma_T = - \frac{m}{8\pi\nu\ell(\ell+1)} \int \rho \; \frac{|\vec{\xi}|^2}{r^2} \; \frac{d^2 T}{d(\cos\theta)^2} \; d^3 \vec{x} \tag{6}$$

$$\gamma_{vis} = - \frac{\ell(\ell+1)}{2} \int \rho\eta \; \frac{N}{2\pi\nu} \; \frac{|\vec{\xi}|^2}{r^2} \; d^3 \vec{x} \tag{7}$$

where $\vec{\xi}$ is the displacement eigenvector, N the Brunt-Väisälä frequency, and $\eta \simeq \nu_T\lambda$ the turbulent viscosity. If we eliminate ν_T by using equation (4), we can obtain an order of magnitude constraint for instability by comparing the expressions (6) and (7) in order to have $\gamma_T \gtrsim |\gamma_{vis}|$:

$$(Tr)^{1/2} \gtrsim \frac{2[\ell(\ell+1)]^2}{m} \; \frac{N}{2\pi\nu} \; N\lambda^{3/2} \tag{8}$$

Since $N/2\pi\nu$ is of order unity for the excited modes, if we use expression (8) for $\ell=m=1$, we finally obtain an order of magnitude of the magnetic field for driving the modes:

$$B^2 \gtrsim 4\pi\rho r^2 \; N^2(4\lambda/r)^3 \tag{9}$$

If the typical scale of turbulence is $\lambda \ll r$, then expression (9) indicates that instability may exist even for dynamically unimportant magnetic fields such that $B^2/4\pi\rho r^2 \ll N^2$ in the whole radiative interior. If we assume $\lambda/r \simeq 0.1$ and insert into equation (9) values appropriate for the solar core, we obtain $B \simeq 5 \; 10^6$ Gauss.

3. CONCLUSIONS

If the energies of the modes calculated in the present work or any other estimate approaching our values are correct, then it seems impossible to connect the excitation mechanism with the effects of convection in the solar outer envelope, including direct energy input, overshooting into radiative interior or five minute mode indirect excitation. On the other hand, overstability of gravity modes due to nuclear or non-adiabatic effects in the solar core may also be excluded.

Therefore, if the turbulence we have postulated is produced by some magnetic instability and it has the properties previously described, we can consider the proposed mechanism as plausible for exciting and driving gravity modes in the core of the Sun.

We thus allow for the existence of some kind of activity in the solar core. Both observational and theoretical evidence for an active Sun's interior are very preliminary. Nevertheless, we point out that recent determinations of the Sun's internal rotation (Duvall et al., 1984; Brown 1986) show that the dependence of Ω on r and θ cannot be understood only in terms of hydrodynamical instabilities, but that magnetic field should play a dominant role especially in the innermost layers. This should support largely our conclusion on the presence of an active magnetic core.

REFERENCES

Boury, A., Gabriel, M., Noels, A., Scuflaire, R., Ledoux, P.: 1975
    Astron. Astrophys. 41, 279.
Brown, T.M.: 1986, Theese Proceedings.
Christensen-Dalsgaard, J., Gough, D.O.: 1982, Montly Notices Roy.
    Astron. Soc. 198, 141.
Christensen-Dalsgaard, J., Dilke, F.W.W., Gough, D.O.: 1974, Montly
    Notices Roy. Astron. Soc. 169, 429.
Delache, P., Scherrer, P.H.: 1983, Nature 306, 651.
Duvall, T.L. jr., Dziembowski, W.A., Goode, P.R., Gough, D.O., Harvey,
    J.W., Leibacher, J.W.: 1984, Nature 310, 22.
Dziembowski, W.A.: 1984, in Theoretical Problems in Stellar Stability
    and Oscillations, eds. M. Gabriel and A. Noels, Université de
    Liège, p. 346.
Dziembowski, W.A., Paternò, L., Ventura, R.: 1985, Astron. Astrophys,
    in press.
Goldreich, P., Keeley, D.A.: 1977, Astrophys.J. 212, 243.
Keeley, D.A.: 1980, in Nonradial and Nonlinear Stellar Pulsation, eds.
    H.A. Hill and W.A. Dziembowski, Springer-Verlag, New York, p. 245
Shibahaski, H., Osaki, Y., Unno, W.: 1975, Publ. Astron. Soc. Japan.
    27, 401.

PROGRESS REPORT ON HELIUM ABUNDANCE DETERMINATION

W. Däppen

High Altitude Observatory
National Center for Atmospheric Research
Boulder, Colorado 80307, U.S.A.

D. O. Gough

Institute of Astronomy and Department of Applied Mathematics
and Theoretical Physics, University of Cambridge, England

Abstract: Intermediate results from a programme to determine the helium abundance of the solar convection zone are presented. The method uses the influence of helium ionization on the local sound speed, which can be gauged from inversions of solar oscillation frequencies. The resolution of the principal diagnostic function by the data justifies optimism for an abundance determination in a future step.

INTRODUCTION

One way of determining the Sun's initial helium abundance is by calibrating solar *evolution* theory (Christensen-Dalsgaard and Gough, 1980, 1981; Claverie *et al.*, 1979; Shibahashi *et al.*, 1983). This method is somewhat unreliable, since it is subject to all uncertainties of that theory (e.g. in the opacities, the treatment of convection, the equation of state, the nuclear physics, and the presumed absence of large-scale material motion). Another way of proceeding (Gough, 1983) is to use the helioseismological sound-speed determination in the solar interior (Christensen-Dalsgaard *et al.*, 1985) to extract information about the bulk modulus of the material in the convection zone. Only weak assumptions concerning the solar *structure* are required by this method. The sound speed c $[c^2 = \gamma p/\rho,\ \gamma = (\partial \ln p/\partial \ln \rho)_S]$ is modulated by the ionization of hydrogen and helium because γ is lowered (due to increased specific heats). Therefore one might hope to detect, in particular, a modulation at the depth of the helium ionization zone that is roughly proportional to the helium abundance Y. The raw sound speed

275

varies too rapidly with depth to reveal this modulation but its derivative with respect to radius r can be related directly to γ and its thermodynamical derivatives (Gough, 1983):

$$W \equiv \frac{r^2}{Gm} \frac{dc^2}{dr} = \frac{1 - \gamma_\rho - \gamma}{1 - \gamma_{c^2}} \equiv \Theta. \tag{1}$$

Here, $\gamma_\rho = (\partial \ln\gamma / \partial \ln\rho)_{c^2}$, $\gamma_{c^2} = (\partial \ln\gamma / \partial \ln c^2)_\rho$, m is the mass in the spherical shell of radius r and G is the gravitational constant. In deriving this equation one uses the equation of hydrostatic support, ignoring Reynolds stresses and one assumes adiabatic stratification: $d \ln T / d \ln p = \nabla_{ad}$. [A special case of this equation (γ = const.) has been used independently by Christensen-Dalsgaard (these proceedings) to discuss the structure of the solar convection zone]. Where the stratification is only approximately adiabatic, Equation (1) holds only approximately.

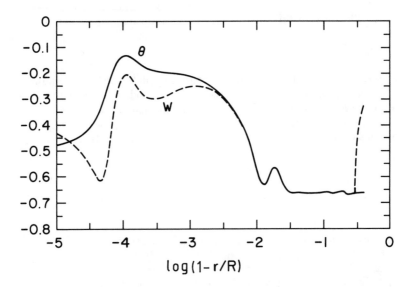

Figure 1. The functions W and Θ defined by Equation (1) plotted against $\varsigma = \log_{10}(1 - r/R)$ for a model of the solar envelope with $\alpha=1.9$, $Y=0.22$ and a heavy-element abundance $Z=0.02$.

The accuracy of Equation (1) according to a theoretical solar model was discussed by Däppen and Gough (1984). Figure 1 shows Θ and W for that model, plotted against $\varsigma = \log(1 - r/R)$, where R is the radius of the Sun. The ionization zones are manifest as humps; the H I and He I zones produce a composite hump in the range $-4 \lesssim \varsigma \lesssim -2$

and the He II zone produces a hump centred near $\varsigma = -1.75$. The hump associated with the ionization of H I and He I is higher than that associated with He II because in the former γ is lowered from its ideal value $5/3$ to about 1.21, whereas in the He II zone γ falls only to about 1.58. However, above $\varsigma \approx -2.5$ the stratification is insufficiently close to being adiabatic for Equation (1) to be usable. Therefore it is only the He II hump that can be of direct value for inferring the value of Y.

To test the potential utility of the diagnostic, Däppen and Gough (1984) computed a grid of envelope models with different mixing-length parameters α and different values of Y ($\alpha = l/H_p$, l being the mixing length and H_p the pressure scale height). The result of that analysis was that the *position* of the hump depended principally on α, and the *height* of the hump on Y. (The influence of α is basically a changed thickness of the superadiabatic layer on top of the convection zone, which leads to a different depth scale.). It was found to be possible to recover the values of α and Y of a given 'unknown' model using the helium humps of 9 models ($\alpha = 1.4, 1.9, 2.4$; $Y = 0.17, 0.22, 0.27$) computed directly from the stratification. The natural extension of that analysis is to use estimates of W computed from oscillation eigenfrequencies of the models. For this purpose, a suitable set of observable modes must be selected that yields maximal resolution in the vicinity of the helium hump (see Gough, 1985). We now present the first results from sound-speed inversions.

W FROM 551 MODES

The first data chosen for inversion was the set of oscillation frequencies of all the p modes of degree between 40 and 1000, in steps of 5, with frequencies between 2500 and 4000 μHz. The equilibrium envelope model was calculated in the manner described by Däppen and Gough (1984), with $\alpha = 1.4$ and $Y = 0.17$; its 551 oscillation frequencies that constitute the chosen set were computed from the programme described by Christensen-Dalsgaard (1982). From those frequencies c^2 was computed by the asymptotic inversion technique used by Gough (these proceedings), which is essentially the same as that described by Christensen-Dalsgaard *et al.* (1985) save that account was taken of the frequency dependence of the phase jump of the wave at the upper turning point. The function W was then computed after numerical differentiation of c^2. In practice it is unnecessary to integrate the equations of hydrostatic support to determine the variation of m with depth, because only about 10^{-5} of the solar mass lies above the location of the peak in the He II hump. The result in the vicinity of the helium hump is displayed in Figure 2. Plotted also (as a dashed line) are the actual values of W computed directly from the model. It is evident that the scatter in the inferred W is much too great to resolve the helium hump by the asymptotic inversion of this selection of modes.

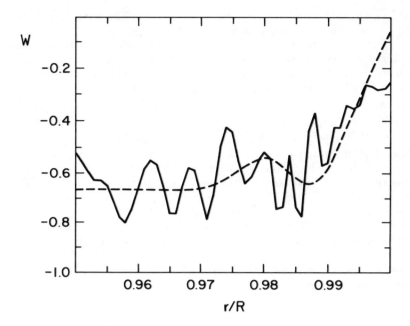

Figure 2. The continuous curve is the inferred function W computed from the set of 551 modes (defined in the text) of an envelope model with $\alpha=1.4$, $Y=0.17$ and $Z=0.02$. The dashed line is the actual function W for that model.

W FROM 2820 MODES

The procedure has been repeated for the 2820 modes used by Christensen-Dalsgaard *et al.* (1985), many of whose lower turning points lie near or above the He II ionization zone. Here the theoretical eigenfrequencies of Model 1 of Christensen-Dalsgaard (1982) were used. The result is shown in Figure 3. The resolution is much better now, and there is indeed some indication of the helium hump. The inferred function W is again compared with its actual values. Encouraged by this modicum of success, we have also carried out the analysis on the real solar data reported by Duvall (1982), Harvey and Duvall (1984) and Christensen-Dalsgaard *et al.* (1985), obtaining results of apparently similar quality.

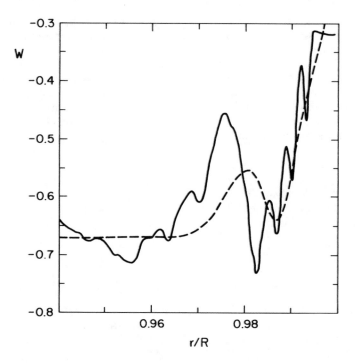

Figure 3. Same as Figure 2, but computed from Model 1 of Christensen-Dalsgaard (1982), using the frequencies considered by Christensen-Dalsgaard *et al.* (1985) in their determination of the solar sound speed.

OUTLOOK

It is fortunate that the He II ionization hump is conveniently located. It is sufficiently deep to be within the nearly adiabatically stratified region of the convection zone, well below the superadiabatic peak (from mixing-length theory one concludes that $\nabla - \nabla_{ad} \approx 2 \times 10^{-4} \nabla$ at $r/R = 0.98$). But it is also sufficiently high for there to be enough modes to obtain a fine resolution of the sound speed. We have not yet attempted to use the results to determine the helium abundance, partly because we think it likely that we can improve the asymptotic analysis to obtain a better representation of W. The analysis must certainly be carried out properly by comparing humps computed from sets of corresponding modes. In that way systematic errors in the inversion procedure largely cancel, thus greatly enhancing the accuracy of the abundance determination.

We stress that the present results give hope for success not only in determining Y but also in probing microphysical properties of the partially ionized matter in the Sun.

The shoulder of the curve in Figure 1 ($-3.5 \lesssim \varsigma \lesssim -2$) can also be seen in the functions W inferred from both the artificial and the real solar data. It contains information about hydrogen (and helium) ionization (see Däppen and Gough, 1984). Finer effects in the statistical physics, such as the action of plasma polarization on the energy levels of bound states or a possible Planck-Larkin cut-off in the partition function [for the Planck-Larkin controversy, see Däppen et al. (1986)], all influence the location and the structure of the ionization zones, and change the values of γ and its thermodynamical derivatives from the ideal-gas values. The possibility exists that these deviations from Saha-type thermodynamics can be measured through the observable function W.

Acknowledgements: We thank T. Bogdan for carefully reading and commenting on the manuscript. The National Center for Atmospheric Research is sponsored by the National Science Foundation.

REFERENCES

Christensen-Dalsgaard, J.: 1982, Mon. Not. R. astr. Soc., **199**, 735-761
Christensen-Dalsgaard, J., Gough, D.O.: 1980, Nature, **288**, 544-547
Christensen-Dalsgaard, J., Gough, D.O.: 1981, Astron. Astrophys., **104**, 173-176
Christensen-Dalsgaard, J., Duvall, T.L. Jr, Gough, D.O., Harvey, J.W.,
    Rhodes, E.J. Jr: 1985, Nature, **315**, 378-382
Claverie, A., Isaak, G.R., McLeod, C.P., van der Raay, H.B.,
    Roca Cortez, T.: 1979, Nature, **282**, 591-594
Däppen, W., Gough, D.O.: 1984, Theoretical Problems in Stellar Stability and Oscilla-
    tions (eds. M. Gabriel and A. Noels, Institut d'Astrophysique, Liège), 264-269
Däppen, W., Anderson, L.S., Mihalas, D.M.: 1986 (in preparation)
Duvall, T.L. Jr: 1982, Nature, **300**, 242-243
Gough, D.O.: 1983, Mem. Soc. astr. Italiana, **55**, 13-35
Gough, D.O.: 1985, Proc. ESA Workshop on Future Missions in Solar, Heliospheric and
    Space Plasma Physics (eds. E. Rolfe and B. Battrick, ESA SP-235, ESTEC,
    Noordwijk), 183-197
Harvey, J.W., Duvall, T.L. Jr: 1984, Proc. Conf. solar Seismol. Space (eds. R.K. Ulrich
    et al., Jet Propulsion Laboratory Publication), 165-172
Shibahashi, H., Noels, A., Gabriel, M.: 1983, Astron. Astrophys., **123**, 283-288

ON WAVE REFLECTION OR ABSORPTION AT HYDROMAGNETIC CRITICAL LEVELS

L.M.B.C. Campos[†] and J.C.G.C. Leitão[†]
† Instituto Superior Técnico, 1096 Lisboa, Portugal[‡]
‡ Work supported by C.A.U.T.L., Instituto de Física-Matemática,
I.N.I.C.

ABSTRACT. We consider the effects (§1) of magnetic field on acoustic
waves, first in an homogeneous medium (§2) and then in an (§3)
atmospheric slab. We choose the case of an horizontal magnetic field,
which is the only one (§4) leading to a wave equation of low (second-)
order, but allows a discussion of the effects of the magnetic field on
cut-off frequencies (§4-5) and amplitude, phase and energy flux (§6-9).
The role of the critical level (§6) in reflecting oblique modes (§7)
and partially absorbing (§8) vertical waves, is associated with
respectively evanescent fields and mode conversion. The latter process
(§9) is illustrated graphically (Figure 1) by plotting wave amplitudes
and phases at and near the critical level. The latter exists only in an
horizontal field, as occurs (§10) at the top of a loop, arch or other
closed coronal structures. It is shown that the velocity perturbations
predicted above a critical level are consistent (Figure 2) with non-
-thermal velocities observed in the transition region.

§1. INTRODUCTION

It is well-known that, in the presence of a magnetic field, acoustic
waves in a compressible, homogeneous fluid, give rise to two modes[1,2],
slow and fast, generally coupled. In the case of an atmosphere, the
presence of gravity and stratification, complicates wave propagation in
two ways: (i) it introduces a third restoring force, namely, buoyancy,
in addition to gas pressure and magnetic forces[3,4]; (ii) it causes at
least one of the wave speeds, sound and Alfvén, to become non-uniform,
leading to non-sinusoidal waveforms[5,6]. The wave equation is of second-
-order in the velocity vector[7,8], and elimination, e.g., for the
vertical velocity, usually decouples the Alfvén wave[9,10], leaving a
fourth-order scalar equation[11], in the general case of vertical[12] or
oblique[13] magnetic field. In the case of an horizontal magnetic field,
the wave equation reduces to the second-order[14,15], as for acoustic-
-gravity waves[16,17], with the important difference that the magnetic
field introduces a singularity[18,19], or critical level[20,21] for these
magnetosonic-gravity waves. There is a considerable controversy in the
281

D. O. Gough (ed.), Seismology of the Sun and the Distant Stars, 281–292.

literature concerning these waves, the incompatible statements including that: (i) the wave amplitude has a logarithmic singularity[20,22], or is finite[13,21] at the critical level; (ii) the energy flux has a discontinuous 'jump'[23] or varies rapidly but continuously[24]; (iii) the cut-off frequencies are affected[25,26] or unaffected[17,27] by the magnetic field. We will show that in each of the issues (i) to (iii) the second is the correct assertion, and also, concerning the polemical issue of whether critical levels are[20] or are not[21] relevant to solar physics, we will argue that they explain the difference between[28] closed and open magnetic structures in the solar corona, adding some observational evidence.

§2. SLOW AND FAST MODES IN AN HOMOGENEOUS FLUID

It is well-known that[1], in the presence of an external magnetic field, acoustic waves in a compressible fluid decouple into slow and fast modes, with phase speeds[2]:

$$u_\pm = \frac{1}{2} \{|c\,\vec{n} + a\,\vec{\ell}| \pm |c\,\vec{n} - a\,\vec{\ell}|\}, \tag{1}$$

where c denotes the sound speed along the wavenormal direction $\vec{n} \equiv \vec{k}/k$, and a the Alfvén speed along magnetic field lines $\vec{\ell} \equiv \vec{B}/B$. Denoting by θ the angle between the wavevector $\vec{k}$ and the magnetic field $\vec{B}$, it follows from (1) that slow u_- and fast u_+ modes are generally coupled, viz.:

$$2u_\pm = \sqrt{c^2 + a^2 + 2ac\,\cos\theta} \pm \sqrt{c^2 + a^2 - 2ac\,\cos\theta}, \tag{2}$$

except in two cases: (1) for propagation along the magnetic field $\theta = o$, the Alfvén a and sound c waves decouple $u_\pm = c,a$; (2) for propagation transverse to the magnetic field $\theta = \pi/2$, they couple into a single magnetosonic mode $u_+ = \sqrt{c^2 + a^2}$, $u_- = o$. Considering the oscillations of a fluid sphere, e.g., the global modes of the sun, the simplest way to include the effects of magnetic field, would be to replace the sound c by fast u_+ speed, viz., in the JWKB or asymptotic formulas for wave-periods[29,30]. Assuming a nearly radial magnetic field, the angle with the direction of propagation θ would be determined roughly by the ratio of radial $k_\perp$ and tangential $k_{//}$ wavenumbers,

$$\tan\theta = k_{//}/k_\perp = n^{-1}\sqrt{\ell(\ell+1)}, \tag{3}$$

i.e., by the order n and degree ℓ of the mode. Thus high(low) $-\ell$ modes, with $\ell \gg n$ ($\ell \ll n$), which are nearly tangential (radial), are most (least) affected by the magnetic field, because they propagate accross (along) magnetic field lines.

§3. MAGNETO–ATMOSPHERIC WAVES IN A COMPRESSIBLE SLAB

The preceding method, i.e., substitution into acoustic–gravity dispersion relations, of the sound speed c by the fast wave speed u_+ (2,3), is successfull in accounting for magnetic field effects, only in the J.W.K.B. approximation, i.e., if the wave speed is nearly constant on a wavelength scale. The Alfvén speed $a = B\sqrt{\mu/4\pi\rho}$, for a nearly uniform magnetic field, and a medium of constant magnetic permeability μ, varies with the inverse square root of the mass density ρ, so that the W.K.B.J. method is limited by the condition $k^2L^2 = (2\pi L/\lambda)^2 \gg 1$ to wavelengths λ much shorter than the density scale height L. This restriction is particularly unnaceptable in the outer layers of the sun, where the scale height is a few hundred kilometers and the reference wavelength, based on wave period, may be a few thousand kilometers. If we consider a region of vertical extent small compared with the radius of the sun, we may use the model of a plane parallel 'slab'; the vertical velocity perturbation:

$$v_z(\vec{x},z,t) = w(z) \exp\{i(\vec{k}.\vec{x} - \omega t)\}, \tag{4}$$

for a wave of frequency ω, has a wavevector $\vec{k}$ in the horizontal plane $\vec{x}$, leaving only the dependence w(z) with altitude z, to be determined. In the case of an isothermal layer, of scale height $L \equiv RT/g$, the density decays as $\rho \sim e^{-z/L}$, and for an uniform magnetic field, the Alfvén speed grows as $a^2 \sim \rho-1 \sim e^{z/L}$, leading to linear wave equations with exponential coefficients, i.e., of the form:

$$\{R(L\ d/dz) - e^{z/L}\ S(L\ d/dz)\}\ W(z) = o, \tag{5}$$

where R,S are polynomials of derivatives, which may also depend on $\vec{k},\omega$, wave speeds c,a, and other atmospheric parameters. There is a general method[13,31] of solving wave equations of type (5) with polynomials of any order, e.g., the fourth-order for vertical[5,12,32,33] or oblique[11, 13,34] magnetic fields. The solutions appear as generalized hypergeometric[13] or Meijer G-functions[11], and only the usual Gauss hypergeometric function has to be used for second-order equations; this is the case for an horizontal magnetic field, which is sufficient to demonstrate several of the physical effects of the magnetic field.

§4. ACOUSTIC AND GRAVITY CUT-OFF FREQUENCIES

An oblique wave (4) in an horizontal magnetic field $\vec{B}$, has an altitude dependence W(z) satisfying[14,20,21] an equation of the form (5), with the polynomials R,S given by:

$$R(\xi) = \omega^2 c^2\{\xi^2 - \xi + \omega^2 L^2/c^2 - k^2L^2 + (\gamma-1)k^2c^2/\gamma^2\}, \tag{6a}$$

$$S(\xi) = (\omega^2 - c^2k^2)\ a^2(\xi^2 - k^2L^2), \tag{6b}$$

where ω is the frequency, k the wavenumber, L the scale height, c the sound speed, and a the Alfvén speed (at altitude z=o). For acoustic-gravity waves, in the absence of a magnetic field a=o, and $S(\xi) = o$, so that the wave equation (5) has constant coefficients. In the case of vertical k=o magnetosonic-gravity waves, (6a,b) simplify to:

$$R(\xi) = c^2(\xi^2 - \xi + \omega^2 L^2/c^2), \quad S(\xi) = a^2\xi^2. \tag{7a,b}$$

In the absence of a magnetic field a=o, the wave equation $\{R(L\,d/dz)\}$ $W(z) = o$, with constant coefficients, has exponential solutions $W(z) \sim \exp(\xi z/L)$, where ξ is a root of (6a), viz:

$$\xi = \frac{1}{2} \pm iKL, \quad K \equiv \sqrt{\{(\omega/\omega_2)^2 - 1\}/4L^2 + k^2\{(\omega_1/\omega)^2 - 1\}}, \tag{8a,b}$$

where ω_1, ω_2 are respectively the gravity and acoustic cut-off frequencies:

$$\omega_1 \equiv \sqrt{\gamma-1}\ (c/\gamma L) = \sqrt{\gamma-1}\ (g/c), \quad \omega_2 \equiv c/2L = \gamma g/2c. \tag{9a,b}$$

For frequencies between the cut-offs $\omega_1 < \omega < \omega_2$, the quantity K in (8b) is imaginary, hence ξ is real, and the wave field $\exp(\xi z/L)$ has no phase, i.e., we have standing modes. Phases, i.e., propagating waves, exist only below the lower cut-off ($\omega < \omega_1$, gravity modes) or above the upper cut-off ($\omega > \omega_2$, acoustic modes), in which case K is real (8b), and plays the role of vertical wavenumber $\exp(\xi z/L) = \exp(z/2L)\exp(\pm iKz)$. For vertical waves k=o, only acoustic modes can propagate, above the cut-off ω_2, as can be shown from (7a). The presence of a magnetic field $a \neq o$, adds the polynomial (6b), which has real roots $\xi = \pm kL$, and cannot introduce phases, i.e., does not affect the preceding reasonings. Thus the acoustic (9b) and gravity (9a) cut-off frequencies are the same[13], [24,27] for (i) acoustic-gravity waves in the absence of a magnetic field and (ii) magnetosonic-gravity waves in the presence of an horizontal field.

§5. FILTERING IN NON-ISOTHERMAL AND MAGNETIC ATMOSPHERES

The cut-off frequencies corresponds to the limit of infinitely spaced nodes $\lambda \to \infty$, and thus can be derived only from exact solutions of the wave equation. If the latter has constant coefficients, e.g., acoustic-gravity waves in an isothermal atmosphere, the dispersion relation is an exact solution, and leads to the cut-off frequencies (9a,b). In a non-isothermal atmosphere, the 'local' form os the dispersion relation would lead to cut-off frequencies varying with altitude[16], suggesting the altitudes at which substantial wave reflection occurs, for each frequency. The exact cut-off frequencies, are strictly speaking always constant, global properties of the atmospheric profiles, which can only be determined from exact solutions of the relevant wave equation. For example, the family of three-parameter temperature profiles:

$$T(z) = T_\infty + (T_o - T_\infty) \exp(-z/h), \tag{10}$$

allowing[17] independent choice of initial T_o and asymptotic T_∞ temperature, and maximum temperature gradient $dT/dz\big|_{z=o} = (T_\infty - T_o)/h$, has global cut-off frequencies given by (9a,b), calculated for the asymptotic temperature T_∞. The presence of an uniform magnetic field, in an isothermal atmosphere, causes the Alfvén speed to increase with altitude as $a \sim e^{z/2L}$; thus local forms of the dispersion relation are valid only for short waves $\lambda < 2L$, and cannot be used to derive cut-off frequencies in the limit $\lambda \to \infty$. Thus the cut-off frequencies which are deduced in the literature, by explicit or implicit use of 'local' dispersion relations for hydromagnetic waves in atmospheres are suspect. For example, the often repeated statement[7,25,26,35,36] that the cut-off frequency for vertical magnetosonic-gravity waves in an horizontal magnetic field is $\Omega_2 = \gamma g/2 \sqrt{c^2 + a^2}$ is incorrect; the correct value[24,27] is $\omega_2 = \gamma g/2c$, the same as for acoustic-gravity waves, as shown before. The 'justification' that $\Omega_2 = \gamma g/2u_+$, where $u_+ = \sqrt{c^2 + a^2}$ is the fast wave speed and replaces the sound speed in $\omega_2 = \gamma g/2c$ is no good, since it supposes (§2) and homogeneous medium rather than atmosphere, and the former has no cut-offs. The conclusion that an horizontal magnetic field, does not affect the cut-off frequencies of atmospheric waves, fails to[13] extend to oblique fields.

§6. CRITICAL LEVEL FOR MAGNETOSONIC-GRAVITY WAVES

The solution of the wave equation (5), is[13] generally:

$$\Phi_\pm(z) = e^{\sigma z/L} \left\{ 1 + \sum_{n=1}^\infty (-)^n e^{-nz/L} \prod_{m=1}^n \{S(\sigma+m-1)/R(\sigma+m)\} \right\}, \tag{11}$$

where σ is a root $R(\sigma) = o$, of (6a), i.e., $\sigma = 1/2 \pm iKL$, with K the effective vertical wavenumber (8b). Substituting (6a,b) in (11), we obtain upward and downward propagating acoustic-gravity waves (first two factors):

$$\Phi_\pm(z) = e^{z/2L} e^{\pm iKz} F(\tfrac{1}{2} - kL \pm iKL, \tfrac{1}{2} + kL \pm iKL; 1 \pm 2iKL; 1/\zeta), \tag{12}$$

modified (third factor) by the magnetic field or Alfvén speed a, which appears in the hypergeometric function, through the variable:

$$\zeta \equiv -(c/a)^2 \{1 - (ck/\omega)^2\}^{-1} e^{-z/L}. \tag{13}$$

The wave field is a linear combination of the particular integrals (12), i.e., a superposition of upward and downward propagating waves:

$$v_z(\vec{x},z,t) = \exp\{i(\vec{k}.\vec{x}-\omega t)\}\{A_+ \, \Phi_+(z) + A_- \, \Phi_-(z)\}, \tag{14}$$

with amplitudes $A_\pm$ which are determined by boundary conditions. The solution (14) is valid only for $|\zeta| \leqslant 1$, i.e., for altitudes below and up to the critical level:

$$z \leqslant z_c \equiv L \, \log \Lambda \,, \qquad \Lambda \equiv (c/a)^2/(1 - c^2 k^2/\omega^2). \tag{15a,b}$$

For $z > z_c$, we have $|\zeta| < 1$ in (13), and the hypergeometric series (12) have variable greater than unity $|1/\zeta| > 1$, so that they do not converge, and fail to provide a valid solution of the wave equation (5).

§7. REFLECTION AND EVANESCENCE OF OBLIQUE MODES

In order to obtain a solution valid above the critical level $z > z_c$, we start by writing the wave equation (5) in inverse form:

$$\{S(L \, d/dz) - e^{-z/L} R(L \, d/dz)\} \, W(z) = 0, \tag{16}$$

which has solutions analogous to (11), interchanging the polynomials R,S, and reversing the sign of z/L, viz.:

$$\psi_\pm(z) = e^{-\sigma z/L} \left\{1 + \sum_{n=1}^{\infty} (-)^n \, e^{-nz/L} \prod_{m=1}^{n} \{R(\sigma+m-1)/S(\sigma+m)\}\right\}, \tag{17}$$

where σ is a root of $S(\sigma) = 0$, i.e., by (6b) we have $\sigma = \pm kL$. Substituting (6a,b) into (17) we obtain two particular integrals:

$$\psi_\pm(z) = e^{\pm kz} \quad F(1/2+iKL \pm kL, 1/2-iKL \pm kL; 1 \pm 2kL; \zeta), \tag{18}$$

again involving hypergeometric functions, but now of the variable ζ defined by (13), so that the solution holds for $|\zeta| < 1$, i.e., at and above the critical level (15a,b), viz., for $z \geqslant z_c$. The general integral is a linear combination of the particular integrals (18), viz.:

$$v_z(\vec{x},z,t) = \exp\{i(\vec{k}.\vec{x}-\omega t)\} \, \{B_+ \, \psi_+(z) + B_- \, \psi_-(z)\}, \tag{19}$$

with $\psi_\pm$ linearly independent for oblique waves $k \neq 0$. By matching the lower (14) and upper (19) solutions at the critical level $z = z_c$, the constants $A_\pm$ and $B_\pm$ are related, viz.:

$$B_\pm = \{\Gamma(1 + 2iKL) \, \Gamma(\mp 2kL)/(\Gamma(1/2 \mp kL + iKL))^2\} \, A_\pm +$$

$$+ \; \Gamma(1 - 2iKL) \, \Gamma(\mp 2kL)/(\Gamma(1/2 \mp kL - iKL))^2\} \, A_\mp \,. \tag{20}$$

In order for the wave field to be bounded above the critical level, we must supress the divergent solution ψ_+ (18), by setting $B_+ = 0$ in (19), leaving only an evanescent field ψ_- (18); the condition $B_+ = 0$ implies $A_\pm \neq 0$ in (20), so that below the critical level (14), we have both upward Φ_+ and downward Φ_- propagating waves (12). It follows that the critical level acts as a reflector of oblique magnetosonic-gravity waves, causing the existence of upward and downward propagating waves below, and only evanescent modes above.

§8. ABSORPTION AND TRANSFORMATION FOR VERTICAL WAVES

The solution (14) below the critical level, remains valid for vertical waves $k = 0$, since the particular integrals (12) remain linearly independent in this case. The solutions (18) above the critical level coincide $\psi_+ = \psi_-$ for $k = 0$, and a new particular integral appears as an hypergeometric function of the second kind G, leading to the solution:

$$v_z(z,t) = \exp(-i\omega t) \ \{B_+ \ F(\zeta) + B_- \ G(\zeta)\}, \qquad (21a)$$

$$F,G(\zeta) \equiv F(1/2 + iKL, 1/2 - iKL; 1; \zeta). \qquad (21b)$$

Asymptoticaly at high-altitude, as $z \to \infty$ and $\zeta \to 0$ by (13), the hypergeometric function of first kind tends to unity $F(0) = 1$, and that of second kind has a logarithmic singularity $G(\zeta) \wedge \log \zeta = \log(-\Lambda \ e^{-z/L}) = \log \Lambda - z/L - i\pi$, leading to a wave:

$$v_z(z,t) \wedge e^{-i\omega t} \ \{B_+ + (\log \Lambda - i\pi) \ B_- - (B_+/L) \ z\}, \qquad (22)$$

which grows linearly with altitude and has bounded phase. The reduction of the rate of growth of the magnetosonic-gravity wave, from (12) the exponential law $\wedge \exp(z/2L)$ 'typical' of acoustic modes, to the (22) linear law characteristic of magnetic modes, shows that the process of mode conversion near the critical level, is accompanied by absorption, i.e., loss of energy to the medium. Since both the low (14) and high (19) altitude solutions converge at $\zeta = -1$, i.e., at the critical level $z = z_c$, they prove that wave amplitude and phase are finite there. The statement in the literature[18,20] that magnetosonic-gravity waves have a logarithmic singularity at the critical level, is erroneous, since it is based on a solution in terms of hypergeometric functions of variable $(1 - \zeta) > 1$, which are divergent; in fact, the logarithmic singularity in the leading term of the expansion in powers of $(1 - \zeta)$, is cancelled[13] by the divergent series that follows. The statement that energy flux is discontinuous at the critical level[23], is associated with the logarithmic 'singularity'; in the absence of the latter, the finite amplitude and phase of the wave, imply a continuous, although possibly rapid reduction of the energy flux, in the vicinity of the critical level. A detailed description of the process of wave absorption would require an inclusion of dissipation mechanisms[37], but it should be

emphasized that, even in the absence of damping (and non-linear
effects), the wave amplitude remains finite at the critical level.

§9. WAVE GROWTH AND EVOLUTION OF PHASES

All of the preceding points can be illustrated by a solution of the
wave equation valid at all altitudes, including the critical level $\zeta = -1$,
e.g., obtained by choosing as variable $1/(1-\zeta) < 1$, where $\zeta < 0$ in (13)
for non-evanescent waves ($\omega > c/k$). The solution is obtained by
transforming[38] from the hypergeometric function (12) of variable $1/\zeta$ to
that of variable $1/(1-\zeta)$, and yields:

$$v_z(\vec{x},z,t) = v_o \exp\{i(\vec{k}.\vec{x}-\omega t)\} \; \{(1+\Lambda)/1+\Lambda \; e^{-z/L})\}^{1/2+kL\pm iKL}$$

$$\{F((1 + \Lambda \; e^{-z/L})^{-1})/F(1/(1+\Lambda))\}, \qquad (23a)$$

$$F(\xi) \equiv F(1/2+kL\pm iKL, 1/2+kL\pm iKL; 1\pm 2iKL; \xi), \qquad (23b)$$

respectively for an upward/downward propagating wave, of initial
amplitude v_o. The amplitude and phase at the critical level $z = z_c$,
$\zeta = -1$, $\Lambda = \exp(z_c/L)$, are determined by hypergeometric functions   of
argument $1/2$, viz.:

$$v_c(\vec{x},t) \equiv v_z(\vec{x},z_c,t) = v_o \exp\{i(\vec{k}.\vec{x}-\omega t)\}$$

$$\{(1+\Lambda)/2\}^{1/2+kL\pm iKL} \; \{F(1/2)/F((1/(1+\Lambda))\}. \qquad (24)$$

The hypergeometric function (23b) simplifies to (21b), in the case of
vertical waves $k = o$. We choose the latter for graphic illustration
(Figure 1), since it includes the important features of mode conversion
and wave absorption. The plasma-β, which we denote by $\Lambda \equiv c^2/a^2$,
specifies the altitude of the critical level by $z_c/L = \log \Lambda$; we plot
the velocity perturbation at the critical level v_c, normalized   to the
initial value v_o, with temporal factor $e^{-i\omega t}$ supressed, i.e., $V_c \equiv e^{i\omega t}$
v_c/v_o, versus Λ, at the top of Figure 1. The amplitude $|v_c|$ at the
critical level increases with Λ, as the wave has to propagate a longer
distance $z_c/L = \log \Lambda$ to reach it; the amplitude $|v_c|$ is (l.h.s.) almost
independent of the dimensionless frequency $\Omega \equiv 2\omega L/c = \omega/\omega_2$, but the
phase $\arg(v_c)$ varies more rapidly (r.h.s) for higher frequency waves.
Having determined the wave field at the critical level $v_c \equiv v(z_c)$, we
may use it to normalize the wave field $v = v(z)$ at arbitrary altitude,
viz.:

$$U(Y) \equiv v(z,t)/v_c(z,t), \qquad Y \equiv (z - z_c)/L, \qquad (25a,b)$$

using as dimensionless altitude Y (25b), the distance from the critical
level divided by the scale height L. It is seen from the bottom of
Figure 1, that, below the critical level $Y < 0$, in the 'acoustic' regime,

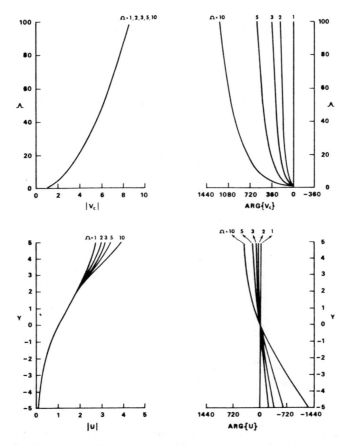

FIGURE 1 — Top: the amplitude (l.h.s.) and phase (r.h.s.) of the dimensionless velocity perturbation $V_c \equiv e^{+i\omega t} v(z_c,t)/v_o$ at the critical level is plotted, versus the plasma-β denoted by $\Lambda \equiv c^2/a^2$ (ratio of sound and Alfvén speeds at altitude $z = o$), for different values of the dimensionless frequency $\Omega \equiv 2\omega L/c = \omega/\omega_2$ (which is the ratio of wave ω to acoustic cut-off ω_2 frequency). The amplitude at the critical level (l.h.s.) increases as the waves have to propagate a longer distance $z_c = L \log \Lambda$ to reach it, but is practically independent of frequency Ω; the phase (r.h.s.) at the critical level increases with both the altitude of the latter (z_c, or Λ) and frequency. *Bottom:* the velocity perturbation normalized to the value at the critical level $U \equiv v(z,t)/v(z_c,t)$, is plotted versus dimensionless distance $Y \equiv (z-z_c)/L$ from the critical level, where L denotes the scale height. The amplitude (l.h.s.) grows in a manner independent of frequency, in the 'acoustic' regime, below the critical level ($Y < o$), and different spectral components are 'split', with faster growth for higher frequencies, in the 'magnetic' regime, above the critical level ($Y > o$). The phase (r.h.s.) grows rapidly, initially almost linearly, in the 'acoustic' regime ($\Lambda e^{-z/L} > 1$), below the critical level, and is bounded in the magnetic regime ($\Lambda e^{-z/L} < 1$) above the critical level, the constant, asymptotic phase being larger for higher frequencies, as a consequence of faster growth at lower altitudes.

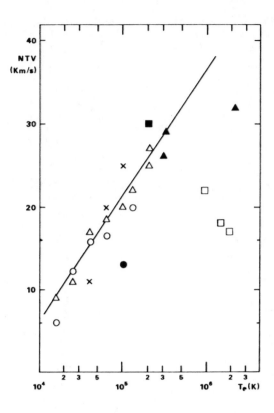

FIGURE 2 – The plots in Figure 1 refer to magnetosonic-gravity waves, propagating vertically in an isothermal atmosphere, under an uniform, horizontal magnetic field; they are calculated from an exact expression of the wave field, valid for all frequencies and distances of propagation, including at the critical level. It can be shown generally, for a non-isothermal atmosphere with bounded temperature, under a non--uniform non-vanishing magnetic field, that the velocity perturbation tends to grow linearly with altitude, above the critical level. This implies a wave velocity perturbation varying approximately linearly with the logarithm of temperature, as is observed for non-thermal velocities in the transition region of the sun, according to the following references: crosses[43], hollow blobs[44] and triangles[45], full blobs[46] and squares[47], hollow squares[48] and full triangles[49]. The location of the critical level at the transition region, agrees with the magnetic field being horizontal at the top of closed structures, such as loops and arches; the intense wave absorption at the critical level, explains why these structures are bright, and dominate the coronal energy balance.

the amplitude (l.h.s.) growth is independent of frequency, and phase
changes (r.h.s.) are strongly affected by it; the 'magnetic' regime,
above the critical level $Y > o$, implies amplitude (l.h.s.) and phase
(r.h.s.) growth that depend both on frequency. There is a general
tendency towards smaller rates of growth beyond the critical level $Y=o$,
due to the magnetic field constraining gas motion.

§10. DISCUSSION

The example of horizontal magnetic fields quoted most often in solar
physics, is that of the sunspot penumbra. Another instance, of no less
importance, is the case of closed magnetic structures in the corona, e.
g., loops and arches, which have an horizontal magnetic field at the
top. The critical level for atmospheric waves only occurs for a purely
horizontal magnetic field, since the presence of a non-zero vertical
component, in an oblique field, would allow[13,18,21] the wave to
propagate through, instead of being trapped, reflected or absorbed. Thus
open magnetic structures in the corona, such as holes, for which the
magnetic field is nowhere horizontal, have no critical levels. The
contrast between closed (open) magnetic structures, which have (do not
have) critical levels, implies the presence (lack of) intense wave
absorption and heating, and explains why the former (latter), e.g.,
loops (holes) are bright (dark). Since it is beyond the scope of the
present work to present detailed models of mass[39] or energy[40] balances,
we note that: (i) the absence of critical levels in coronal holes,
implies that hydromagnetic waves are not absorbed, and can propagate
out with the solar wind, accelerating[41,42] high-speed particle streams;
(ii) the presence of critical levels in coronal loops and arches,
implies intense wave absorption and heating, and should be observed as
non-thermal velocities. We discuss elsewhere the evidence for (i), and
conclude with (ii). The wave velocity perturbation grows linearly (22)
above the critical level, so that $v \sim bz$, and $e^v \sim e^{bz} \sim \rho^{-b}$; the mass
density ρ scales $\rho \sim 1/T$ on the inverse of the temperature T, in the
transition region, leading to $v \sim -b \log \rho \sim b \log T$. The linear
variation of wave velocity perturbation v with the logarithm of
temperature, above a critical level, is consistent (Figure 2), with the
observed[43-49] variation of non-thermal velocities in the transition
region.

REFERENCES

1* Cowling, T.G. 1960 *Magnetohydrodynamics*. Interscience, 2nd ed. 1980.
2* Campos, L.M.B.C. 1977 *J. Fluid Mech.* 81, 529.
3* McLellan, A. & Winterberg, F. 1968 *Solar Phys.* 4, 401.
4* Campos, L.M.B.C. 1983a *Port. Phys.* 14, 145.
5* Ferraro, V.C.A. & Plumpton, C. 1958 *Astrophys. J.* 129, 459.
6* Campos, L.M.B.C. 1983b *Port. Phys.* 14, 179.
7* Bray, R.J. & Loughhead, R.E. 1974 *The solar chromosphere*. D. Reidel
8* Campos, L.M.B.C. 1982 *Port. Math.* 41, 13.
9* Hollweg, J.V. 1972 *Cosmical Electr.* 2, 423.

10* Campos, L.M.B.C.   1983c *J. Méc. Théor. Appl.* 2, 861.

11* Zhugzhda, Y.D. & Dzhalilov, N.S.   1984 *Astron. Astrophys.* 132,45.

12* Leroy, B. & Schwartz, S.J.   1982 *Astron. Astrophys.* 112, 84.

13* Campos, L.M.B.C.   1985 *Geophys. Astrophys. Fluid Dyn.* (to appear).

14* Nye, A.H. & Thomas, J.H.   1976 *Astrophys. J.* 204, 573.

15* Campos, L.M.B.C.   1983d *Solar Phys.* 82, 355.

16* Moore, D.W. & Spiegel, E.A.   1964 *Astrophys. J.* 139, 48.

17* Campos, L.M.B.C.   1983c *Wave Motion* 5, 1.

18* McKenzie, J.F.   1973 *J. Fluid Mech.* 58, 709.

19* Rae, I.C. & Roberts, B.   1981 *Geophys. Astrophys. Fluid Dyn.* 18,197.

20* Adam, J.A.   1977 *Solar Phys.* 52, 293.

21* Schwartz, S.J. et. al.   1984 *Solar Phys.* 92, 81.

22* Thomas, J.H.   1983 *Ann. Rev. Fluid Mech.* 15, 321.

23* El Mekki, O. et. al.   1978 *Solar Phys.* 57, 261.

24* Campos, L.M.B.C.   1983f *J. Phys.* A16, 417.

25* Stein, R.F. & Leibacher, J.W.   1974 *Ann. Rev. Astron. Astrophys.* 12, 407.

26* Priest, E.R.   1982 *Solar magnetohydrodynamics.* D. Reidel.

27* Thomas, J.H.   1982 *Astrophys. J.* 262, 760.

28* Campos, L.M.B.C.   1984a *Proc. 25th Liège Astrophys. Conf.* 393.

29* Tassoul, M.   1982 *Astrophys. J. Suppl.* 43, 469.

30* Deubner, F.L. & Gough, D.O.   1984 *Ann. Rev. Astron. Astrophys.* 22, 593.

31* Campos, L.M.B.C.   1984b *Proc. 25th Liège Astrophys. Conf.* 387.

32* Weymann, R. & Howard, B.   1958 *Astrophys. J.* 128, 142.

33* Lust, R. & Scholer, M.   1966 *Zeit. Naturf.* A21, 1098.

34* Schwartz, S.J. & Bel, N.   1984 *Astron. Astrophys.* 112, 93.

35* Michalitsanos, A.G.   1973 *Solar Phys.* 30, 47.

36* Athay, R.G.   1976 *The quiet sun.* D. Reidel.

37* Cally, P.S.   1985 *Astron. Astrophys.* 136, 121.

38* Caratheodory, C.   1952 *Theory of functions.* Birkhauser V., repr. Chelsea Publ. 1967.

39* Campos, L.M.B.C.   1984c *Month. Not. Roy. Astron. Soc.* 207, 547.

40* Campos, L.M.B.C.   1984d *Mem. Soc. Astron. Ital.* 55, 267.

41* Belcher, J.L. & Davis, L.   1971 *J. Geophys. Res.* 76, 3534.

42* Denskat, K.U. & Burlaga, L.F.   1977 *J. Geophys. Res.* 82, 2683.

43* Boland, B.C. et. al.   1975 *Month. Not. Roy. Astron. Soc.* 171, 697.

44* Doschek, G.A. et. al.   1976 *Astrophys. J. Suppl.* 31, 417.

45* Mariska, J.T. et. al.   1978 *Astrophys. J.* 226, 698.

46* Roussel-Dupré, R. et. al.   1979 *Month. Not. Roy. Astron. Soc.* 187,9.

47* Kjeldseth-Moe, O.K. et. al.   1979 *Solar Phys.* 61, 319.

48* Cheng, C.C. et. al.   1979 *Astrophys. J.* 227, 1037.

49* Vial, J.C. et. al.   1980 *Solar Phys.* 68, 187.

ON UMBRAL OSCILLATIONS AS A SUNSPOT DIAGNOSTIC

L.M.B.C. Campos
Instituto Superior Técnico
1096 Lisboa
Portugal*

ABSTRACT. The observation that p-modes are attenuated in sunspots[1], i.e. sunspots act as 'sinks' of acoustic wave energy, is explained by the transfer of energy to motions along the magnetic field lines. The latter appear as Alfvén waves (Figures 1-2), reflected between the atmosphere, and anchoring points in the convection zone, and have periods (Table I) and amplitudes (Table II) consistent with observations [2-10] of umbral oscillations. The depth of generation of these waves, as inferred from a crude inversion of periods, is about 10 Mm, i.e., consistent with the depth of origin[11] of ephemeral magnetic regions in the solar surface. The oscillation model is used to calibrate a general dimensional law, for the intensity of radiation of MHD waves, which gives reasonable results for the sun, and may apply to other stars.

1. INTRODUCTION

The global p-modes are observed, in attenuated form[1], in sunspots, suggesting that the acoustic waves could shake magnetic field lines, thereby transfering some of the energy to the magnetic field. The sunspots thus acts a 'sink' of p-mode energy, viz., more precisely as a 'transformer' of acoustic into magnetic energy. The shaking of magnetic field lines is a process of excitation of Alfvénic oscillations, that could form standing modes, if trapped between two reflecting layers: (i) the upper reflector is the rapid but gradual increase in Alfvén speed in the chromosphere, or the temperature 'jump' across the transition region to the corona; (ii) the lower reflector is the sub-photospheric level at which magnetic field is anchored in the convection zone. This outline physical picture of wave-type phenomena in sunspots, is discussed in more detail, in comparison with observations of umbral oscillations. We start by a number of checks of a simple theoretical model, and then use it to infer physical conditions in sub-photospheric layers of sunspots. The discussion of wave generation mechanisms, which includes estimates of radiation intensity, could be relevant to the sun as a whole, and to other stars.

D. O. Gough (ed.), Seismology of the Sun and the Distant Stars, 293–301.
© *1986 by D. Reidel Publishing Company.*

2. OBSERVATIONAL DATA AND CURRENT THEORIES

There are several sets of observations of oscillations in the umbra of
sunspots, with some variation in the frequencies reported [2-10],
respectively: 300-470s;300,180,128-140s;450,300,180,100,65s;300-470,
164-196,100-123s;173,136s;180,145s;197±4,171±3,155±2,138±3s;300,170,
130s;300,165s. The frequencies appear to be independent of sunspots size
and magnetic field strength, i.e., are probably a diagnostic of some
general solar properties; they also exhibit a variability in time, about
10%, which exceeds observational scatter. We interpret this empirical
data as evidence for the existence of periods of oscillation of about
300,180,120,100,65s, suggesting a fundamental and four harmonics, of
some type of wave.
        The fundamental period of 300s, is the response[1] of sunspots to
the global five-minute oscillations of the sun[12], which are attenuated
relative to non-spot regions. The fact that the five-minute oscillations
are acoustic modes, does *not* imply necessarily that the 300s period in
sunspots is a purely hydrodynamic mode; the acoustic waves can 'shake'
the magnetic field lines, and thus generate MHD waves, such as Alfvèn
modes, which propagate readily [13] in the near vertical magnetic fields
of sunspots umbras. The 180s period corresponds to the most widely
observed chromospheric (as distinct from global) oscillation; the
various theories of the 'three-minute' oscillation, e.g., those based
on acoustic waves[14], slow[15], fast[16] and Alfvèn[17] MHD modes, and flux
tube[18] oscillations, have concentrated on explaining the 180s period.
This leaves unacounted for the various shorter periods, which are
reported in most observations[2-10] of umbral oscillations, although with
less certainty since resolution is more critical.

3. RATIOS OF PERIODS OF HARMONICS TO FUNDAMENTAL

The sequence of periods does not appear fit into a sequence of
submultiples, such as, 300,150,100,75s, implying that the speed of
propagation is *not* nearly constant, e.g., this argues against acoustic
modes, since the sound speed, which depends mainly on temperature (and
molecular mass), does not vary too much in sunspots. In contrast the
Alfvèn speed $A = B/\sqrt{4\pi\rho}$, even in a gradually varying magnetic field B,
will increase rapidly with altitude, as the mass density decreases, e.g.,
in a isothermal atmosphere $\rho \sim e^{-z/L}$, where L is the scale height, and
thus $A = a\ e^{z/2L}$ for constant B. The velocity v and magnetic field h
perturbations, for an Alfvèn wave of frequency ω, in an isothermal
atmosphere, permeated by an uniform vertical magnetic field, are given
[19-20] respectively by:

$$v(z,t) = C\ e^{-i\omega t}\ J_0((2\omega L/a)\ e^{-z/2L}), \tag{1a}$$

$$h(z,t) = i(B/a)\ e^{-i\omega t}\ e^{-z/2L}\ J_1((2\omega L/a)\ e^{-z/2L}), \tag{1b}$$

where C is a constant and J_0, J_1 Bessel functions; we have assumed perfect reflection from a level z=d, several scale heights away $d/L \gg 2 \log(2\omega L/a)$, from the level of generation z=0.

Thus we model umbral oscillations as Alfvēn modes, travelling along magnetic field lines, generated in the convection zone, propagating through the photosphere, and reflected in the atmosphere[21-22], e.g., in the transition region from the chromosphere to the corona, where these oscillations are also observed[5]. The 'footpoints' of magnetic field lines are shaken by convection, in sub-photospheric layers, up to a certain depth, beyond which they are 'anchored'[11]. We may expect the excitation of waves to occur just above the level of 'anchoring' of magnetic field lines; the condition of vanishing magnetic field perturbation h(0,t)=0 at this level z=0, implies, by (1b), that $2\omega L/a = j_n$, where j_n are the positive roots of the Bessel function J_1, and specify the eigenfrequencies ω_n, or periods T_n, of resonant modes:

$$T_n = 2\pi/\omega_n = 4\pi L/j_n a; \qquad T_n/T_1 = j_n/j_1, \qquad (2a,b)$$

shows that the ratios of periods (2b), are fixed numbers, independent of atmospheric properties, and they are consistent (Table I) with the observed[3-10] values, which are also reported[7] to be independent of particular sunspot characteristics.

4. VARIABILITY OF PERIODS AND DEPTH OF GENERATION

Since the periods of umbral oscillations exhibit some variability in time, viz., about 10%, not yet fully resolved, the comparison of theory and observation cannot be taken anywhere near the levels of accuracy now demanded for the five-minute oscillation. Our model, is in fact, oversimplified, at least in two respects: (i) it assumes isothermal conditions, which is not too unreasonable for sunspot atmospheres, but sub-photospheric layers are more nearly polytropic; (ii) it assumes an uniform, averaged magnetic field, instead of expanding magnetic flux tubes[24]. Although an exact analysis of Alfvēn waves in polytropic atmospheres does not appear to be available, a multi-layer model can be used[25] to construct any temperature and magnetic field[26] profile, in a stepped approximation. These refinements do not appear, however, to be necessary, for a theoretical estimate of resonant frequencies, to within the observed variability; the more important issues, are to estimate the order of magnitude of the frequencies, and to explain their variability.

Having discussed the *ratios* of periods, we have still to consider the absolute valve of *one* of them, e.g., the fundamental. Using L=0.15 Mm for the scale height at $5 \times 10^3 K$, the period $T_1 = 300s$ implies, by (2a), an Alfvēn speed $a = 4\omega L/j_1 T_1 = 1.66$ Km/s at the level of generation, in the sub-photospheric layers of the sunspot. Assuming that the sunspot radius[24] decreases by a factor f=1.75 relative to the surface[1], and taking $B_0 = 2$ kG, for the surface magnetic field[27], we obtain, by flux conservation $B = f^2 B_0 = 6.4$ kG, corresponding to a density $\rho = B^2/4\pi a^2 =$

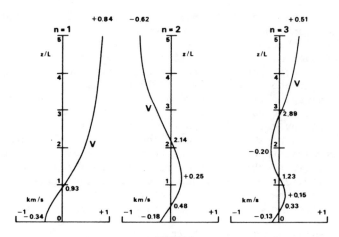

FIGURE 1

FIGURE 1 - Velocity perturbation of the first three Alfvẽn modes, versus
altitude divided by scale height. Note that: (i) the n-th mode has n
nodes, and the nodes of succeeding modes are interlaced; (ii) assuming
the same excitation amplitudes, the initial and asymptotic amplitude
decrease with the order of the mode; (iii) the velocity perturbation
tends to a constant value, at high-altitude, alternating with the order
of the mode n. These profiles assume a standing Alfvẽn-gravity wave, in
an isothermal atmosphere, under an uniform, vertical magnetic field, in
the absence of damping.

TABLE I
PERIODS OF UMBRAL OSCILLATIONS IN SUNSPOTS

Order of mode	Root $J_1(j_n)=0$	Periods (s) observed	predicted	Ratio to fundamental theoretical	empirical
n=1	3.83	300	300	1.00	1.00
2	7.02	180	164	0.55	0.60
3	10.3	120	113	0.38	0.40
4	13.3	100	86	0.29	0.33
5	16.5	65	71	0.23	0.22

TABLE I - The roots j_n (second column) of the Bessel function J_1,
specify, through their ratios, the theoretical estimates of the ratio
(fifth column) of the periods of harmonics to the fundamental, of
umbral oscillations; these ratios are consistent with empirical values
(sixth column), taking into account the observed ±10% variability of
periods. The absolute values of periods, theoretical (fourth column) and
empirical (third column), correspond to the preceding ratios, for a
fundamental period of 300s. The latter period specifies the depth of
generation of Alfvẽn waves as 10 Mm, and the 10% variability of periods
implies a fluctuation of ±0.1 Mm in this depth.

$=1.2 \times 10^{-4}$ g cm^{-3} at generation level. A photospheric density $\rho_0 = 3.0 \times 10^{-7}$ g cm^{-3}, and a polytropic law $\rho/\rho_0 = (z/\lambda)^2$ with scale $\lambda = 0.53$ Mm, place the depth of generation of Alfvén waves at $\Delta z = 10.4$ Mm below the photosphere; this is consistent[11] with the estimated $\sim 10^4$ Km depth of origin of ephemeral magnetic regions on the solar surface. The variability of periods (±10%), corresponds to changes of Alfvén speed (±10%) or mass density (±20%), which imply moderate displacements (< 0.1 Mm) of the sub-photospheric level of excitation of Alfvén waves.

5. AMPLITUDES, WAVEFORMS AND ACOUSTIC COUPLING

The periods of Alfvén waves, and their variability, are determined mainly by physical conditions in the deeper layers, because there the mass density is higher and Alfvén speed lower, i.e., it is in these layers that Alfvén waves spend most of the time. Another consequence, is that the velocity (1a) and magnetic field (1b) perturbations, viz.:

$$|v(z,t)| = C_n \, J_0(j_n \, e^{-z/2L}), \tag{3a}$$

$$|h_n(z,t)| = (B/a) \, C_n \, e^{-z/2L} \, J_1(j_n \, e^{-z/2L}), \tag{3b}$$

for the n-th mode, oscillate mostly in the first few scale heights, as illustrated respectively in Figures 1 and 2, for the first three modes $n=1,2,3$. The rapid increase of Alfvén speed with altitude, not only causes an increasing spacing of modes, but also, together with reflections in the atmosphere, leads to a bounded amplitude C_n of velocity oscillations, and exclusion of the magnetic field perturbation from upper layers. The amplitude C_n appearing in (3a,b) can be evaluated[28] by Fourier integration of (1a,b), as $C_n = \pi h_0 (a/B) j_n q_n$, with $q_n \equiv J_1'(j_n) = J_0(j_n)$ a constant, and h_0 the initial magnetic field perturbation. The ratio of amplitudes is $C_n/C_1 = j_1 q_1/j_n q_n$, for uniform forcing in the spectral band ω_1 to ω_n; the ratio C_2/C_1 is consistent (Table II) with the limited observational data available[5] on amplitudes, and other ratios are listed in the expectation of more empirical results for future comparison.

The amplitude of fundamental $C_1 = 0.84$ Km/s, implies an excitation magnetic field $h_0 = B \, j_1 C_1/\pi a = 3.9$ kG, corresponding, by equipartition of kinetic and magnetic energies, to a reference velocity $V_0 = (B/a) h_0 = 1.0$ Km/s, which is higher than the velocities 0.13–0.34 Km/s at the generation level. In these deep sub-photospheric layers, the oscillation is a purely alfvénic perturbation, travelling along a nearly vertical magnetic field, with a nearly horizontal velocity perturbation; the latter would produce almost no Doppler shift at disk center, adding to the difficulties of observation due to the optical depth. In the atmosphere of a sunspot, the strong magnetic field and rapidly decreasing density, imply that the Alfvén speed exceeds the sound speed, so that non-linear coupling[29] with acoustic waves can occur; the latter are longitudinal modes, and have a vertical velocity component,

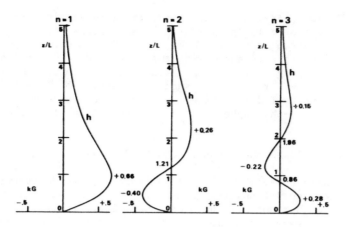

FIGURE 2

FIGURE 2 - As in figure 1, for the magnetic field perturbation of the first three Alfvén modes, versus altitude divided by scale height. Note that: (i) the initial magnetic field perturbation is zero, at the depth where magnetic field lines are 'anchored'; (ii) asymptoticaly at high altitude, the magnetic field perturbation decays, because the magnetic (and acoustic) energy must decay for a standing mode; (iii) the n-th mode has (n-1) nodes (besides zero and infinity) and n local maxima, which decrease with the order of the mode, and with altitude.

TABLE II
PEAK-TO-PEAK AMPLITUDE OF VELOCITY PERTURBATION

Order of mode	slope $q_n = J_0(j_n)$	Amplitude (Km/s) observed	Amplitude (Km/s) predicted	Ratio to fundamental theoretical	Ratio to fundamental empirical
n=1	−0.403	0.84	0.84	1.00	1.00
2	+0.300	0.66	0.62	0.73	0.79
3	−0.250	−	0.51	0.61	−
4	+0.218	−	0.45	0.53	−
5	−0.196	−	0.40	0.48	−

TABLE II - The slope q_n (second column) of the Bessel function J_1, specifies the theoretical ratio of asymptotic amplitudes (fifth column) of the velocity perturbation, at a layer several scale heights above the level of generation; the absolute values (third column) depend on atmospheric properties, and assume the same excitation magnetic field for all modes. The observational data available for comparison is limited to two amplitudes (fourth column) and one ratio (sixth column).

Note that theoretical estimates refer to the horizontal velocity for an Alfvén wave, and observations to the vertical velocity for an acoustic wave, so that there is a non-linear coupling coefficient to relate the two of them, which need not be unity or the same for all modes (as assumed in the table).

producing Doppler shifts in disk observations, and tending to form
shocks[14]. The apparent alternation of the 300s and 180s oscillations in
sunspots[4], could be explained (Figure 1), by the fundamental n = 1 and
first harmonic n = 2 of Alfvén modes, being mostly out-of-phase, in the
upper, atmospheric layers, and the phase shifts, corresponding to a
phase speed of 10-25 Km/s in the chromosphere[30], are associated with
the non-linearly coupled acoustic waves.

6. WAVE EXCITATION AND RADIATION INTENSITY

The coexistence of acoustic and Alfvén waves applies, not only to   [31]
sunspot atmospheres, but to other regions of the sun, such as spicules
and the chromosphere[32]. Satellite data, from the SMM mission, suggest[33]
that Alfvén waves in the solar wind, associated with particle events,
originate from sunspot umbras. The presence of Alfvén waves in the solar
wind has been reported for some time[34], but these more recent
observations[33] imply their existence in sunspot atmospheres. It may
presumed that the Alfvén waves are able to propagate away from sunspots,
during flare events[35], since then the magnetic field lines 'open-up' to
eject matter, due to the rise of a flux tube or an MHD instability. In
the quiet sun, the magnetic 'arcades' would confine the Alfvén waves,
which appear as standing modes, associated with oscillations;
alternatively, the tilting of the magnetic field towards the horizontal,
could transform Alfvén into magnetosonic modes[36], viz., running
penumbral waves. The theories of umbral oscillations based on slow or
fast MHD waves[6], can account for the observed 10% variability of periods,
only by assuming a corresponding variation in the height of the
'photospheric'[16] or 'chromospheric'[15] resonant cavities, which is
difficult to reconcile with observation; the theories based on acoustic
modes[14,18] would lead to a sequence of periods consisting of submultiples,
which does not appear to be consistent with the observations. In the
present model, as Alfvénic oscillations, the uneven spacing of periods
is explained by the non-uniform wave speed, and their variability is
accounted for by a small displacement of the sub-photospheric layers of
the resonanting cavity. The atmospheric layers, where the Alfvén modes
become coupled, non-linearity with acoustic waves, have little effect
on wave periods, since there the Alfvén speed is large, and the crossing
time (less than 20s), is a small fraction of the total period.
    The preceding arguments on the anchoring of magnetic field lines,
and the variability of periods, suggest that the wave excitation region,
may be less than one-scale height deep. In this case, the intensity of
radiation, may be predicted[37-38] from the theory of generation of   MHD
waves, viz:

$$I_m = k \ \rho(\ell^2/u) \ (v/u)^{2m} \ (v^2 + \mu \ b^2/4\pi\rho)^2 , \tag{4a}$$

For a wave of phase speed u, produced by r.m.s fluctuations of velocity
v and magnetic field b; the integer m specifies[39] the source type, viz.,

a monopole (m=0), dipole (m=1) or quadrupole (m=2), e.g., respectively[39] volume changes, ionized inhomogeneities and hydromagnetic turbulence. The formula (4a) includes, as particular, acoustic case, the laws of aerodynamic acoustics[40], and is extended, to account for the presence of a magnetic field[41], by adding to the magnetic $\mu b^2/8\pi$ to the dynamic $1/2\ \rho v^2$ pressure, or equivalently, multiplying by $2/\rho$, adding to the velocity perturbation squared v^2 the Alfvén speed squared $a^2 \equiv \mu b^2/4\pi\rho$.

7. DISCUSSION

In a stratified medium, such as an atmosphere, wave sources induce volume changes, i.e., their efficiency is enhanced[42] to monopole level, and we obtain from (4a) with m=0, for the intensity of radiation of Alfvén waves (u=a), the formula:

$$I_o = K\ \rho(\ell^2/a)\ (v^2 + a^2)^2; \qquad\qquad\qquad (4b)$$

both in (4b) and (4a), K is a constant of order of magnitude unity, independent of flow conditions or magnetic field, and typical of the wave excitation mechanism, e.g., overstable convection. We can 'calibrate' the formula, i.e., estimate the constant K, by comparing with the energy flux in Alfvén waves $F = \rho\ v^2 a$, assuming radiation $I_o = F\ell^2$ through an area determined by the scale ℓ of the source region; comparing $I_o = \rho\ell^2 v^2 a$ with (4b), we obtain $K \sim (v^2 + a^2)^2/v^2 a^2 = 5.1$, using the preceding values of v,a. As an example, we apply the formula (4b) to the solar photosphere, where the mass density is $\rho = 3\times10^{-7}$ g cm^{-3}, taking v = 1 Km/s for the velocity perturbation and a = 10 Km/s for the Alfvén speed, so that the total intensity of radiation, from the solar disk $\ell^2 \sim \pi R^2$ of radius R = 7×10^{10} cm, may be estimated as $I_o \simeq 2.2\times10^{33}$ erg cm^{-2} s^{-1}, which is close to the total radiative loss of the sun (3×10^{33} erg s^{-1}); although the theoretical estimate depends on the assumed magnitude of velocity and magnetic field perturbations, it indicates that Alfvén waves are a significant energy carrier. The general formula (4a) could also be used to estimate the energy flux of the various MHD modes in stars other than the sun.

REFERENCES

1. Thomas, J.H. & Cram, L.E. & Nye, A.H., 1982 *Nature* 297, 485.
2. Bhatnagar, A. & Livingston, W.C. & Harvey, J.W. 1972 *Solar Phys.*27,80
3. Rice, J.B. & Gaizauskas, V. 1973 *Solar Phys.* 32, 421.
4. Schroter, E.H. & Soltau, D. 1976 *Astron. Astrophys.* 49, 463.
5. Soltau, D. & Schroter, E.H. & Wohl, H. 1976 *Astron. Astrophys.* 50,367
6. Gurman, J.B. et al. 1982 *Astrophys. J.* 253, 939.
7. Lites, B.W. 1984 *Astrophys. J.* 277, 874.
8. Thomas, J.H. & Cram, L.E. & Nye, A.H. 1984 *Astrophys. J.* 285, 368.
9. Soltau, D. & Wiehr, E. 1985 *Astron. Astrophys.* (to appear).
10. Balthasar, H. & Wiehr, E. 1985 *Astron. Astrophys.* (to appear).
11. Parker, E.N. 1984 *Astrophys. J.* 280,422.

12. Christensen-Dalsgard, J. & Gough, D.O. 1976 *Nature* 259,89.
13. Campos, L.M.B.C. 1983 *Solar Phys.* 82, 355.
14. Leibacher, J.W. & Gouttebroze, P. & Stein, R.F. 1982 *Astrophys. J.*258,383
15. Zhugzhda, Y.D. & Locans, V.A. 1981 *Astron. J.* USSR 7, 144.
16. Scheuer, M.A. & Thomas, J.H. 1981 *Solar Phys.* 71, 21.
17. Uchida, Y. & Sakurai, T. 1975 *Publ. Astron. Soc. Japan* 27, 259.
18. Hollweg, J.V. & Roberts, B. 1981 *Astrophys. J.* 250, 398.
19. Ferraro, V.C.A. & Plumpton, C. 1957 *Astrophys. J.* 129, 459.
20. Campos, L.M.B.C. 1983 *J. Méc. Theor. Appl.* 2, 861.
21 Thomas, J.H. 1978 *Astrophys. J.* 225, 275.
22. Leroy, B. 1982 *Astron. Astrophys.* 97, 245.
23. Staude, J. 1981 *Astron. Astrophys.* 100, 284.
24. Deinzer, W. 1965 *Astrophys. J.* 141, 548.
25. Hollweg, J.V. 1972 *Cosmical Electrod.* 2, 423.
26. Nye, A.H. & Hollweg, J.V. 1980 *Solar Phys.* 68, 279.
27. Golub, L. et al. 1981 *Astrophys. J.* 243, 309.
28. Campos, L.M.B.C. 1983 *J. Phys.* A16, 417.
29. Uchida, Y. & Kaburaki, O. 1975 *Solar Phys.* 35, 451.
30. Uexkull, M. & Kneer, F. & Mattig, W. 1983 *Astron. Astrophys.* 123, 263.
31. Campos, L.M.B.C. 1984 *Month. Not. Roy. Astron. Soc.* 207, 547.
32. Campos, L.M.B.C. 1984 *Mem. Soc. Astrop. Ital.* 55, 267.
33. Mullan, D.J. & Owens, A.J. 1984 *Astrophys. J.* 280, 346.
34. Belcher, J.L. & Davis, L. 1971 *J. Geophys. Res.* 76, 3534.
35. Priest, E. 1984 *Solar Flare Magnetohydrodynamics*. Gordon & Breach.
36. Nye, A.H. & Thomas, J.H. 1976 *Astrophys. J.* 204, 582.
37. Campos, L.M.B.C. 1977 *J. Fluid Mech.* 81, 529.
38. Stein, R.F. 1981 *Astrophys. J.* 246, 966.
39. Campos, L.M.B.C. 1978 *Proc. Roy. Soc.* A351, 65.
40. Lighthill, M.J. 1952 *Proc. Roy. Soc.* A211, 564.
41. Campos, L.M.B.C. 1983 *Port. Phys.* 14, 121.
42. Parker, E.N. 1964 *Astrophys. J.* 140, 1170.

CHANGE OF SOLAR OSCILLATION EIGENFREQUENCIES WITH THE SOLAR
CYCLE

Martin F. Woodard[*] and Robert W. Noyes
Mail Code 16
Harvard-Smithsonian Center for
Astrophysics, Cambridge, MA 02138 U.S.A.

ABSTRACT.   Solar p-mode eigenfrequencies depend on the radial stratification of the solar interior and this stratification may be
perturbed by the mechanism of the 11-year magnetic activity cycle.
Observations of long-term changes in p-mode oscillation frequencies may therefore help to identify the cause of the solar cycle.
A comparative analysis of two separate years of data (1980, 1984)
from the ACRIM instrument on the Solar Maximum Mission satellite
was performed to access the level of frequency variation over the
declining phase (~ 1980-1984) of the activity cycle.   The difference for the two epochs (1980 minus 1984) in the mean frequency
of nine ℓ =0 and ℓ =1 oscillation peaks in the ACRIM power
spectrum is

$$\Delta\nu = 0.42 \pm 0.14 \ \mu \ Hz, \ that \ is,$$

the mean frequency of these modes was approximately 1.3 parts in
10^4 higher at the epoch of sunspot maximum (~ 1980) than at
the later epoch. Analysis of synthetic oscillation data seems to
rule out possible frequency pulling effects due to differences
between the 1980 and 1984 observing windows as the cause of the
observed variation.

A variety of phenomena associated with the solar cycle may
influence the p-mode eigenfrequencies (see reference below and
articles mentioned therein).   To leading order the frequencies
of the low-degree p-modes are inversely proportional to the
sound travel time between the inner and outer reflection points
of the solar p-mode 'cavity'. Thus, the observed frequency increase at sunspot maximum could be explained by a fractional
decrease in the radius of the outer reflection point (essentially

*Present address: Jet Propulsion Laboratory, Mail Code 169-506
4800 Oak Grove Drive, Pasadena, CA 91109 U.S.A.

D. O. Gough (ed.), Seismology of the Sun and the Distant Stars, 303–304.

the radius of the photosphere) of order 10^{-4}, by a relative increase (of the same order) in the mean sound speed of the solar interior, or by some combination of such changes. From a frequency-variation analysis of a much wider selection of solar p-modes it may be possible to infer the depth dependence of structural perturbations associated with the solar cycle. We expect that observations of low – ℓ p-modes in distant stars will similarly help to clarify the nature of stellar activity.

Support for this work was provided by NASA contract NAG 5-506.

Reference

"Change of Solar Oscillation Eigenfrequencies with the Solar Cycle", M.F. Woodard and R.W. Noyes (submitted to Nature).

LONG PERIOD VARIATIONS IN THE SOLAR DIAMETER

Philippe DELACHE
Observatoire de Nice
Laboratoire associé au C.N.R.S. 128
B.P. 139
F-06003 Nice-Cedex
France

ABSTRACT. Observations of the solar diameter with a Danjon-type visual astrolabe have been recorded at the C.E.R.G.A. Observatory since 1975. An harmonic analysis of these measurements shows the existence of very long solar periods. In order to assess the significance of peaks in the power spectrum we present four different methods. Their common conclusion is that we are entitled to identify a few long periodicities in the solar diameter fluctuations. Most of the periods, although not all of them, appear to be correlated with the solar activity monitored by the Zürich sunspot number. We present a discussion of these correlations.

1. INTRODUCTION

Most of the present analysis and results has already been described (Delache et al.,1985). The present contribution, with the exception of figure 2 below, will thus be essentially a summary of the presentation.
    The primary goal of the C.E.R.G.A. astrolabe observations is the determination of the orbital elements of the Earth around the Sun. However they provide, as a by-product, very good quality, homogeneous values of the solar diameter. Measurements are obtained up to sixteen times per day. We consider all these measurements on equal footing. We present here the analysis of the low frequency part of their power spectrum.

2. FOURIER ANALYSIS

Figure 1 shows the low frequencies in the power spectrum. It has been computed from 504 daily averages of the solar diameter observed from 1975 to 1984. In order to assess the significance of peaks in this spectrum, we start with the three following methods:

2.1. Independant sets

First, we divide our data set into two parts. One contains the morning

305

D. O. Gough (ed.), Seismology of the Sun and the Distant Stars, 305–308.

observations, and the other, the afternoon ones. We may then obtain two
independant spectra on which the three major low frequency peaks, C, P,
W, remain significant. Along a similar idea, see also paragraph 2.4.

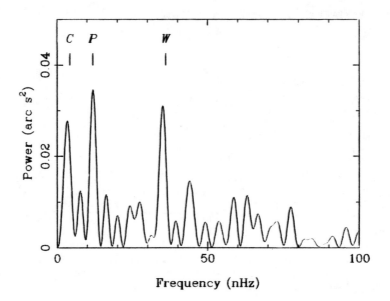

Figure 1. Power spectrum of the solar diameter variations. Peaks C, P,
W, are found to be significant.

2.2. Distribution of spectral estimates

A more quantitative analysis makes use of the observed statistical
distribution of spectral estimates. We compare the distribution derived
from the actual spectrum with the distribution obtained from a similar
spectrum computed with the same data taken at random and distributed on
the actual observing days. We thus keep the same window and error distribution in the measurements, but we destroy coherent signal. One observes that the real data spectral estimates distribution has a significantly
large excess of high values, indicating the existence of peaks well over
the noise level.

2.3. Comparison with solar activity

One may expect that magnetic activity will modify the visual definition
of the solar limb. It is then in order to compare our diameter spectrum
with some global solar activity index, such as the daily Zürich sunspot
number (ZSN) spectrum, computed from observations recorded during the
same years. One finds that peaks labeled C, and W, are present too in
the ZSN spectrum. We conclude that features shared by both spectra, at
least, are intrinsic solar periods.

2.4. Random windows

This fourth test has been suggested to us by J. R. Kuhn during the meeting, and we are glad to report here the results. It consists in making
a random separation of the original daily data set into two parts, in
order to get two different window functions. The result is shown in
figure 2 a,b where the two spectra are plotted. Of course they have a
lower quality as compared to the spectrum of figure 1, since they rely
upon only 252 independant values. One observes readily that the three
peaks C, P, W continue to dominate both spectra.

      We thus feel confident that we are observing true solar diame-
ter variations.

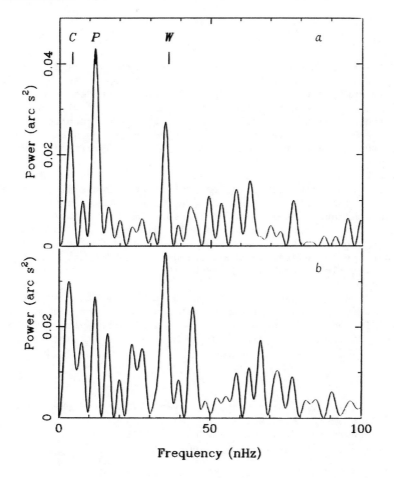

Figure 2. Power spectra of the solar diameter variations, selecting
randomly : a, one half of the original data, b, the other half.

Table I gives amplitudes and frequencies for the three main peaks.

TABLE I

Peak	frequency (from diameter) (nanoHertz)	amplitude (arc s)	frequency (from ZSN) (nanoHertz)
C	3.3 ± 1.5	0.16	4.4 ± 1.5
P	11.9 ± 1.5	0.18	-
W	35.0 ± 1.5	0.17	36.0 ± 1.5

3. DISCUSSION

Peak C is the signature of the solar cycle variations. Inside this peak, the complex Fourier spectra appear to have nearly opposite phases. One can also apply a very low band pass filter to our data sets and verify the strong anticorrelation between the solar activity and the diameter: the active Sun is smaller. Notice that such an effect, with the same order of magnitude is also suggested by the changes in the periods of the 5 min oscillations from SMM/ACRIM observations (Noyes and Woodard, 1986).

Peak W has already been identified in the ZSN spectrum (Wolff, 1983). At this period, of order 320 days, the variations in the diameter and in the activity are nearly *in phase*, as if the emergence of active regions were lifting the apparent level of the solar limb.

Peak P, the strongest one in the solar diameter, has no counterpart in the ZSN spectrum. Its period, of order 1000 days, is very intriguing. It might perhaps be related to the marginally significant periods found in the neutrino flux (Haubold and Gerth, 1983).

Further work is in progress, and incudes a study of the heliographic latitude dependance of our diameter measurements, a comparison with SMM/ACRIM radiometer observations, and with other global Sun properties such as the 10 cm radio flux.

4. REFERENCES

Delache, P., Laclare, F. & Sadsaoud, H.,1985, *Nature*, in the press
Haubold, H.J. & Gerth, E.,1983, *Astron. Nach.*, 304, 229
Noyes, R.W. & Woodard, M.,1986, *these proceedings*
Wolff, C.L.,1983, *Astrophys. J.*, 264, 667

THE 1984 SOLAR OSCILLATION PROGRAM OF THE MT. WILSON 60-FOOT TOWER

Edward J. Rhodes, Jr.
Department of Astronomy
University of Southern California
Los Angeles California   90089
    and
Space Physics and Astrophysics Section
Jet Propulsion Laboratory
California Institute of Technology
Pasadena, California 91109

Alessandro Cacciani
Department of Physics
University of Rome
Rome, Italy

Steven Tomczyk and Roger K. Ulrich
Department of Astronomy
University of California at Los Angeles
Los Angeles, California 90024

ABSTRACT. We describe here the instrumentation, data, and preliminary results from the summer, 1984, solar oscillation observing program which was carried out using the 60-foot tower telescope of the Mt. Wilson Observatory. This program was carried out with a dedicated solar oscillation observing system and obtained full-disk Dopplergrams every 40 seconds for up to 11 hours per day. Between June and September, 1984, observations were obtained with a Na magneto-optical filter on 90 different days. The data analysis has progressed to the point that spherical harmonic filter functions have been employed to generate a few one-dimensional power spectra from a single day's observations.

1. INTRODUCTION

    Shortly after the initial observations of high-degree solar p-mode oscillations (Deubner, 1975; Rhodes, Ulrich, and Simon, 1977; Deubner, Ulrich, and Rhodes, 1979) two of us (EJR and RKU) recognized the need for a dedicated solar oscillation observing system. Accordingly, we began to develop such a system jointly with Dr. Robert F. Howard at the Mt. Wilson Observatory. The initial design of this system was developed during 1978 and major modifications were made to

D. O. Gough (ed.), Seismology of the Sun and the Distant Stars, 309–332.

this design during 1980 and 1981. The two most significant improvements were the replacement of one-dimensional Reticon detectors with a CID camera system, and the use of a magneto-optical filter instead of a spectrograph. The developmental status of this system was first described by Rhodes et al. (1981). More recent reports about this system were given by Rhodes et al. (1983 and 1984a). The first solar oscillation power spectrum obtained with this system during July, 1983, was presented in Rhodes et al. (1984a and 1984b).

Here we wish to report on the first extended operation of this oscillation system in the manner for which it was orginally designed: the acquisition of daily full-disk Dopplergrams with high time resolution. During the fall of 1983 and the spring of 1984 the full-disk imaging capability of the system was installed and tested with the goal of obtaining such observations throughout the 1984 observing season. We will first report on the observations which were obtained and then finish with a progress report on the status of the reduction of that data. The analysis of a larger quantity of this data is currently in progress with our primary goals being an improvement in the published accuracy of low- and intermediate-degree p-mode frequencies and the probing of solar internal rotation as functions of depth and latitude as Brown has reported earlier at this meeting (Brown, 1985).

2. OBSERVATIONS

The observations obtained during the summer of 1984 consisted of pairs of narrow-band filtergrams which were obtained in the two Na D lines. The filtergrams were obtained with a two-cell version of the magneto-optical filter (MOF) which was originally invented by Cacciani. This filter was first described in Cimino et al. (1968a and 1968b), and in Cacciani et al. (1970). Its uses for solar observations were later described in Agnelli, Cacciani, and Fofi (1975); in Cacciani and Fofi (1978); in Cacciani, Fortini, and Torelli (1980); and in Cacciani et al. (1981). More recently, its principles of operation were reviewed by Cacciani and Rhodes (1984), who described both one- and two-cell versions of it.

Figure 1 shows the two-cell configuration of the MOF which was employed during May through September, 1984, to obtain the data described here. During that time the filter employed two 1000-Gauss permanent magnets. (Subsequent to September, 1984, the first of these two magnets was replaced with a 6000-Gauss permanent magnet and this combination of 6000- and 1000-Gauss magnets has been adopted for all subsequent Na observations.)

At the left of the magnets in Figure 1 we show the mechanism which controlled the moveable quarter-wave plate. This mechanism consisted of a reversible stepper motor and a Geneva movement. When operated under the control of an LSI-11 microcomputer, this mechanism allowed us to rotate the wave plate through angles of +45° and -45°

Figure 1. Magneto-Optical Filter. This figure shows the two-cell Na MOF as it was installed at Mt. Wilson during 1984. At that time the filter employed two 1000-Gauss magnets.

with respect to the crossed polarizers which were also part of the MOF.

When the wave plate was positioned in the +45° position, for example, we were able to record filtergrams in the bandpass located on the red side of the NaD lines, while a rotation of the wave plate to the -45° angle tuned the filter to the blue bandpass.

The exposures of the Photometrics, Ltd. CID camera which we employed were regulated by a rotating disk shutter. The rotation of this shutter was governed by a series of uniformly-spaced stepper motor pulses which were supplied to the shutter by a Superior Electronic Buffered Translator. Each filtergram took 5 seconds for both exposure and camera read-out and so a pair of red- and blue-wing filtergrams could have been obtained once every 10 seconds. In fact, this sequence was only repeated once every 40 seconds for up to 11.4 hours per day. Therefore, a single day's observing run consisted of a maximum of 1024 red filtergrams and 1024 blue filtergrams.

The transmission profiles of the two-cell MOF were similar to those shown in Figure 2. Here we show two more-recent prfiles which were obtained in the NaD1 line with the 6000-Gauss/1000-Gauss combination of magnets mentioned earlier. These wavelength scans were obtained with the spectrograph of the 150-foot tower telescope at Mt. Wilson. The two profiles corresponding to the +45° and -45° positions of the quarter-wave plate are superimposed upon each other in this Figure. The degree of separation of the two bandpasses is illustrated by the low level of the transmission of each curve at the location of the peak of the other curve.

The cumulative amount of data obtained during our 1984 campaign is shown in Figure 3. Here the number of hours of cloud-free observations obtained on each day is plotted as a function of time. The initial observations were obtained on May 8, while the last observations were obtained on September 9. The stipled portion of the diagram shows that most of the May data was subsequently destroyed when all but the first one of each day's tapes were later re-written with newer data in early August. A single tape per day from that period was saved in order to have some Dopplergrams for eventual comparison with those produced at the 150-foot tower on the same days. On some of the days between May 20 and June 8 several other tests were run with the MOF and with the observing hardware; hence, the absence of data during that interval was not due to a period of bad weather.

Figure 3 shows that on several occasions during the campaign we did indeed obtain 7 to 11 hours of cloud-free data per day for intervals exceeding one week in duration. On only 12 days after with June 24 was the weather bad enough to prevent the recording of any data at all. The fact that this much data was obtained during 1984 was due solely to the diligence of S. Tomczyk who observed and archived

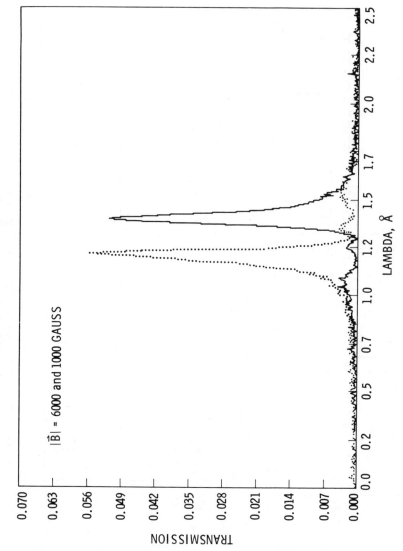

Figure 2. Magneto-Optical Filter Transmission Profiles. The red- and bluewing Na D1 transmission profiles for a two-cell MOF are shown. The magnetic field strengths were 6000 Gauss and 1000 Gauss, in cells 1 and 2, respectively. The input power was 25.0 watts in cell 1 and 14.5 watts in cell 2.

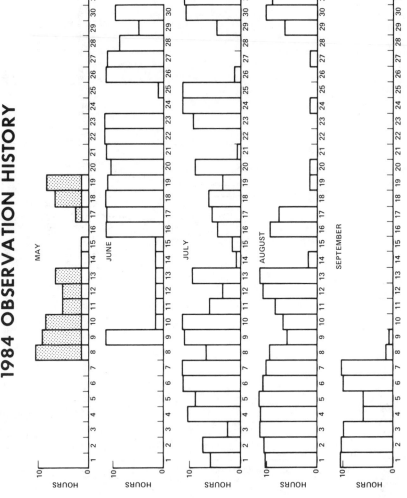

Figure 3. Observational History. The number of hours of data obtained each day is plotted as a function of time during May–September 1984.

the data almost single-handedly for the entire three months.

A histogram of the number of days having a given length run is shown in Figure 4. Inspection of this figure shows that on 45 of the 90 days at least 8 hours of data was obtained. Since this histogram only includes the number of cloud-free hours on each day, the total amount of data that was obtained actually exceeded the amount shown in the histogram. The large number of days having between one and two hours of data was the result of our effort to record at least a single tape's worth of data even when it became obvious early in a given day that the weather would not permit the full-day run to be completed.

3. CONVERSION INTO DOPPLERGRAMS

At the conclusion of each day's observations the time series of alternating red and blue filtergrams was archived from disk to tape. The first steps in the analysis of this data were the computation of the location of the disk center of each filtergram and the translational registration of each filtergram to a common center location. The registration process was necessary to correct for the image motion which occurred between the acquisition of each red filtergram and its corresponding blue companion. It also allowed us to correct for slow drifts in the solar image introduced by the guider on the time scale of several hours.

After the filtergrams were registered in the computer, each pair could then be converted into a Dopplergram with the following simple formula:

$$\text{velocity }(x,y) = \text{constant} * \frac{I_{blue}(x,y) - I_{red}(x,y)}{I_{blue}(x,y) + I_{red}(x,y)} \tag{1}$$

where I_{blue} and I_{red} referred to the intensities of the pixel at location (x,y) in the blue and red filtergrams and the constant was a calibration constant which had to be determined in some way.

A Dopplergram similar to those employed with the 244x248 pixel CID camera and the Na MOF is shown here in Figure 5. The figure shows a Dopplergram which was obtained with a JPL 800x800 pixel CCD camera (although only an area 500 pixels on a side was employed in this instance). This Dopplergram was obtained on June 1, 1984. It was obtained by differencing two filtergrams which were taken 15 seconds apart; hence, the appearance of the solar limb is due to slight image distortions caused by the Earth's atmosphere during and between the exposure of the two filtergrams. Solar rotation is visible as the shading from white to black across the visible disk.

By comparison we show the Dopplergram obtained in λ5250 with

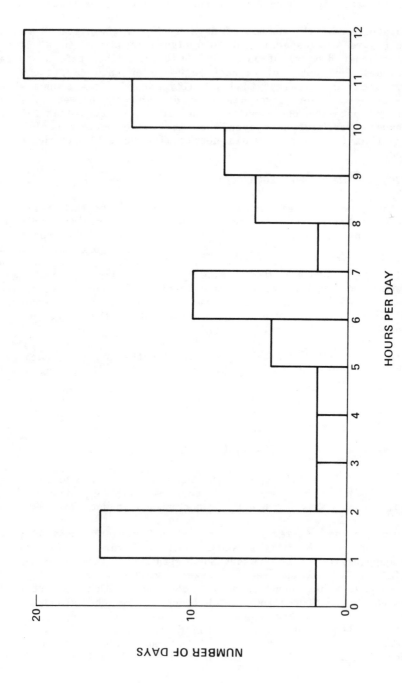

Figure 4. Mean Acquisition Histogram. The number of days having a given run length are plotted as a function of observing run length.

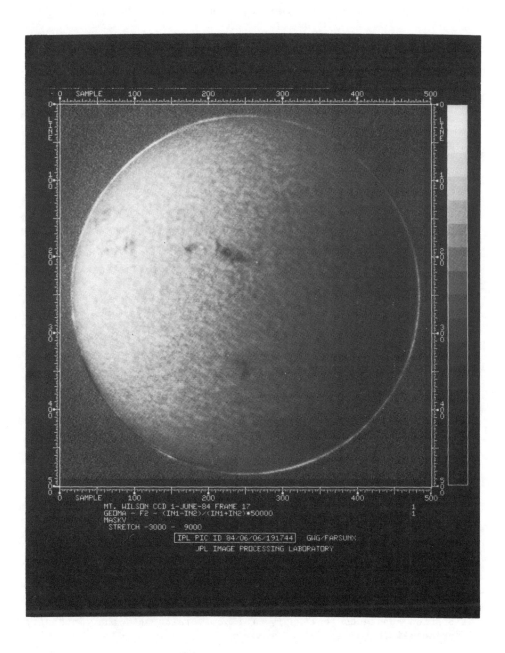

Figure 5. MOF Dopplergram. Na Dopplergram obtained with MOF and 800x800 pixel CCD camera on 1 June 1984. Solar rotation is visible as the shading across the disk.

the spectrograph of the 150-foot tower on the same day in Figure 6.

Again, solar rotation is visible across the disk as are several active regions.The velocity flows around the active regions are less pronounced in the $\lambda 5250$ Dopplergram than they were in Figure 5. Some of this difference may be due to the fact that the NaD lines are chromospheric lines while $\lambda 5250$ is a photospheric line. Additional tests at Mt. Wilson will be needed to clarify this possibility.

The constant given in equation (1) can in principle be determined in two different ways and hence confirmed independently. First, it is possible to use the rate of solar rotation as measured in a particular latitude zone on a given day at the 150-foot tower to provide the radial velocities against which the observed intensity differences can be compared. The idea is illustrated schematically in Figure 7. Here the scaled Doppler signal (defined for now as $50000 \cdot [I_B - I_R]/[I_B + I_R]$) from a full-disk Dopplergram obtained with the CID camera is shown as a function of position along a swath that was located parallel to the solar equator. If the MOF were responding linearly to the solar radial velocity over a range of ± 200 m/sec, then we would expect to see a perfectly straight line running from one end of the equator to the other, while the slope of this straight line would be related to the solar rotational velocity. That is, for the 150-foot tower Dopplergram of Figure 6, a slope of roughly 2 m/sec/arcsecond would be obtained near the equator.

A regression analysis of the so-called Doppler signal against the measured solar velocities along the equator would then yield a straight line whose slope could later be used to convert a particular intensity difference anywhere on the disk into an equivalent radial velocity. The degree of constancy of this calibration constant over the solar image would then be a measure of the presence or absence of systematic velocity errors introduced by the MOF. The principal objection to employing this calibration method is that it makes use of the 150-foot tower Dopplergrams to provide measurements of the very quantity which the MOF is trying to measure, namely, the solar radial velocity at each point on the visible disk.

An alternative calibration procedure is to compute an average Doppler signal for every single Dopplergram and to then compute what the predicted Earth-sun radial velocity was for the moment in time. This later calculation can be made with an ephemeris program and subsequently a regression analysis can be carried out of these Doppler signals as a function of the predicted velocity. The slope and intercept of the regression line that is so obtained will then be the calibration constants necessary for conversion of the intensity differences into radial velocities.

Two different applications of this latter approach are illustrated here in Figures 8 and 9. In the upper left panel of Figure 8 we have plotted the mean Doppler signal ($10000 [I_B - I_R \cdot [I_B + I_R]$) which

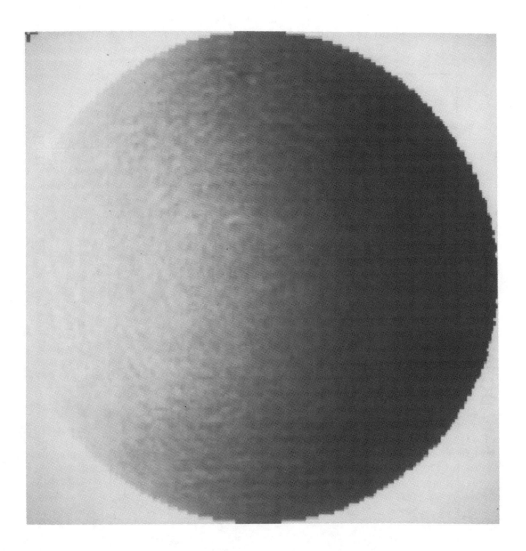

Figure 6. Comparison Dopplergram. A Dopplergram obtained in λ5250
with the spectrograph at the Mt. Wilson 150' tower telescope on 1
June 1984.

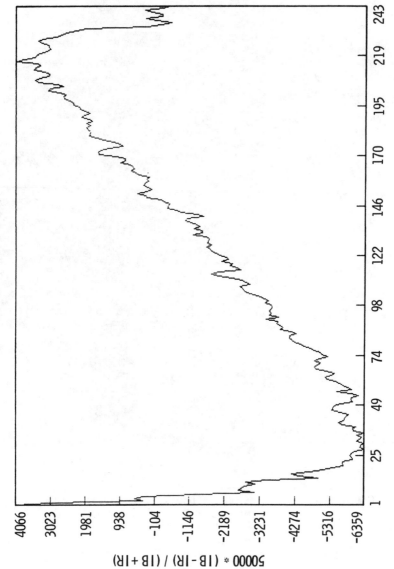

Figure 7. East-West Doppler Signal Profile. The Doppler signal (50000 • [IB-IR]/[IB+IR]) from a full-disk Dopplergram obtained with the CID camera is shown as a function of position along a swath parallel to the solar equator. Solar rotation is responsible for the slope of the profile.

we obtained by averaging the raw Doppler signal from individual pixels within a $512'' \times 512''$ area at the center of the solar disk. Here this mean Doppler signal is shown as a function of the Universal time at which each frame of data was recorded.

In the upper right panel of Figure 8 we have plotted the radial velocity of the Earth relative to the sun as a function of Universal Time. The radial velocities shown in the panel were computed with an emphemeris program which was originally developed at JPL for the interplanetary program, and which includes the effects of planetary perturbations on the Earth. The output from this JPL program was combined with computations of the Earth's diurnal motion and of the Earth's lunar barycentric motion in order to provide the total radial velocity between Mt. Wilson and the sun at every moment of time. The rotation of the Earth is clearly visible in this panel.

In the bottom panel we have plotted the mean Doppler signal versus the corresponding computed sun-Earth velocity for that moment of time. A linear regression analysis was performed upon the data shown in this panel and we obtained the regression coefficients listed at the top of the Figure. The slope of 1.222 m/sec per Doppler unit is what would later be used to convert individual pixel difference signals into radial velocities. The intercept of the regression line includes the effects of: 1) the gravitational redshift, 2) asymmetries in the solar line profile, and 3) asymmetries in the transmission profile of the MOF, as were seen in Figure 2. The correlation coefficient of 0.993 from the regression analysis shows that the Doppler/velocity conversion was indeed quite linear for this particular data.

The MOF data used to generate Figure 8 were obtained in July 1983. When we repeated this analysis with the 1984 full-disk data described in Section 2, we got the results shown in Figure 9. Here the mean Doppler signal shown in the upper left panel was obtained by averaging the Doppler signal over the entire visible disk of the sun. Again the computed sun-Earth radial velocities for the observing times of each frame are plotted at the upper right. And again, at the bottom, we show the results of plotting the mean Doppler signal against the computed radial velocity. Here the relationship is not nearly as linear as was that shown in Figure 8. While we are not yet certain of the cause of this departure from linearity, we have noticed that the discrepancy is worst at large hour angles early in the morning and late in the afternoon. Thus, we currently believe that telluric H_2O contamination in the wings of the NaD2 line, which would have the greatest effect at the largest air masses, may be introducing the time-dependent departure from linearity which we are seeing in Figure 9. We will repeat this type of analysis for every day's worth of data in order to see whether or not this departure from linearity is always present in the 1984 data. (In the meantime, we have observed only in the D2 line during 1985 since that line has much less telluric contamination.)

REGRESSION ANALYSIS

Y-INT: -. 11695D+04 +/-   .42635D+01 M/SEC
SLOPE: .12222D+01 +/-   .48351D-02 M/SEC / VEL UNITS
R: .99304D+00

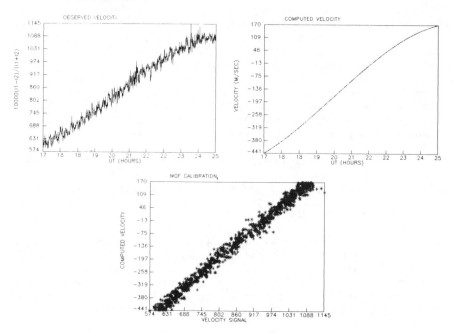

Figure 8.   Calibration of Doppler Signal. This figure shows how the
rotation and orbital motion of the Earth can be used to calibrate the
MOF's Doppler signal. In the upper left the Doppler signal obtained by
averaging over a 512" x 512" area at the center of the solar disk is
displayed as a function of Universal Time. In the upper right the radial
velocity of the Earth relative to the sun as computed from an ephemeris
program is also displayed versus versus Universal Time. At the bottom
the computed velocity is plotted against the average Doppler signal for
the corresponding moments of time. The linear regression analysis shows
that the two quantities are linearly related with a correlation coefficient of 0.993.

REGRESSION ANALYSIS

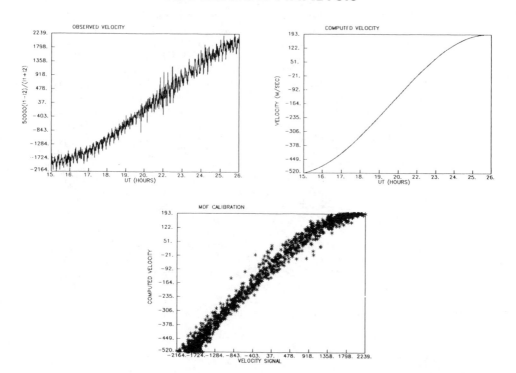

Figure 9. Re-Calibration Using Full-Disk Averages. The calibration procedure shown in Figure 8 is repeated here with Doppler signals obtained
from full-disk averages of Dopplergrams taken on 27 July 1984. Here the
departure from linearity is believed to be due to time-varying telluric
contamination of the Na D2 line.

4. INITIAL DATA ANALYSIS

4.1 Spherical Harmonic Filtering

The next step in the data reduction sequence was the spherical harmonic filtering of the various time series of Dopplergrams. In order to carry out this filtering operation, we developed a computer program with which we could project spherical harmonic functions onto the planar 240 x 240 pixel grid of the CID camera images. Since the image registration program which translated all of the Dopplergrams from a given run to a common center also computed the radius of each solar image in pixels, we needed a program which was general enough to correct for the varying size of the solar image. In addition, we wanted a program which would allow for the position angle of the solar rotation axis and for the non-zero heliographic latitude of the sub-earth point on the solar disk. We also felt that it would be desirable for the harmonic generation scheme to be accurate up to ℓ-values of 1000 or more. Thus, the program which we developed can now generate fore-shortened spherical harmonics of the appropriate size and of any orientation.

Once the spherical harmonic generator had been developed, we next used its output to perform the spatial filtering. This was accomplished by dot-multiplying each Dopplergram with a given spherical harmonic on a pixel-by-pixel basis. This dot-multiplication corresponded to a double integration of the product of the Dopplergram and the harmonic. The output of this double integration was a separate filtered time series for each individual harmonic.

An example of such a filtered time series is shown here in Figure 10. In this Figure we show the detrended amplitude of the $\ell = 10$, m = 10 sectoral spherical harmonic as a function of time on July 27, 1984. Because of the non-linearity of the time-dependent calibration analysis for this July 27 observing run, as was shown in Figure 9, we did not in fact subtract off the computed sun-Earth velocity from each individual filtered Dopplergram. Instead, we decided to employ a simple third-order polynomial detrending analysis to remove the diurnal motion of the Earth from the filtered time series until we could correctly determine how to calibrate the July 27 dataset. Therefore, the time series displayed in Figure 10 has been both filtered and detrended.

Once the detrending analysis was complete, the information which remained in the filtered time series shown in Figure 10 was the amplitude which resulted from the beating together of all of the various solar oscillation modes having the spatial pattern given by the $\ell = 10$ sectoral harmonic, as well as some cross-talk from other, nearby sectoral and near-sectoral harmonics. (An analysis is now underway to determine more precisely the amount of cross-talk expected from such adjacent spherical harmonics.)

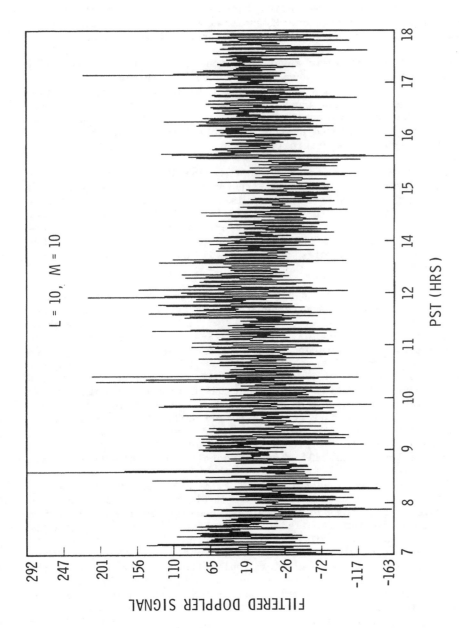

Figure 10. Filtered Doppler Signal. Here we show the result of filtering each Dopplergram obtained on 27 July 1984 with the sectoral harmonic having $\ell = 10$ and $m = 10$. The filtered time series shows the amplitude of this particular harmonic over the entire 11-hour observing day.

4.2 Power Spectral Computations

Once a filtered time series had been generated it was then a simple matter to employ a one-dimensional FFT program and compute the corresponding one-dimensional power spectrum. Three such power spectra from the July 27, 1984 run are shown in Figures 11, 12, and 13. Figure 11 shows the power spectrum which resulted from transforming the filtered time series shown in Figure 10. Therefore, the relative power levels of several $\ell = 10$ sectoral harmonics (along with some cross-talk from adjacent harmonics) are illustrated in this Figure. The power levels shown on the vertical axis are in arbitrary units since we had not employed the correct normalization constants at the time this spectrum was originally computed.

In Figure 11 the frequency unit employed is the circular frequency, ω instead of the linear frequency, ν. In these units the peak of the "5-minute" p-mode band is near $\omega = 0.020$ sec^{-1}. Due to the limited frequency resolution available from a single 11-hour observing run, the individual p-modes are not fully resolved in this spectrum. Nevertheless, the relatively low noise level of the MOF can be seen at frequencies between 0.004 and 0.016 sec^{l} and above 0.060 sec^{-1}.

At this time we should point out that the filters were operated without the benefit of any temperature stabilization scheme whatsoever. Furthermore, they were operated within the spectrograph chamber without the presence of any insulating material to shield them from ambient air temperature fluctuations. Lastly, the power supplies which controlled the heaters on the MOF cells, and hence the optical depth of the Na clouds within the cells, were not operated with any power regulation beyond that built into the supplies themselves. The rise in power level at frequencies below $w = 0.004$ sec^{-1} could be due in part to uncontrolled drifts in the power inputs to the cell heaters. (A closed-loop servo system which will control the temperature of the Na pits within the MOF cells to $\pm 0.1°C$ is now nearing completion and will soon be installed to test these ideas.)

Two more power spectral slices from the same day's data are illustrated here in Figures 12 and 13. In Figure 12 the power spectrum was completed from the dataset that had been filtered with the $\ell = m = 20$ sectoral harmonic, while in Figure 13 the $\ell = m = 40$ was used. The power levels are again not expressed in (m/sec)^2, although both power spectra should be normalized consistently with the spectrum of Figure 11. (That is to say that, while the absolute value of the normalization coefficients may not be correct, their relative values are. All of the factors which are ℓ- and m-dependent have been correctly included in the normalization coefficients already.) Note that while the noise level looks higher in Figures 12 and 13 than in Figure 11, the vertical scales are different. In fact, the absolute noise levels are quite similar in all three cases.

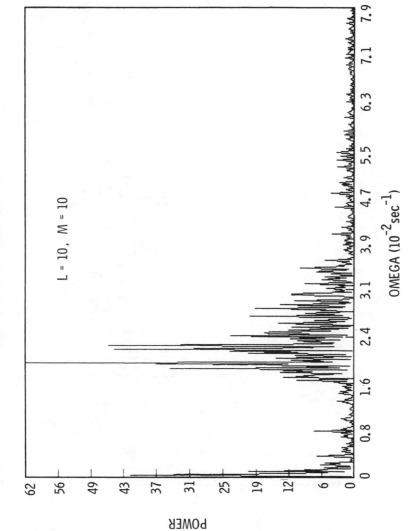

Figure 11. Power Spectral Slice. The one-dimensional power spectrum obtained by Fourier transforming the filtered time series displayed from Figure 10 is shown here. The power is plotted as a function of the circular frequency, ω, measured in units of 10^{-2} radians/sec.

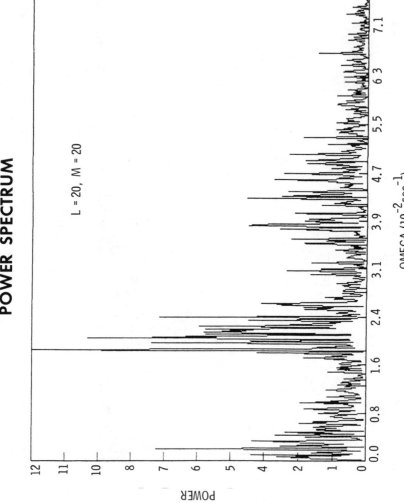

Figure 12.   Same as Figure 11 except that this spectrum corresponds to the ℓ = 20, m = 20 sectoral harmonic.

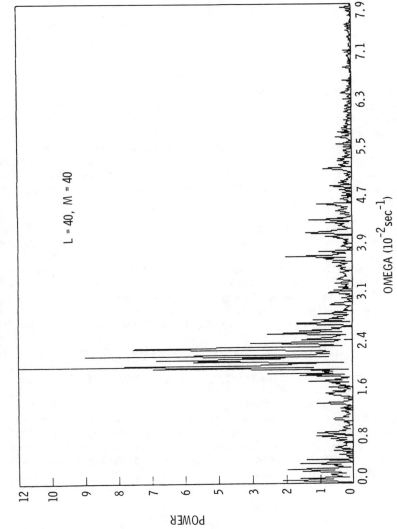

Figure 13. Same as Figures 11 and 12 except that this spectrum corresponds to the ℓ =40, m = 40 sectoral harmonic.

5.   TASKS REMAINING

    The tasks which remain in our processing of the summer 1984 data
can be devided into three major areas. First, there are several
difficulties with the preliminary processing of the data which we
have already outlined above and which we must resolve before we can
proceed. Second, there is the normous task of actually carrying out
the analysis on the large number of data tapes which we have accumu-
lated. Third, there is the scientific interpretation of the frequencies and frequency splittings which we will obtain from the power
spectra.

    There are four principal difficulties which comprise the first
group of tasks. These are: 1) the problem with the non-linearity of
the time-dependent velocity calibration analysis, 2) the search for
possible spatial non-uniformities in the calibration coefficients, 3)
the determination of the correct spherical harmonic normalization
scheme, and 4) the re-coding of the spherical harmonic generation
program and the spatial filtering program for our CSPI MAP 300 and
Mini-MAP array processors in order to cut their execution times down
to a manageable level.

    The actual production runs to be carried out in the second phase
will include: 1) the registration and calibration of the individual
Dopplergrams for many more days of observations, 2) the spatial
filtering analysis, 3) the time-synchronization of the filtered time
series for multi-day runs, 4) the application of gap-filling techniques for the nighttime data gaps, 5) the computation of the power
spectra from these multi-day filtered time series, and 6) the determination of the modal frequencies from the resulting power spectra.

    The key scientific analyses to be carried out include: 1) the
computation of the absolute frequencies for all of the modes observed
as a function of harmonic degree, 2) the comparison of these frequencies with the predictions of various solar models, 3) the computation
of the modal frequency splittings introduced by solar rotation, 4)
the inversion of the frequency splittings in order to measure internal solar rotation as a function of both depth and latitude, and 5) a
comparison of the absolute frequencies and the frequency splittings
with those from earlier epochs in order to search for possible time-
dependent changes in the sun.

6.   SUMMARY

    We have presented a review of the acquisition and preliminary
analysis of the most extensive set of imaging data yet obtained
anywhere for the study of solar oscillations. We have outlined the
unique aspects of the instrumentation which have made the acquisition of this much data possible. We have gone on to describe some
currently-unresolved difficulties with the data (as well as methods

which will hopefully allow us to resolve these difficulties). And we have presented a few preliminary results from our analyses of only a small amount of the available data. Finally, we have outlined the tasks which remain to be completed in our analysis.

ACKNOWLEDGEMENTS

The research reported herein could not have been accomplished without the assistance of the following individuals: Thomas Andrews, Thomas Bursch, Douglas Clay, and Thomas Thorpe, all of the Jet Propulsion Laboratory; John Boyden, Maynard Clark, and Harvey Crist of the Mount Wilson Observatory; Nick Magnone, Daniel McKenna, James Wilkie, and Bradley Wood of the U.C.L.A. Astronomy Department; and Jeffrey Mannan of the Orbis Corporation. We also wish to acknowledge the assistance of Mr. Enzo Zante of the Italian National Laboratory for Alternative Energy (ENEA) in Frascati who helped with the fabrication of the MOF cells. The portion of the work carried out at U.S.C. was supported in part by NASA Grant NAGW-13. The work at U.C.L.A. was supported in part by NASA Grant NAGW-472. The work carried out at the Jet Propulsion Laboratory of the California Institute of Technology, Pasadena, California, was conducted under contract with the National Aeronautics and Space Administration. The portion of the work conducted in Italy was supported by the Italian Consiglio Nazionale delle Ricerche (CNR) and by the Ministero della Pubblica Istruzione.

REFERENCES

Agnelli, G., Cacciani, A., and Fofi, M., "The MOF I: Preliminary Observations in the Na D Lines," Solar Physics, 44, 509, 1975.

Brown, T. M., these proceedings; also HAO preprint submitted to Nature, 1985.

Cacciani, A., Cimino, M., and Fofi, M., "Some Developments of the Magnetic Beam Absorption Filter," Solar Physics, 11, 319, 1970.

Cacciani, A., Croce, U., Fortini, T., and Torelli, M., "Searching for ℓ=1 Modes of Solar Oscillation," Solar Physics, 74, 543, 1981.

Cacciani, A., and Fofi, M., "The MOF II: Velocity Field Measurements," Solar Physics, 59, 179, 1978.

Cacciani, A., Fortini, T., and Torelli, M., "Na-light Flare Observations: McMath 13043-July 1974," Solar Physics, 67, 311, 1980.

Cacciani, A., and Rhodes, E. J., Jr., "The Magneto-Optical Filter, Working Principles and Recent Progress," in Solar Seismology from Space, R. K. Ulrich, ed., JPL Publication 84-84, 115, 1984.

Cimino, M. Cacciani, A., and Sopranzi, N., "A Sharp Band Resonance Absorption Filter," Applied Optics, 7, 1654, 1968a.

Cimino, M., Cacciani, A., and Sopranzi, N., "An Instrument to Measure Solar Magnetic Fields by an Atomic-Beam Method," Solar Physics, 3, 618, 1968b.

Deubner, F.-L., "Observations of Low Wavenumber Nonradial Elgenmodes of the Sun," Astronomy and Astrophysics, 44, 371, 1975.

Deubner, F.-L., Ulrich, R. K., and Rhodes, E. J., Jr., "Solar p-Mode Oscillations as a Tracer of Radial Differential Rotation," Astronomy and Astrophysics, 72, 177, 1979.

Rhodes, E. J., Jr., Cacciani, A., Blamont, J., Tomczyk, S., Ulrich, R. K., and Howard, R. F., "Evaluation of a Magneto-optical Filter and a Fabry-Perot Interferometer for the Measurement of Solar Velocity Fields from Space," in Solar Seismology From Space, R. K. Ulrich, ed., JPL Publication 84-84, 125, 1985.

Rhodes, E. J., Jr., Cacciani, A., Tomczyk, S., Ulrich, R. K., Blamont, J., Howard, R. F., Dumont, P., and Smith, E. J., "A Compact Dopplergraph/Magnetograph Suitable for Space-Based Measurements of Solar Oscillations and Magnetic Fields," in Advances in Space Research, 4, 103, 1984.

Rhodes, E. J., Jr., Howard, R. F., Ulrich, R. K., and Smith, E. J., "New Instrumentation for Solar Oscillation Measurements at Mt. Wilson Observatory," in Solar Instrumentation: What's Next?, R. B. Dunn, ed., Sacramento Peak Observatory Publication, p. 102, 1981.

Rhodes, E. J., Jr., Howard, R. F., Ulrich, R. K., and Smith, E. J., "A New System for Observing Solar Oscillations at the Mount Wilson Observatory, I. Design and Installation," Solar Physics, 82, 245, 1983.

Rhodes, E. J., Jr., Ulrich, R. K., and Simon, G. W., "Observations of Nonradial p-Mode Oscillations on the Sun," Ap.J., 218, 901, 1977.

THE K 769.9 nm LINE PROFILE

H.B. VAN DER RAAY
Physics Department, Birmingham University, U.K.

P.L. PALLE, T. ROCA CORTES
I.A.C. Universidad de La Laguna, Tenerife

ABSTRACT. The observation of a solar absorption line by means of resonant scattering over several months in the study of solar oscillations, yields data which may be analysed to provide details of the line shape over the region being studied. The asymmetry and non-linearity of the K 769.9 nm line are found and the effect on the overall response of the resonant scattering spectrometer is illustrated. Finally a simple second order correction term to the response function of the system is empirically found which greatly improves overall linearity.

INTRODUCTION

In the study of solar oscillations by optical resonance scattering, the mean line of sight velocity between the observer and the solar surface was determined continuously over $\sim$ 12 hours per day for $\sim$ 5 months. The measured line of sight velocity, V_m, may be expressed as

$$V_m = V_{spin} + V_{orb} + V_{grs} + V(t) \qquad (1)$$

Amplitude (ms^{-1}) 400   500   640   1

Period   day   year   -   minutes

where the corresponding amplitude and periods of the terms are indicated. The earth's rotational motion is represented by V_{spin}, its orbital motion by V_{orb}, V_{grs} represents the gravitational red shift and $V(t)$ denotes the solar oscillations of mean amplitude 1 ms^{-1} which are the primary motivation of the investigations.

During the present analysis V_{grs} will be assumed constant over the time period of $\sim$ 165 days considered and $V(t)$ will be neglected due to its small amplitude in comparison with the other terms in equation (1).

333

D. O. Gough (ed.), Seismology of the Sun and the Distant Stars, 333–337.
© *1986 by D. Reidel Publishing Company.*

The resonant scattering technique depends essentially on a
comparison of the light intensity at two points (a,b) on the limbs of
the absorption line, as illustrated diagramatically in fig. 1. During
any one day these points of operation
move over each limb as indicated by
the arrows in fig. 1, each scanning
a region $\sim$ 800 ms^{-1} wide. These
regions are shifted from day to day
due to the earths orbital motion
(V_{orb}) thus enabling a scan of
$\sim$ 1800 ms^{-1} to be obtained for each
limb during a 6 month season.

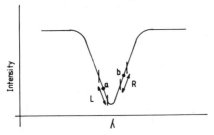

Fig. 1. Absorption line showing
region of operation during one day.

METHOD

The resonant scattering spectrometer[1] was used to obtain intensity
measurements alternately of the right and lefthand portions of the
K 769.9 nm line at 1 sec intervals for $\sim$ 12 hours per day from April
18th to September 14th 1984. These data were collected at the
observatory at Izana (2400 m) on the island of Tenerife. The total
light intensity, as defined by a 1.5 nm passband interference filter
centred on 769.9 nm, passing through the resonant scattering cell was
monitored by a separate detector. All intensity measurements were
obtained from fast photon counting, the mean count rate being 2 x 10^6
counts/sec. The number of counts recorded in each second for both the
transmitted and scattered intensities are transferred to magnetic tape
for subsequent analysis.

ANALYSIS

Considering the lefthand limb of the line, the scattered intensity
(S_L) and transmitted intensity (T_L) are both corrected for deadtimes in
the counting system and then the ratio $R_L = S_L/T_L$ is found for every two
second observation. A period of 40 seconds is then chosen and the mean
and standard deviation of R_L is found. Any individual ratio differing
by more than 2 standard deviations is discarded and a new mean found.
The alternate data corresponding to the righthand limb are similarly
treated. The time as recorded by a crystal oscillator, checked daily
against a radio time signal, is also transferred to the magnetic tape
at 40 sec intervals. This is used to calculate the mean line of sight
velocity as defined in equation 1. Any one day hence gives a portion
of the line profile extending over $\sim$ 800 ms^{-1}. By combining 4 days,
suitably spaced over the observing season the sections of line profile
shown in fig. 2a and b were obtained. The straight lines merely serve
to emphasise the non-linearity of the line at the operating points. The
greater linearity of the righthand limb results largely from the fact
that the operating point is further removed from the minimum in the
absorption line.

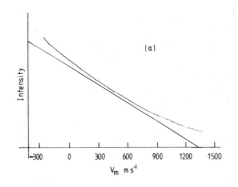 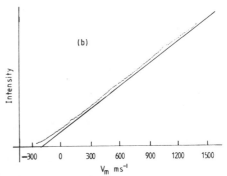

Fig. 2. The variation of intensity as a function of doppler velocity for a) lefthand and b) righthand limbs of the K 769.9 nm absorption line.

As the strength of the magnetic field applied to the vapour cell is known and hence the Zeeman splitting may be found, the composite curve illustrated in fig. 3 may be obtained. It should be noted that the line is asymmetric, there being an approximate 10% difference in the slopes of the two sides. This asymetry has been previously detected [2].

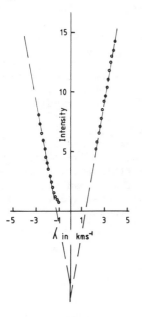

Fig. 3. Experimentally determined K 769.9 nm line profile illustrating asymmetry.

In using the spectrometer to measure velocity rather than the line profile, the ratio

$$R = \frac{S_L - S_R}{S_L + S_R} \qquad (2)$$

is used. This will only yield a linear relationship between R and velocity if the line is a) symmetric and b) linear over the operating region. The actual response function of the spectrometer is found by evaluating R as defined in equation 2 and plotting this against V_m as given in equation 1. Again the data for any one day will provide a range of V_m of ~ 800 ms^{-1} but by considering four days suitably spaced over the observing season the plot shown in fig. 4 is obtained. The dotted line is drawn merely to emphasise the non-linear response function.

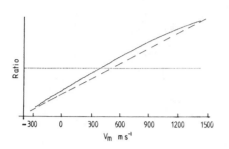

Fig. 4. Response function of
resonant scattering
spectrometer.

Although the mean slope of the
response function does not alter
greatly, when measuring solar
oscillations (V(t)) it is the
incremental slope of the response
function that must be considered.
This is illustrated in fig. 5 by
curve (a). Clearly in the early part
of the observing season when V_m is
large, the daily change of $800 ms^{-1}$
in V_m can cause changes in
sensitivity of $\sim 50\%$ during the day.
Naturally during the late summer
observing season these daily
variations in sensitivity are greatly
reduced ($\sim 10\%$).

The effect may however be largely removed by fitting the
response function by a second order polynomial and defining the ratio R'
as

$$R' = (1 + kV_m^2)R \qquad (3)$$

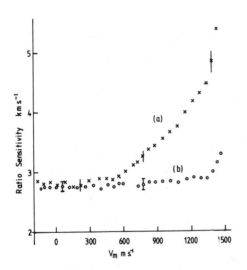

Fig. 5. Instrumental senstivity
a) normal response, b) empirically
corrected response.

With this definition of R' and an
empirically determined value of
$K = 8 \times 10^{-8}$, the modified response
shown as curve (b) in fig. 5 is
obtained. Clearly as long as values
of $V_m < 1400$ only are considered
a fairly constant instrumental
response is obtained. The
departures seen for values
$V_m > 1400$ arise since the operating
point on the lefthand limb is now
approaching the minimum in the
absorption line and a greater
magnetic field would be required to
increase the Zeeman splitting if
these large negative V_m data are
to be utilized.

CONCLUSION

The data normally obtained during the study of solar oscillations by
optical resonant scattering, may be used to give a detailed study of the
line profile over the region of operation of the instrument. This has
shown in the case of the K 769.9 nm line, that the line is both
asymmetric and non-linear over the region of interest. This in turn
leads to a non-linear response function for the instrument with
consequent sensitivity changes of up to 50% during the day in early
parts of the observing season. By fitting a simple second order term
to the instrumental response function expressed in terms of the line of
sight velocity a satisfactorily linear response is obtained provided the
line of sight velocity is limited to $V_m < 1400$ ms^{-1} with the present
instrument. Alternatively a greater magnetic field could be applied to
the resonant scattering vapour cell.

ACKNOWLEDGEMENTS

The hospitality of Professor F. Sanchez is gratefully acknowledged.
We would like to thank A. Jimenez, J. Perez and C. Regulo for assistance
in the data collection and the technical staff of the Birmingham solar
group for their invaluable contribution. This work was financially
supplied by a SERC grant.

REFERENCES

1) J. R. Brookes, G. R. Isaak and H. B. van der Raay, Mon Not R astr
    Soc 185 (1978) 1.

2) J. L. Snider, Solar Phys 12 (1970) 352.

GROUND BASED MEASUREMENTS OF GLOBAL SOLAR INTENSITY OSCILLATIONS

B.N. Andersen
Institute of Theoretical Astrophysics
University of Oslo
Norway

ABSTRACT

An instrument designed to observe the low l-value solar intensity oscillations from ground based sites is described. This instrument measures
the relative change in the line equivalent width. By taking the ratio
between the integrated line intensity and a nearby continuum the effect
of the variations in the atmospheric transmission is to a large extent
removed. The instrument is expected to have an accuracy of 2-5 ppm.

1. INTRODUCTION

The last ten years a large amount of knowledge of the solar interior has
been acquired through the measurement of solar velocity oscillations
(see references in Belevedere and Paterno 1984). These oscillations have
been measured in detail for modes of degree from zero to several
hundred. Especially the modes of low degree have yielded information of
the regions close to the solar core. For the intensity oscillations the
low degree observations have been scarce. This is mainly due to the
influence of the variations in the transparency of the terrestial atmosphere.

Up until now the most reliable observations of models with l=0,1,
and 2 have been aquired with the ACRIM instrument on the SMM satellite
(Woodard and Hudson, 1983). Other attempts have been carried out with
baloon observations (Frölich 1984), but these observations consist of
only a few measurement points. The results are therefore not very detailed with respect to amplitude and frequency information.

The informations of intensity oscillations may apart from the information of the solar interior give information of the interaction
between the oscillations and the solar atmosphere (Andersen 1984,
Frandsen 1984). They find that the amplitude of the oscillation decreases strongly from the line center to the continuum.

At Oslo Solar Observatory we have observed the amplitude dependence of the intensity oscillations through several spectral lines. Only
a small portion (30"x30") near the disc center was observed thus the

D. O. Gough (ed.), Seismology of the Sun and the Distant Stars, 339–342.
© *1986 by D. Reidel Publishing Company.*

main contributions to the signal were from modes with high l-values. In
Figure la and lb we have shown the results for two g=0 Fe I lines, 557.6
and 543.4 nm. Also shown are the theoretical calculations carried out by
Frandsen (private communication). We see that the general form is reproduced in the observations. The secondary peak seen both in the observations and in the theoretical data for the 557.6 line is not clearly
understood. The normalization of the theoretical points to the line
center is undertaken.

It is however clear that by observing the ratio of the oscillations in the line and the continuum, an amplification of the intensity
oscillations may be achieved. Also, if the continuum is close to the
line the effects of the transmission variations in the atmosphere may be
reduced. The applications of the signal is of great help since the
signal in the continuum is expected to be of the order of 1-5 ppm. The
reduction of the terrestial effects is crucial since these variations
are of the order of 0.1%.

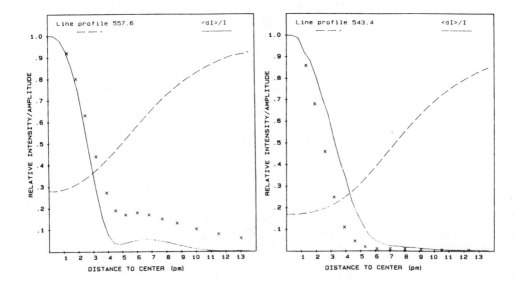

Fig. 1. The variation of the ratio of the amplitude of the intensity
oscillation in spectral line and continuum. Data for the Fe I 557.6 and
543.4 nm lines are shown. The crosses indicate the theoretical values by
Frandsen.

2. INSTRUMENTATION

We have developed an instrument to measure accurately the ratio between
two channels of photon input in the large Czerny-Turner spectrograph at
Oslo Solar Observatory.

A high astigmatic image of the sun (0.15x12 mm) is formed by two
ortogonally oriented cylindrical lenses. This image is used as entrance
slit to the spectrograph.

At the focus of the spectrograph two variable width slits are
placed 50 mm from each other along the spectrum. In the red and near
infrared this separation corresponds to a difference in wavelength of
0.8-5 nm, depending on the order used.

One of the slits is positioned such that a spectral line free of
blends is situated wholly within the slit. The other is located in a
local continuum. The continuum entrance slit is adjusted to make the
average intensity of the two channels equal. A rotating chopper (50 HZ)
placed in front of the exit slits let the single detector alternatively
detect the line and continuum signal. The DC signal from the detector is
the average of the two signals, while the AC component is a measure of
the difference between the channels. Both the AC and DC signals are
separately amplifiend and registered by a 10 bit ADC in a mini computer.
The mini-computer keeps track of the sampling and telescope guiding and
writes the two signals continously to magnetic the tape.

For the current version the total noise of the darkened detector
and amplifier system is about 5 ppm. When lighted this noise increases
by a factor of 5 and long term drifts of 50 ppm/hour are introduced.
This noise and instability is probably caused by a combination of phase
errors in the chopper and the difficulty of maintaining mechanical/
thermical stability in the optical components. These problems are currently being investigated.

The principles of operation are shown in Figure 2.

3. OBSERVATIONAL EXPECTATIONS

From the data presented Figures 1,2 we expect an amplification of the
oscillation signal from the continuum to the line core of a factor of
5-10, depending on the line parameters. Our instrument integrates the
total line intensity, therefore the expected amplification is reduced to
1.5-3.0. Taking the ACRIM data as reference, we would expect an oscillation signal of 3-6 ppm for each of the l=0, 1,2 modes of the 5 minutes
oscillations.

The geographical location of our observatory allows us to observe
up to 18 hours continously during the summer months. For a few days a
duty cycle of more than 70% may be obtained.

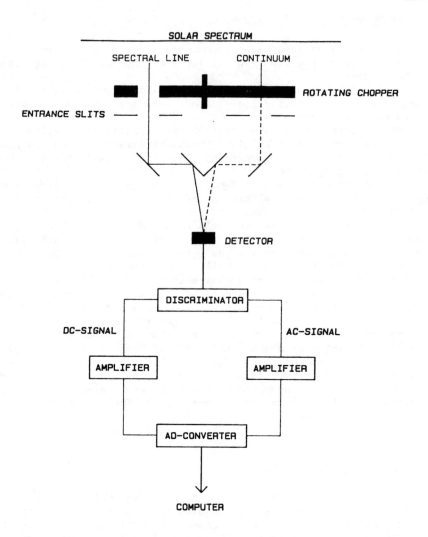

Fig. 2. The working consept of the intensity oscillation instrument at
Oslo slar Observatory

References

Andersen, B.N., 1984, Proceedings 25. Liege Astroph.Coll., 220.
Belevedere, G. and Paterno, L., 1984, "Oscillations as a Probe of the
    Sun's Interior", Mem.Soc.Astr.I. 55.
Frandsen, S., 1984, Proceedings 25. Liege Astroph.Coll., 303.
Frölich, C., 1984, Mem.Soc.Astr.I. 55, 237.

A LIQUID CRYSTAL MODULATOR

D.G. ECCLES, Y. ELSWORTH, H.B. VAN DER RAAY
Physics Department, Birmingham University, U.K.

P.L. PALLE, T. ROCA CORTES
I.A.C., Universidad de La Laguna, Tenerife

ABSTRACT. The use of a liquid crystal as an optical modulator for use with a resonant scattering spectrometer has been investigated. Preliminary data indicate that using this simple device velocity signals corresponding to 5 minute oscillations have been detected.

Introduction

The operation of resonant scattering optical spectrometers[1] depends on the alternate response of the two Zeeman components of a magnetically split line. As these two components have opposite circular polarizations, they are normally selected by either alternating the input or the scattered light circular polarization. This may be achieved mechanically by rotating a $\lambda/4$ plate, electrically by using the Pockels effect (electro optical light modulator, EOLM) or by making use of the photo elastic effect. Each of the above methods have disadvantages, mechanical motion ($\lambda/4$), high voltages (EOLM) and rapid switching (photo elastic). A far simpler, and less costly, device is based on the liquid crystal used in conjunction with a polaroid and $\lambda/4$ plate.

Method

The liquid crystal commonly used in electronic displays (LCD) has the property of rotating linear polarization by $90°$. By the application of a low voltage this rotation is removed and hence the device becomes a convenient rotator of linear polarization. When used in conjunction with two crossed polaroids the device becomes a light intensity modulator, LCM.

However by using a polaroid, liquid crystal, $\lambda/4$ plate combination as illustrated in fig. 1, the device performs

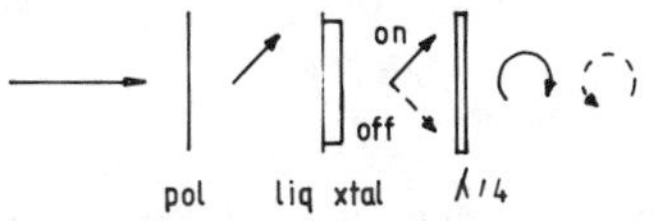

pol   liq xtal $\lambda/4$

Fig. 1. Schematic arrangement of liquid crystal modulator

343

D. O. Gough (ed.), Seismology of the Sun and the Distant Stars, 343–345.
© 1986 by D. Reidel Publishing Company.

the same function as that of the three optical modulators discussed
previously. Hence by the application of a low voltage the liquid crystal
modulator (LCM) can switch from left to right handed circular
polarization. The voltage applied to the crystal should be a.c. since
d.c. signals cause electrolytic
effects which destroy the surface
electrodes. However the optical
response does not follow the a.c.
signal if the frequency is>50 Hertz.
The measured optical response as
determined by measuring the transmission
with the crystal placed between two
crossed polaroids is shown in fig. 2.
The LCM responds in 10 ms after the
application of a 300 Hz 10^V_{pp} signal.
The device is insensitive to both
frequency in the range 50 – 5 KHz
and applied voltage 25<V>8. The decay
time, after removal of the 300 Hz

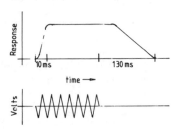

Fig. 2. Response of LCM to
applied a.c. voltage

signal is 130 ms. Hence the device may be cycled at 1 sec say, with a
dead time of 200 ms giving a 80% duty cycle. The duty cycle may of
course be increased by switching less frequently.

The transmission of the liquid crystal was measured to be in excess
of 90% and its angular acceptance, a limiting factor for many modulators
is given by a cosine squared dependence.

The EOLM with its associated 3KV EHT supplies was replaced by a
LCM and the normal resonant scattering spectrometer operating on the
K 769.9 nm line was used to record solar data. A typical velocity curve
showing the effect of the earth's rotational velocity is illustrated in
fig. 3. The fine structure on this curve corresponds to the 5 minute
solar oscillations.

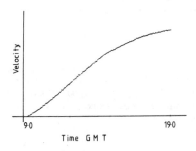

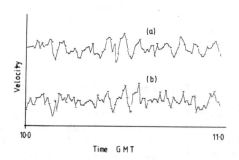

Fig. 3. Measured line of sight
velocity

Fig. 4. Comparison of 5 minute solar
oscillations from a) standard
spectrometer with b) spectrometer
with LCM

Simultaneously, another K 769.9 nm resonant scattering spectrometer was operated in an adjacent room and the direct raw data from the two spectrometers are compared in fig. 4. The correlation of the two data sets is evident.

Tests performed with a Na 589.3 nm resonant scattering spectrometer have also proved satisfactory.

Conclusion

A simple circular polarization modulator has been constructed from a commercially available liquid crystal in conjunction with a polaroid and $\lambda/4$ plate. Initial tests have shown that the device is capable of detecting 5 minute solar oscillations. The reduction in applied voltage, complexity, and cost should make the LCM an attractive alternative to the EOLM. Further tests are being carried out to establish the long term stability of the device.

Acknowledgements

The hospitality of Professor F. Sanchez at the observatory at Izana is gratefully acknowledged. The technical assistance of C.P. McLeod and the financial support of the Paul Fund of the Royal Society and the SERC made this work possible.

References

1) J. R. Brookes, G. R. Isaak and H. B. van der Raay, Mon Not R astr Soc 185 (1978) 1.

2) E. B. Priestly, P. J. Wojtowicz and D. Sheng, Introduction to liquid crystals, Plenum Press NY (1974).

A FABRY-PEROT BASED STELLAR SEISMOMETER

H.R. Butcher
Kapteyn Observatory
NL-9301 KA   Roden

T.R. Hicks
Queensgate Instruments, Ltd
112 Windmill Road
Middlesex   TW16 7HB,   UK

ABSTRACT. A simple stellar seismometer for velocity or line strength observations is described, based on a stabilized tunable Fabry-Perot interferometer. Preliminary studies indicate that a stability of near 30 cm/s over many hours should be attainable. The techniques for achieving such stability are described, and a discussion is given of the observational program possible with such an instrument on existing large telescopes.

1. INTRODUCTION

The discovery that the normal mode oscillation spectrum of the sun is observable and exhibits large numbers of very sharp lines, even in integrated light, opens the possibility that the interior structures of other stars may be inferrable directly from observation (cf Christensen-Dalsgaard 1984). If the sun is representative of near solar-mass stars generally, a new generation of instruments will be required to make these observations, devices stable to well below 1 m/s equivalent, tunable over tens of km/s to accommodate a range of stellar sources, and optically very efficient. The extreme sharpness of the modes will also make it desirable, after an initial exploratory phase of observation with existing public telescopes, to construct moderate to large diameter telescopes which are dedicated, or almost dedicated, to stellar seismology.

The only instrument currently known to be stable enough for velocity observations of the sort desired, is the alkali vapor atomic resonance cell (discussed extensively elsewhere in this workshop). The technique is a stong candidate for any program in seismology, but does suffer from restricted tunability.

During the course of a project to construct an imaging Fabry-Perot

D. O. Gough (ed.), Seismology of the Sun and the Distant Stars, 347–357.
© *1986 by D. Reidel Publishing Company.*

instrument (the TAURUS II) for the new 4.2-m William Herschel telescope
on La Palma, we have examined closely the ultimate stability likely to
be achievable at reasonable cost with the Queensgate Instruments F-P
servo system, which is based on capacitance micrometery (cf. Hicks et al
1974). That examination indicated that modest improvements in the
existing servo electronics plus very stringent control of the optical
and electronic environments would very probably lead to a velocity
stability of about 30 cm/s, for spectral resolutions near 50000. The
resulting velocity analyzer could be an efficient, versatile component
of a stellar seismometer to be used for exploring the nature of the
oscillation spectra of bright stars.

It is the purpose of this contribution to outline our ideas on the
design of a seismometer incorporating such a tunable, stabilized Fabry-
Perot, and to comment on the possibilities and limitations this kind of
instrument has for seismology when used on existing large telescopes.

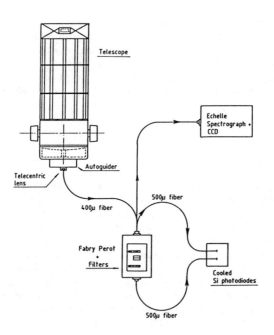

Figure 1.   Schematic layout of the seismometer.

2. THE INSTRUMENT

2.1 Operational Modes

A schematic layout of the instrument is shown in Figure 1.

Stellar light is conducted via a fiber optic cable from the telescope, to a room below the observing floor and to a thermostat containing collimating optics, one of several possible F-P etalons, and band limiting filters. The stellar light is either reflected from the front of the F-P, re-focused and inserted into one of two output fibers immediately adjacent to the input fiber, or it is transmitted through the F-P and focused into a fiber at the opposite end of the thermostat. The output fibers in turn lead either directly to discrete Si photodiodes or to a resident spectrograph with multichannel detector. This arrangement allows operation of the device in any of four modes, each of which is designed to optimize some aspect of the instrument and/or study some particular observational facet of stellar seismology. These four modes are as follows:

2.1.1 Chopped narrow passband

In this mode, the F-P transmission passband alternates rapidly between the red and blue sides of the chosen line profile. A single detector is operated synchronously with the F-P to provide a difference-divided-by-sum signal. If the chopping is rapid enough, guiding and atmospheric scintillation and seeing fluctuations will be compensated, and slow electronic drifts will be naturally removed. A typical strong line of Fe I will produce a signal of $dI/I = 5 \times 10^{**-4}$ for 1 m/s velocity amplitude. While this is almost two orders of magnitude larger than oscillation signals seen in luminosity, even in the bottoms of strong lines (cf. Andersen 1984, Frandsen 1984), very large numbers of detected photons are nevertheless required. Because of their 60 - 80% quantum efficiency and large dynamic range, we choose Si photodiodes as detectors. To minimize their contribution to system noise the diodes are run in a charge integration mode and with a double correlated sampling circuit to remove 1/f and reset noise. A noise level of below 300 electrons per readout should be readily attainable, and the instrument should remain quantum noise limited chopping at frequencies of 10 Hz or greater on Alpha Cen A or Procyon with a 4-m class telescope.

Chopping across the line profile allows one to work in a null mode, in which the line centroid is continually tracked by the F-P control servo. The mode is in principal therefore the least sensitive to detector instabilities and drifts, non-linearities in the velocity calibration, etc. A serious problem with the mode, however, may occur if the response of the F-P order sorting filter is not extremely flat across the line profile in question. The passband response of this filter to temperature fluctuations is very large, and depending on its detailed spectral response may provide the dominant component of drift.

2.1.2 Fixed narrow passband on line shoulder

This mode consists of a fixed narrow passband transmitted by the F-P and recorded by one detector, and a second much broader passband derived from the light reflected from the front of the F-P and conducted to a second detector. The narrow passband yields a velocity signal channel, and the broad one a monitor channel for signal level fluctuations unrelated to velocity. The F-P order sorting filter may be tilt-tuned in this mode to null its sensitivity to temperature, but the price paid is the necessity for 2 detectors and the resulting likelihood of relative drift at some level. Which of these first two modes of operation will ultimately prove most stable will not be known until the instrument is fully operational, being dependent on the not easily predicted limiting performance of specific components. In both of these modes two pairs of filters are provided to permit observation of at least 2 different spectral lines (e.g. one stellar line which hopefully shows oscillations, and one absorption line from the earth's atmosphere, which presumably does not), without having to disassemble the device.

2.1.3 Fixed broader passband

This mode is similar to mode 2.1.2, in that it uses a broad band off the front of the F-P, a narrower band transmitted by the F-P, and 2 detectors, but is intended to measure the intensity ratio of the bottom of a strong absorption line to the continuum, or of an emission line to the continuum. Very young stars typically exhibit washed out lines, due to rapid rotation, and are unsuitable for study in velocity space. There is evidence however that the Ca II K-line emission from active chromospheres shows enhanced oscillation amplitudes (Noyes et al 1985), which should be observable in even rather rapidly rotating, young stars. The influence and evolution of rotation are important unsolved questions in stellar evolution theory, and a capability to investigate the observability of oscillations in rotating stars is clearly of fundamental importance. Stability requirements for this mode are somewhat relaxed, and for study of the Ca II K-line it will be possible to mount the seismometer next to the telescope on the observing floor and utilize short lengths of UV transmitting fiber.

2.1.4 Multi-channel operation

Connes (1984) and others have shown that enormous gains in limiting magnitude are in principle possible if all velocity information inherent in a stellar spectrum is extracted in an optimal fashion. To permit study of multi-channel velocity measurement techniques, provision is made for conducting light reflected from the front of the F-P to a resident echelle spectrograph with CCD array detector (e.g. the CES at ESO La Silla). This light will have a series of very stable sharp absorption lines from the F-P superposed on its spectrum, which correspond in wavelength to the transmission peaks of the F-P. Spectral templates of the two spectra may be prepared by observing, for the F-P a hot, rapidly rotating OB star (which has almost no lines) or a continuum lamp, and for the star in question its own spectrum secured with an appropriately tilted mirror in the top filter slide. In this way the achievable velocity stability for a multi-channel system may be explored, with the F-P providing the "absolutely stable" component in the system. It is likely that detector instabilities will limit the performance of this mode to several m/s, but even this level of stability may prove interesting for stars in the Ap and Delta Scuti regions of the HR diagram.

2.2   Stabilization program

To achieve a stability equivalent to a few tens of cm/s, strict
attention must be paid to the limiting performance of each component of
the instrument.   In this section is described the measures being taken
to approach stability in the present seismometer.

2.2.1   Optical illumination field

It is in practice an unavoidable feature of F-P etalons that the
spectral response varies over the etalon aperture.   It is therefore of
the utmost importance that the illumination field presented to the F-P
be constant.   In addition to the obvious use of a fiber optic cable to
conduct the stellar light from the telescope to the F-P, in the process
scrambling any image structure completely, two additional tactics are
being employed.   Because fibers scramble less efficiently in input angle
than in input azimuth, it is important to re-image the telescope's pupil
to infinity and remove the angular consequences of image wander in the
input focal plane. This re-imaging is accomplished with a telecentric
beam ratio conversion lens (which converts from the telescope's intrinsic focal ratio to roughly f/3, a value transmitted by most fibers with
little degradation).   And to minimize non-uniform illumination of the
pupil caused by vignetting in the input focal plane, an autoguider is
employed together with a large diameter fiber.   The autoguider keeps the
image centered on the fiber, and the large size of the fiber ensures
that all but a few percent of the total stellar light successfully
enters the fiber.   The maximum fiber diameter will in the first instance
be determined by the size of the Si photodiode detectors, which are to
be coupled with unit magnification to the output fibers.   For a 0.5 x
0.5 mm diode, a 400 micron input fiber may be used, yielding an input
image field some 7 arcsec in diameter (on a 3.5-m telescope).

2.2.2   F-P servo electronics

The versatility of a tunable F-P carries with it the possibility
for severe stability problems.   The F-P servo control system to be used
in this seismometer employs the technique of capacitance micrometry to
monitor the etalon plate separation, which as a measurement technique is
capable of sensing displacements very much more precisely than is required here.   The details of the F-P servo system based on this technique are discussed by Hicks et al (1974).   In this section is summarized the measures being taken to stabilize an upgraded version of the
servo system currently being produced by Queensgate Instruments.

One of us (TRH) has used a combination of laboratory tests on
existing electronics and a computer circuit analysis program to derive
the drift behavior and noise characteristics of the F-P servo to be
built for the seismometer. This study showed that the dominant factor
inducing drift in the servo is temperature, and that stabilizing the
temperature environment of the electronics will be required to achieve
the desired performance. The resulting temperature drift figures are
shown in Table I, which has been prepared assuming an etalon finess of
35 and a spectral resolution of 50000.   It is immediately evident that

TABLE I

Sub-component	Temperature sensitivity	Power dissipation
Bridge drive & scan control	7 m/s/K	4 W
Error signal preamp	0 m/s/K	0.3 W
Servo loop electronics	0.4 m/s/K	2 W
Etalon plates	300 m/s/K	< 1 mW
Order sorting filter	10000 m/s/K	
Servo noise performance:	3 m/s/√Hz	

----

the bridge drive and scan control sub-unit, having 4 Watts of power
dissipation and a substantial drift coefficient with temperature, will
be the most difficult to stabilize. A circuit board design utilizing
space technology techniques and a water circulation temperature regula-
tor system are envisioned for this task.

In addition to drift control, the noise performance of the servo
is of importance. Table I also shows that the noise level of the
current servo design is such that a 1 m/s equivalent rms noise can be
achieved in under 10 sec of integration. This level is just adequate
for the present purposes, but can probably be improved in the future
with additional circuit design effort.

2.2.3 Thermostat

In Table I are also given the drift coefficients to be expected for
a fused silica etalon and a typical order sorting interference filter.
The influence of the latter could be disastrous unless its response with
wavelength is either exceedingly flat, or is carefully nulled by tilting
the filter with respect to the light beam. Etalon servo noise considerations dictate no more than a 50 micron control capacitor gap, while to
achieve the desired spectral resolution an optical gap of many times
that value is needed. Hence the expansion of the etalon plates may be
expected to produce an important component of drift, and the temperature
of the plates must be stabilized -- evidently to a milli-Kelvin, if an
overall stability near 30 cm/s is to be realized. Luckily, power dissipation from the etalon piezo stacks (less than a mW even in the chopping
mode) is the largest source of heat within 20 cm of the etalon. A
relatively straightforward thermostat may therefore be designed which
stabilizes the environment of the F-P to a few mK on a timescale of
minutes, and also seals the etalons against pressure and humidity
variations.

One thermostat design capable of this degree of temperature control
is shown in Fig. 2. In this design, an inner volume containing the F-P
etalon, filters, and collimating and camera lenses, communicates ther-

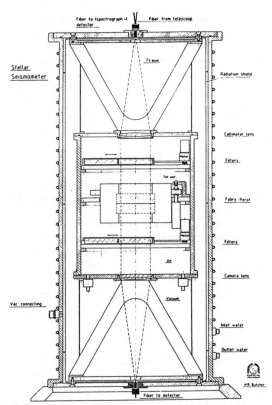

Figure 2. Schematic layout of the vacuum thermostat. Light enters at the top, and may exit either from the bottom or the top. A temperature sensor on the inner vessel controls the water temperature in the coils surrounding the outer vessel.

mally with the outside world primarily via radiation (i.e. is isolated using a vacuum space). A thermistor sensor placed on the inner can controls the temperature of water flowing in a coil encircling the outer can. Preliminary calculations show that variations of a tenth of a degree in the water temperature translate into variations of about a milli-degree inside the inner can, and the long thermal time constant of the etalon plates ensures that they vary in temperature even less.

It may be possible to design a non-vacuum thermostat which can also realize the needed thermal control. Such systems appear always to require rather complex multi-stage servos, but do show promise of being able simultaneously to perform some of the servo electronic temperature control tasks.

2.2.4 Detectors

As discussed in section 2.1.1, cooled Si photodiode detectors are to be employed in the seismometer because their quantum efficiencies are factors of 3 to 5 greater than those of photomultipliers. They are to

be run in a charge integration mode, much like low noise CCD or CID
solid state array detectors. Experience has shown that this kind of
circuit design is much stabler and has lower noise than the phase sensitive detection schemes commonly used with photodiodes. Slow drift
contributions typically do remain, due to the temperature sensitivity of
various of the electronic components, but because these drifts are
primarily in the electronic offset (as opposed to the gain), they may be
defeated by reading the detector twice in quick succession following an
integration, differencing the results to obtain a signal with offset
removed.

Finally, to stabilize the optical input field onto the detectors,
it is planned to attempt to bond the diodes directly onto the fiber
ends. Preliminary tests indicate that this scheme is likely to work
satisfactorily, but at the time of writing it is not known what the
effects of repeated temperature cycling will be. In particular, it is
possible that the core-cladding interface of the fiber itself may
deteriorate upon temperature cycling, thus degrading the coupling
efficiency. If direct bonding to the fiber ends is ultimately excluded,
an ellipsoidal mirror can be used to image the fiber ends onto the
diodes, allowing the components to remain at different temperatures.

3. OBSERVATIONAL PROSPECTS

While stellar seismology holds promise of greatly improving the
theory of stellar evolution, the practical difficulties are substantial.
Observationally, to be confronted include the instrumental problems of
achieving adequate velocity stability, the political/social difficulties
of gaining access to telescopes for long periods, and perhaps most
difficult of all, the simple collection of adequate numbers of photons
during the phase coherence lifetimes of the oscillations. The philosophy
of the present instrument is that enough photons are available with a
single channel device and a 4-m class telescope, over several nights of
observation, to allow a first study of the very brightest stars. It
becomes important in this situation to understand just what may be
learned from observations of the quality likely to obtainable, and from
the sample of stars which can be observed at sufficient S/N.

With a resolution of a micro-Hz or greater in the oscillation
spectrum (i.e. for a contiguous observing period of weeks per object),
it is possible to study the rotational splitting of individual lines,
and even begin to discuss rotation as a function of radius in the star
(cf. Duvall and Harvey 1984). Unfortunately, until a successful multi-
channel seismometer is operating on a dedicated rather large telescope,
there seems little chance of performing this kind of measurement on
stars other than the sun.

For resolutions near 5 micro-Hz (i.e. several nights of observation), however, one has in principle enough photons with existing telescopes and the seismometer design presented here to acquire interesting

observations. By interesting in this context is meant determination of individual frequencies sufficiently accurately to be able to perform an asymptotic fit of the kind described by Christensen-Dalsgaard (1984). The parameters describing the run of frequencies may be used to derive the mass and age of an isolated star, independently of knowledge of its distance -- the $\Delta\nu$o - Do diagram as discussed by Christensen-Dalsgaard. This diagram may ultimately replace the color-magnitude diagram as the pivotal contact between stellar evolution theory and observation, if a way can be found to secure the needed data.

Alpha Cen A and Procyon appear to be bright enough for this quality of data to be obtainable with present and forseeable facilities, and a convincing placement of these two stars in the $\Delta\nu$o - Do diagram is a principal goal of our seismometer program. If the instrument works well, the main difficulty will be accounting for the day-night sidelobes in the spectrum, which are unavoidable with a single, earth-bound instrument. Because this side lobe pattern is well known, however, it should not by itself preclude obtaining an overall fit to the run of frequencies.

With one night of observation, or the incoherent addition of multiple single nights, a resolution of 50 micro-Hz is attainable, which is sufficient to detect the sequence of regularly spaced low degree modes expected from near-solar type dwarfs. The mean spacing of these peaks is a measure of the sound travel time to the stellar core, and hence the radius of the star. If useful S/N must be obtained by reducing the resolution even further, it remains of considerable interest to investigate the total power and bandwidth of the spectrum of excited modes as a function of stellar parameters, to help elucidate excitation and damping mechanisms. In Fig. 3 is shown all near-solar type dwarfs brighter than V = 4.0 mag (data from Eggen 1964) and free from contaminating companion stars within roughly 15 arcsec or 10 mags of the main star. These stars are bright enough to yield data at least on excitation mechanisms, and in some cases also on the stellar radius.

If future efforts towards multi-channel seismometers and dedicated large telescopes bear fruit, this sample of stars will also be adequate for an extensive discussion of most aspects of stellar seismology. Objects of all ages are represented, and the interplay of excitation by the H - He ionization zone and by whatever excites the 5 min p-mode oscillations may be studied. Not plotted in Fig. 3 for technical reasons, but clearly lying somewhere among the stars shnown, are a bright Am star (HR 8278) and a bright Ap star (HR 5463), the latter being known to exhibit short period p-mode oscillations (Kurtz and Balona 1984). The opportunity thus also exists to explore the behavior of oscillations in abnormal situations, which possibility has often in the past been very helpful in understanding normal objects. The most important difficulty in all these discussions, over and above the current worries about how to get adequate S/N in the observed spectra, will likely be an insufficiently accurate knowledge of the stellar composition.

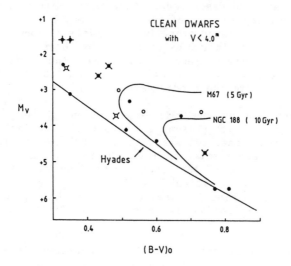

Figure 3. Bright solar type dwarfs having no contaminating companion.
Observed cluster isochrones indicate likely ages. Plus signs indicate
known Delta Scuti stars, crosses suspected ones. Open circles are drawn
for stars for which potential compromising problems have been reported,
but which are not yet clearly unsuitable for seismological observation.

ACKNOWLEDGEMENTS

We wish to thank P. Atherton, K. Reay, and J. Ring for many helpful
discussions concerning the performance characteristics of the Queensgate
Instruments, Ltd, Fabry-Perot systems; and J. Christensen-Dalsgaard and
E. Fossat for explaining much of the reasoning behind current practice
in solar and stellar seismology, and also for communicating results
before publication.

REFERENCES

Andersen, B.N. 1984, in ´Theoretical Problems in Stellar Stability and
        Oscillations,´ Proc. 25th Liege Colloquium, Univ. de Liege, p. 220;
        and this workshop.

Christensen-Dalsgaard, J. 1984, in Proc. Workshop on Space Prospects in
        Stellar Stability and Oscillations, ed. F. Praderie, Meudon; and
        this workshop.

Connes, P. 1984, in Proc. Workshop on Space Prospects in Stellar
    Stability and Oscillations, ed. F. Praderie, Meudon; and this
    workshop.

Duvall, T.L. and Harvey, J.W. 1984, Nature, 310, p. 19.

Eggen, O.J. 1964, Astron. J., 69, p. 570.

Frandsen, S. 1984, in ʻTheoretical Problems in Stellar Stability and
    Oscillations,ʼ Proc. 25th Leige Colloquium, Univ. de Liege, p. 303;
    also this workshop.

Hicks, T.R., Reay, N.K., and Scaddan, R.J. 1974, Journal of Physics E:
    Scientific Instruments, 7, p. 27.

Kurtz, D.W., and Balona, L.A. 1984, M.N.R.A.S., 210, p. 779; also Kurtz,
    this workshop.

Noyes, R.W., et al 1984, Ap.J., 285, p. L23.

APPLICATIONS OF THE MAGNETO-OPTICAL FILTER TO STELLAR PULSATION MEASUREMENTS

Edward J. Rhodes, Jr.
Department of Astronomy
University of Southern California
Los Angeles, California   90089
      and
Space Physics and Astrophysics Section
Jet Propulsion Laboratory
California Institute of Technology
Pasadena, California 91109

Alessandro Cacciani
Department of Physics
University of Rome
Rome, Italy

Steven Tomczyk
Department of Astronomy
University of California at Los Angeles
Los Angeles, California   90024

ABSTRACT. A proposed method of employing the Cacciani magneto-optical filter (MOF) for stellar seismology studies is described which is different from that employed by Fossat and his colleagues in Nice. The method relies on the fact that the separation of the filter bandpasses in the MOF can be changed by varying the level of input power to the filter cells. With the use of a simple servo system the bandpass of a MOF can be tuned to compensate for the changes in the radial velocity of a star introduced by the orbital motion of the earth. Such a tuned filter can then be used to record intensity fluctuations through the MOF bandpass over an extended period of time for each given star. Also, the use of a two cell version of the MOF makes it possible to alternately "chop" between the bandpass located in the stellar line wing and a second bandpass located in the stellar continuum. Rapid interchange between the two channels makes it possible for atmospheric-introduced noise to be removed from the time series.

D. O. Gough (ed.), Seismology of the Sun and the Distant Stars, 359–375.

I. INTRODUCTION

As reported recently (Fossat et al., 1983; Fossat et al., 1984; Fossat, 1986), Fossat has employed a single-cell version of the magneto-optical filter (Agnelli, Cacciani, and Fofi, 1975; Cacciani and Fofi, 1978; and Cacciani and Rhodes, 1984), to search for pulsations in α Centauri and in Procyon. In this paper we will describe an alternative method of employing the magneto-optical filter (hereafter referred to as the MOF) to make such observations which should, in principle, be more sensitive and flexible than the method used by Fossat and his colleagues.

In the method of operation employed by Fossat's group a single bandpass is employed in each of the two NaD lines to transmit starlight to a photomultiplier tube. Each one of these 80 mÅ-wide bandpasses is located at the center of the corresponding D line in the laboratory frame of reference. In order to position these bandpasses at the steepest portion of the stellar Na D line profiles, Fossat and his colleagues then allow changes in the radial velocity of the star under study to shift the stellar line profile relative to the filter profiles. This type of "tuning" of the MOF relative to the stellar line means that a given star can only be observed for a very small number of nights each year. Otherwise, the changes in the earth-stellar radial velocity which are introduced by the changes in the component of the earth's orbital velocity along the direction to that star become large enough to move the two stellar D lines by a substantial amount relative to the corresponding filter bandpasses. This wavelength shift in turn means that each filter bandpass travels over such a wide range of wavelengths relative to its own width that the sensitivity of the filter to the star's radial velocity changes with time. Hence, it is not possible to employ such "orbital motion" tuning to obtain time series of more than a few days in length during which the velocity sensitivity of MOF is kept constant.

An additional problem with the use of a single bandpass at each spectral line is the difficulty of removing the scintillation noise introduced by the earth's atmosphere from the time series of intensity fluctuations transmitted through that bandpass. Even when a second detector is employed to monitor the integral starlight falling on that detector as a function of time, differences in the acquisition times of the two channels or in the location observed (in the case of so-called "dual-beam" photometers which chop against a nearby patch of the sky) can make the removal of the telluric atmospheric noise difficult. Before going on to describe the method in detail, we wish to stress that our ideas are based upon our experience gained with the MOF in the study of solar oscillations and they have not yet been tried out for the stellar case. Nevertheless, we feel that the ideas are sufficiently novel to be described here.

2. PROPOSED METHOD

During the course of evaluating the suitability of the MOF as a Doppler analyzer for the study of solar oscillations, we obtained spectral scans of the transmission profiles of the MOF for varying levels of input power and varying magnetic field strengths. These profile scans were obtained with sodium-filled MOF cells using the spectrograph of the Mt. Wilson 150-foot tower telescope.

As is shown here in Figure 1 these spectral scans indicate that the wavelength separation, the peak transmission, and the halfwidth of each filter bandpass can be altered by altering the amount of power which is input to the heater coils which vaporize the sodium. In Figure 1 we show four different spectral scans for a single Na MOF cell which was embedded in a longitudinal magnetic field having a strength of 1000 Gauss. As the input power level is increased, the two narrow bandpasses located on the sides of the Na D1 line can be seen to separate, to increase in height, and to increase in width.

The sensitivity of the wavelength separation of the two peaks to the amount of input power is summarized in Figure 2 for two different MOF cells. Here we have plotted one-half of the total peak-to-peak separation of the two bandpasses (i.e., the wavelength separation between the line center in the laboratory reference frame and one of the peaks) converted into an equivalent radial velocity measured in km/sec.

Figure 2 illustrates that the two filter bandpasses can be separated by up to 17 km/sec when the field strength is only 1000 gauss. This figure also shows that the dependence of the peak separation upon the input power is nearly linear and is very similar for both the Na D1 and D2 lines.

In Figure 3 we show that the peak transmission of the filter bandpass increases slightly with increasing input power. In Figure 4 we show that the fullwidth at half-maximum (FWHM) of each bandpass also increases as the level of input power is increased. Here the relatively small scatter in the two D1 curves suggests that the bandpasses are actually somewhat broader in D2 than they are in D1.

Figure 5 is a spectral scan of the MOF at the D1 line which is similar to those shown in Figure 1, except that here the magnetic field strength has been increased from 1000 to 6000 Gauss. This profile shows that at some power settings it is possible to get similar-appearing profiles even though the field strengths are considerably different.

Na D1 FILTER PROFILES
IBI = 1000 GAUSS

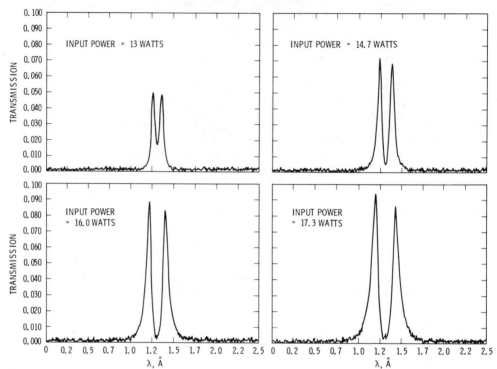

Figure 1:   Transmission profile scans of the MOF in the Na D1 line.
These measurements were made with the pit spectrograph of the Mt.
Wilson 150-foot tower telescope. They were made for four different
input power settings. A longitudinal field strength of 1000 Gauss was
employed for all four scans. A single MOF cell was used in all four cases,
so that one filter bandpass is shown for each wing of the Na D1 line.

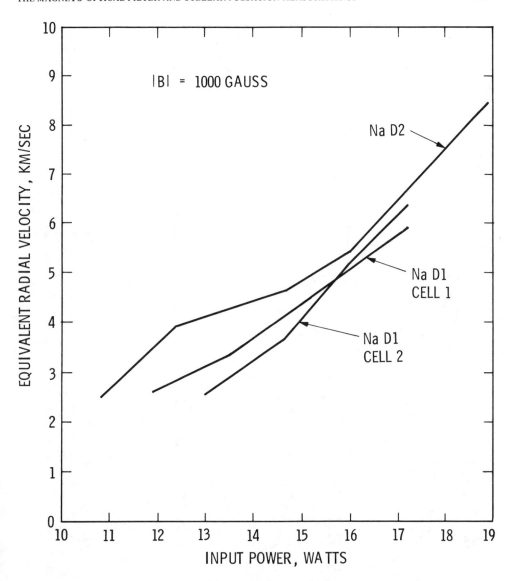

Figure 2: One-half of the peak-to-peak wavelength separations of the two transmission bandpasses is plotted here as a function of the input power applied to the cells. The wavelength separations have been converted into their equivalent radial velocities relative to the line center. The results from two different cells are shown for the D1 line, while the results from one of those two cells is also shown for D2.

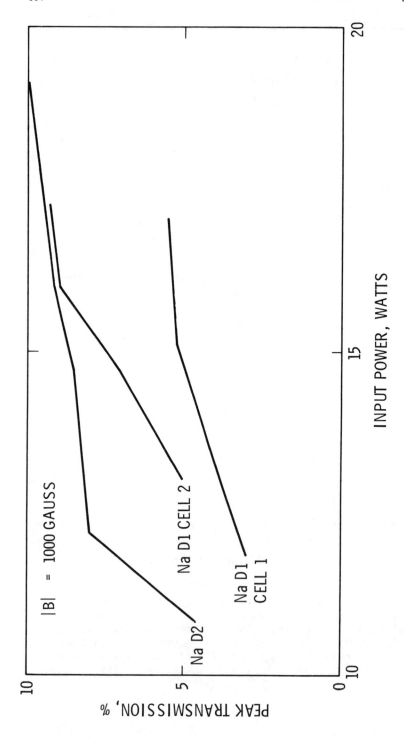

Figure 3: The percentage transmission at the peak of the MOF bandpasses shown in Figure 1 is plotted as a function of the input power applied to the cell. Also shown are similar curves for another cell at the D1 line and for cell #2 at the D2 line. The increase in transmission percentage with increasing power which was evident in Figure 1 is also seen here.

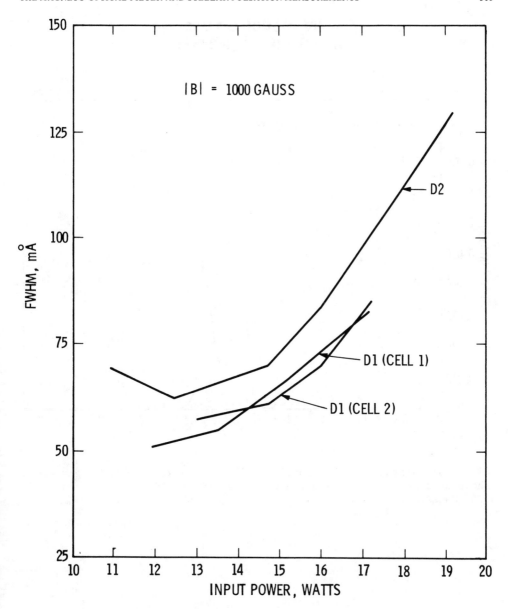

Figure 4: The fullwidth at half maximum of one of the two filter bandpasses shown in Figure 1 is shown here as a function of the input power for a field strength of 1000 Gauss.

|B| = 6000 GAUSS
Na D1 FILTER PROFILE

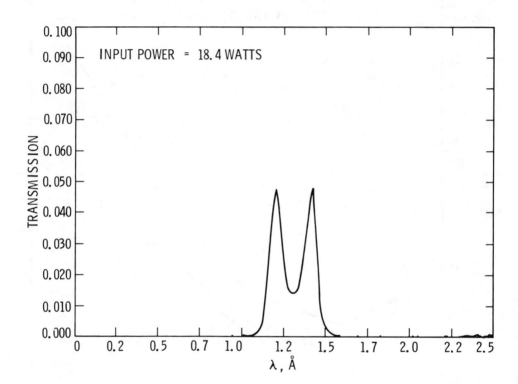

Figure 5:   A transmission profile similar to those shown in Figure 1 is shown here but in this case the peak magnetic field strength was 6000 Gauss.

Figure 6, on the other hand, shows that when the input power is increased beyond a threshold level, the transmission profile of the MOF becomes more complicated. Here the individual local maxima are caused by multiple Faraday rotations of the phase of polarization of the beam within the cell. This is due to the so-called "Macaluso-Corbino Effect" described in the earlier papers on the MOF. (We note here that the end windows of the particular cell employed in the measurements shown in Figure 6 had become contaminated from operation at very high power levels and hence the peak transmission values shown of about 5% are much too low. Consequently, the vertical scale of this Figure should be ignored.)

Figure 7 shows spectral scans for the Na D2 line again at a field strength of 6000 Gauss. The two bandpasses can again be seen to move apart smoothly with increasing power and to be much cleaner than the D1 profiles shown in Figure 6.

The dependence upon input power of the center-to-peak separations of the bandpasses illustrated in Figures 6 and 7, again expressed in terms of the equivalent radial velocity of each peak from the line center, is summarized in Figure 8. Here we note that the D1 profiles can be separated by up to $\pm$ 50 km/sec from the line center by simply varying the level of power input to the cell heaters. Figure 8 shows that the tuning of the filter (at least for Na D1) is quite linear with input power and that the bandpasses can be moved to compensate for the 29 km/sec orbital velocity of the earth around the sun. Figure 8 also shows that the D2 bandpasses are not as widely separated as are the D1 bandpasses for a given level of input power.

For those stars whose heliocentric radial velocity never exceeds about 20 km/sec the data in Figures 6, 7, and 8 suggests that use of Na D2 alone might yield cleaner filter bandpasses and higher peak transmissions. However, before any final choice could be made, additional MOF spectral scans should be obtained to confirm the differences shown in these Figures.

In order to exploit the properties of the MOF just described for stellar pulsation studies, we would propose to employ a two-cell version of the filter (see e.g., Cacciani and Rhodes, 1984; and Rhodes et al. 1984) to "chop" between the red-and blue-wing bandpasses of the filter. The transmission profile from such a two-cell MOF would be similar to that shown in Figure 9. Here the dashed curve represents the filter profile when only the "red" bandpass is transmitted, while the dotted curve shows the profile when only the "blue" bandpass is being transmitted.

Na D1 FILTER PROFILES
IBI = 6000 GAUSS

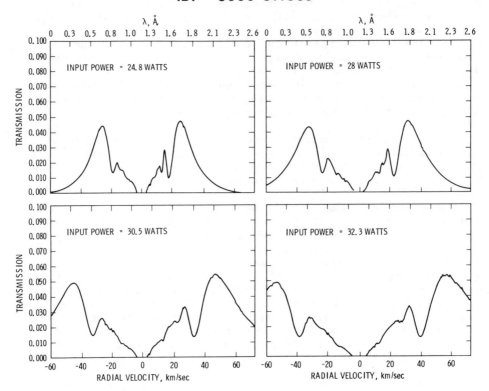

Figure 6:   Same as Figure 5 but for four higher input power levels. The profiles are more complicated here because, at the higher optical depths generated by the higher input power levels, the transmission peaks due to several complete Faraday rotations appear in each wing.

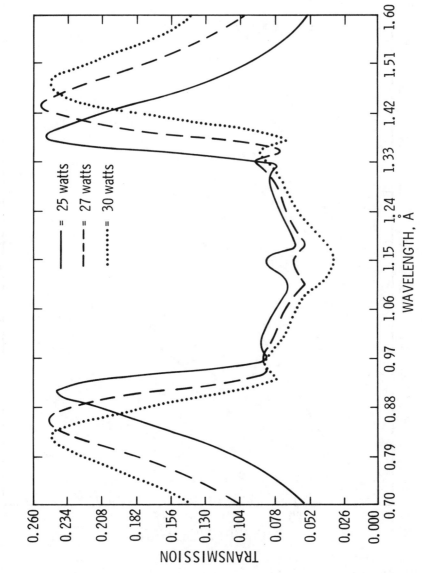

Figure 7: Same as Figure 6 except that the Na D2 line was employed instead of the D1 line as in Figures 5 and 6.

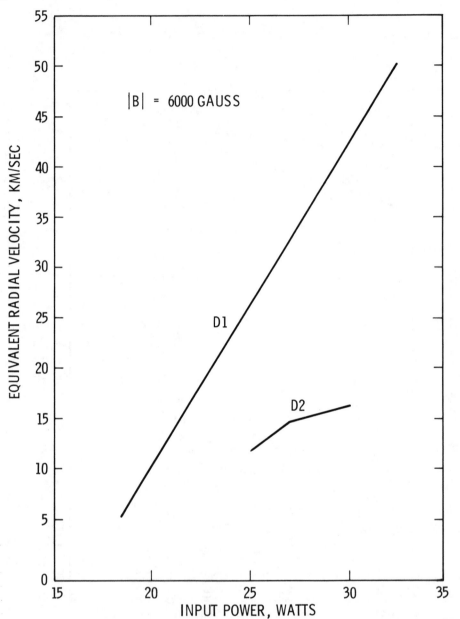

Figure 8:   Same as Figure 2 except that the field strength is 6000 instead of 1000 Gauss. Large stellar radial velocity shifts could indeed be compensated for by adjusting the level of the input power each night and then stabilizing that power level for the duration of that night.

SOLAR AND FILTER PROFILES

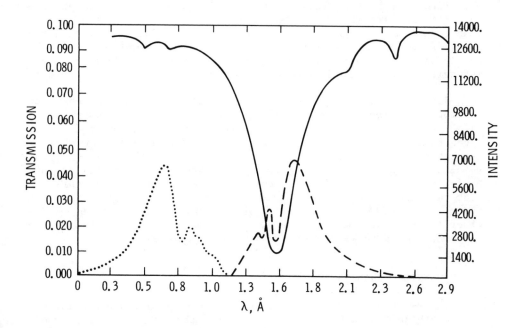

Figure 9:   The result of employing a two-cell filter rather than a one-cell
filter is shown schematically. Here the dashed profile represents the
transmission of the filter at one moment in time, while the dotted profile
shows that only the other bandpass would be transmitted at a later moment.
By electronically or mechanically tuning the two-cell filter, it is
possible to "chop" sequentially between these two bandpasses. The solar D1
line profile is shown at the top for comparison. It has been shifted
redward by 0.37A to simulate a stellar radial velocity of 19 km/sec.

Superimposed upon the MOF transmission profile scans is a red-shifted profile of the solar D1 line. Clearly, a two-cell filter system could chop between one side of the stellar line and the stellar continuum. By recording interspersed intensity time series with the filter alternating between the two bandpasses, we would be able to obtain one signal that is sensitive to the Doppler shift of the Na line and another signal that is a record of telluric atmospheric changes. By simply dividing the first of these two time series by the second, we would then be able to remove the telluric-induced intensity shifts recorded in the chosen bandpass to first order.

Returning once again to the ability of this proposed scheme to put together time series stretching over many days, we note that a slight alteration in the level of input power to the cells from one night to the next would allow us to keep the bandpass fixed with respect to the stellar line profile even though the star's radial velocity was apparently being changed by the earths orbital motion. A simple closed-loop servo system would be employed each night to keep the input power level steady throughout that night, and then the mean power level would be adjusted prior to the next evening's observing. Thus, the input power would make a series of discrete jumps but would be kept stable between each change.

3. MECHANICAL CHANGES REQUIRED

The relatively high power levels necessary to reach the largest wavelength shifts shown in Figure 8 mean that the MOF cells would have to be run at relatively high optical depths. Hence, unless, an alteration is made in the physical construction of the cells themselves, there is a reasonable chance that the end windows of the cells could become clouded in an unacceptably short interval of time. Therefore, we are proposing that the physical arrangement of the MOF cells be altered for the stellar case. Our proposed alterations are illustrated schematically in Figure 10. Hence we sketch the existing solar MOF design at the top and our proposed stellar design at the bottom. In the solar case the need to produce an image of the extended solar disk limits the length-to-diameter ratio of the cells. Thus, the existing metal pole pieces cannot become very long before the solar beam is partially obscured by the sides of the pole pieces.

Since the pole pieces are employed in part to keep the end windows relatively cool and to provide a place for the vapor in the cell to deposit before it reaches the end windows of the cell, it becomes important to move the end windows farther from the heated "pits" (i.e., the sources of the vapor) as the optical depth inside the cells is increased. Thus, with the high optical depths envision-

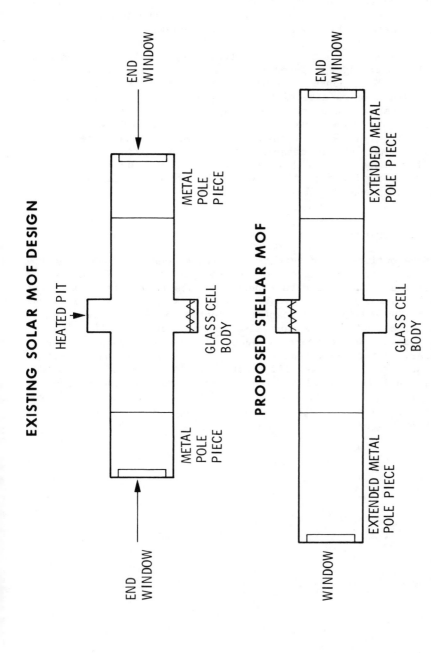

Figure 10: (top) Sketch of existing MOF cell design as is currently employed at Mt. Wilson for solar observations. (bottom) Proposed alteration in the cell design for the stellar application. The pole pieces would be longer in order to keep the end windows further away from the heated pits containing the Na.

ed for the stellar case, we are proposing that extended pole pieces
be employed which would in fact move the end windows well away from
the vapor sources. Since in the stellar case we would not be imaging
an extended object like the sun, the larger length-to-diameter ratio
of such a modified cell should not pose a problem.

We should also point out here that additional photons can be
captured by observing the intensity in both Na D lines simultaneously, albeit at the expense of a more complicated filter profile. By
simply removing the calcite plate which allows us to choose which of
the two D lines we are using at any one moment, we can indeed employ
both lines at the same time, as long as the prefilter we employ is
at least 15 A wide. We have in fact employed this dual-line mode of
operation in many of our measurements of solar oscillations during
the past few years, and we believe that the confusion introduced by
the more complex filter profiles should be offset by the added photons that would be available.

SUMMARY

We have outlined an alternative method of employing the magneto-
optical filter for observations of stellar pulsations. The method we
have described should allow an observer to observe a single star for
more than a few nights each year and should also allow for the monitoring of the telluric atmospheric contamination to the stellar
intensity time series. We have also indicated the modifications to
the filter design which would be necessary to allow us to operate in
this manner. Before we intend to try out the scheme we have described at a large telescope, we will first fabricate a few of the revised
MOF cells and build the closed-loop servo system for the control of
the input power. If laboratory tests with this hardware yield promising results, then we will attempt to test our ideas with actual
stellar observations.

REFERENCES

Agnelli, G., Cacciani, A., and Fofi, M. "The MOF I: Preliminary observations in Na D Lines," Solar Physics, 44, 509, 1975.

Cacciani, A., and Fofi, M., "The MOF II: Velocity Field Measurements," Solar Physics, 59, 179, 1978.

Cacciani, A., and Rhodes, E. J., Jr., "The Magneto-optical Filter, Working Principles and Recent Progress," in Solar Seismology from Space, R. K. Ulrich, ed., Jet Propulsion Laboratory Pub. 84-84, 115-123, 1984.

Fossat, E., these Proceedings, 1986.

Fossat, E., Decanini, Y., and Grec, G., "First Test of the New Spectrometer Devoted to Stellar Seismology," The Messenger, 33, 29-30, 1983.

Fossat, E., Gelly, B., and Grec, G., "Five-Minute p Modes Detected in Doppler Shift Measurement on Alpha Centauri," in Theoretical Problems in Stellar Stability and Oscillations, Proceedings of 25th Liege International Astrophysical Colloquium, A. Noels and M. Gabriel, eds., University de Liege, Cointe-Ongree, Belgium, pp. 229-233, 1984.

Rhodes, E. J., Jr., Cacciani, A., Blamont, J., Tomczyk, S., Ulrich, R. K., and Howard, R. D., "Evaluation of a Magneto-optical Filter and a Fabry-Perot Interferometer for the Measurement of Solar Velocity Fields from Space," in Solar Seismology from Space, R. K. Ulrich, ed., Jet Propulsion Laboratory Publication 84-84, pp. 125-155, 1984.

ACKNOWLEDGEMENTS

The research reported herein could not have been accomplished without the assistance of the following individuals: Thomas Andrews and and Juergen Linke, both of the Jet Propulsion Laboratory; John Boyden, Maynard Clark, and Harvey Crist of the Mount Wilson Observatory; and Nick Magnone and James Wilkie of the U.C.L.A. Astronomy Department. We also wish to acknowledge the assistance of Mr. Enzo Zante of the Itanlian National Laboratory for Alternative Energy (ENEA) in Frascati who helped with the fabrication of the MOF cells. The portion of the work carried out at U.S.C. was supported in part by NASA Grant NAGW-13. The work at U.C.L.A. was supported in part by NASA Grant NAGW-472. The work carried out at the Jet Propulsion Laboratory of the California Institute of Technology, Pasadena, California, was conducted under the contract with the National Aeronautics and Space Administration. The portion of the work conducted in Italy was supported by the Italian Consiglio Nazionale delle Ricerche (CNR) and by the Ministero della Pubblica Instruzione.

A NEW FABRY-PEROT RADIAL VELOCITY SPECTROMETER

K. A. R. B. Pietraszewski, N. K. Reay and J. Ring
Astrophysics Group
Blackett Laboratory
Imperial College
London
SW7 2BZ

ABSTRACT. We have developed a Fabry-Perot Interferometric Stellar Oscillation Spectrometer (FP-ISOS) to measure the radial velocity curves of ultra short-period variable stars and to search for solar-type oscillations in stars. The Fabry-Perot etalon is servo-stabilised and the instrument operates under full microprocessor control.
  Sensitivity to radial velocity is limited ultimately by electronic noise in the servo-control system. For an etalon of 100 micron spacing (resolving power $\sim$12000 in the mid-visible) a noise limited sensitivity of $1ms^{-1}$ requires an integration time of approximately 30 minutes.
  The instrument has been used at the 1.5m IRFC (Tenerife) and on the 1.9m SAAO telescope (Sutherland). A photon shot noise limited sensitivity of a few metres per second was achieved on bright stars. On the basis of these observations we expect to achieve a sensitivity of $1ms^{-1}$ on $m_v = 0$ stars in $\sim$30 hours using a 4m telescope.

1. INTRODUCTION

The correlation mask technique (Griffin 1967) forms the basis of the most widely available instrument for measuring changes in stellar radial velocity. Perhaps the most accurate instrument using this technique is CORAVEL (Baranne et al, 1979) which, on relatively bright stars, has achieved a velocity sensitivity $\sim$125ms^{-1}. The sensitivity appears to be limited by instrumental drifts.
  The development of optical resonant scattering methods for measuring solar oscillations (c.f. Brookes et al, 1976) has demonstrated that velocity sensitivity of millimetres per second can be achieved providing photon flux levels are sufficiently high.
  The resonance scattering technique is not widely applicable to stars because the flux levels are too low and the vapour cell passband can only be tuned over a very limited range. It is suitable only at specific times of the year, when the difference between the Earth's orbital velocity and the stellar radial velocity is within the

377

D. O. Gough (ed.), Seismology of the Sun and the Distant Stars, 377–389.
© 1986 by D. Reidel Publishing Company.

the tuning range of the cell, for bright stars which have alkali
metal lines in their spectra.

In this paper we describe a Fabry-Perot Interferometric Stellar
Oscillation Spectrometer (FP-ISOS) in which the etalon spacing and
parallelism are servo-stabilised using a CS100 (Queensgate Instruments
Ltd.) stabilisation system (Atherton et al, 1981). The method of
measuring radial velocity is similar to the vapour resonance cell
method, but overcomes the luminosity and tunability shortcomings.

The instrument is more stable than the correlation mask
spectrometers described in the literature and more luminous than the
vapour resonance methods used so successfully on the Sun. In its
present form it is comparable in accuracy to the coude-Reticon method
described by Campbell and Walker (1979), but it is distinctly more
luminous. It has an intrinsic sensitivity comparable to that of the
Michelson-ISOS which has achieved sub-metre per second accuracy
(Pietraszewski et al - these proceedings).

2. MEASUREMENT OF RADIAL VELOCITY

FP-ISOS is a scanning spectrometer and as such it samples spectral
points sequentially.

As for the vapour resonance cell, the most effective way of
measuring changes in radial velocity is to monitor the intensities
at the points of inflexion on an absorption line profile.

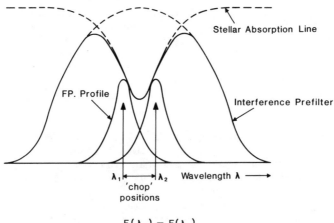

$$R(t) = \frac{F(\lambda_1) - F(\lambda_2)}{F(\lambda_1) + F(\lambda_2)}$$

Figure 1. Mode of operation. The absorption line is isolated by a
narrow-band interference filter and the Fabry-Perot passband is
'chopped' between the points of inflexion.

The technique is illustrated in Figure 1 which shows an idealised
absorption line profile as seen through a narrow band premonochromator

filter. Superimposed is the Fabry-Perot passband which is switched between the points of inflexion in order to alternately sample the intensities at these points. A change in the position of the line will cause an increase in the intensity at one of the points of inflexion and a corresponding decrease in the intensity at the other.

To investigate periodic changes in radial velocity we construct, as a function of time, the ratio:

$$R = (I_a - I_b)/(I_a + I_b)$$

Division by the sum of the intensities $(I_a + I_b)$ compensates for fluctuations in atmospheric transparency and the like which would otherwise be interpreted as changes in radial velocity. Fourier Transform techniques are used to search for periodicities in the ratio 'R'.

The sensitivity to radial velocity is determined by the slope of the absorption line at the point of inflexion - the steeper the line the more sensitive to radial velocity it is. However, the precision of the measurements depends also on the signal at that point, clearly, as the photometric accuracy increases so does the accuracy with which the absorption line wavelength can be determined.

An analytical solution which assumes that the filter and absorption line profiles are Gaussian, and the Fabry-Perot is an Airy function, shows that maximum sensitivity is achieved when the instrumental passband width is comparable to the width of the absorption line. This is not unexpected in view of the 'matched filter' theorem.

Adopting a continuum flux level at the top of the atmosphere of $3.7 \ 10^{-9}$ erg.cm^{-2}.Å^{-1}.s^{-1} for a zero magnitude star, and assuming that the absorption line is 50% deep, 1 Å wide and that the observation is made at a resolving power $\sim$10000 with a 4m telescope, an observing time of $\sim$30 hours will be required to achieve a sensitivity of 1ms^{-1}.

The highest radial velocity periodicity that can be determined depends on the frequency at which the Fabry-Perot passband is switched between the points of inflexion on the line. By the 'sampling theorem' the shortest period recoverable from the data is equal to the time needed to make two determinations of the ratio 'R'. The frequency resolution in the data depends on the length of the data window.

3. THE INSTRUMENT

We adopted the philosophy that the instrument should be as simple as possible, with the minimum of moving parts and optical elements to minimise mechanical flexure, positional non-reproducability, and unnecessary light losses. The effect of environmental changes on the etalon, the filter and the CS100 servo system were also important considerations in evolving the layout shown schematically in Figure 2.

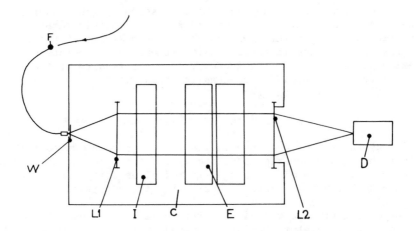

Figure 2. Schematic of the instrument. Light from the telescope is
directed to the instrument by a fibre optic cable. The etalon is
sealed in a constant volume chamber which can be located some way
from the telescope.

3.1 Optical Layout

Light from the telescope or calibration source is directed to the
interferometer by an optical fibre (F) the output end of which is
the entrance aperture for the interferometer. A standard photographic
lens (L1) collimates the light through a narrow-band interference
filter (I) and the Fabry-Perot etalon (E). A simple lens (L2)
focusses the light onto the photomultiplier detector (D). The etalon,
interference filter and collimator are housed in a rigid cylindrical
chamber (C), sealed by a small window (W) after the entrance aperture
and by the simple lens (L2). This eliminates the effect of
environmental changes on the etalon and filter.

3.2 The Etalon and Pre-filter

The choice of etalon gap (i.e. spectral resolving power) depends
largely on the width of the absorption line to be used in making the
observations. For δ-Scuti and Ap stars the 5317.4Å FeI absorption
line has a width in the range 0.7 to 1.3Å and a depth of about 80%.
A computer simulation of the sensitivity of the instrument to radial
velocity changes in this line showed that the optimum resolution is
∿0.5Å, corresponding to an etalon gap of 100µm.
   With a finesse of ∿28 the Fabry-Perot (FP) orders repeat at
intervals of approximately 14Å. To eliminate all but the central
order, a three cavity interference filter of half power bandwidth
10Å was placed in the collimated beam before the etalon.

3.3   The Fibre

There are two important reasons for using fibres. The first is that the instrument need not be mounted on the telescope. It can be permanently located in a stable, vibration free environment remote from the observing floor.

The second is that the fibre scrambles the stellar image to produce an evenly illuminated entrance aperture, eliminating the spurious radial velocity changes that might otherwise be caused by movement of the image.

We chose to use a QSF 400/600 ASW step index fibre. The core has a diameter of 400 μm which defines an entrance aperture size which is well within the Jacquinot criterion (Jacquinot 1954).

3.4   The Detector

An RCA 31034A/02 photomultiplier was used because it has a high quantum efficiency and low dark current when cooled. The photomultiplier was used in photon counting mode and when cooled to $-78°C$ had a dark count rate of 2 to $4s^{-1}$.

4.   STABILITY REQUIREMENTS

4.1   Environmental Effects

Changes in temperature, pressure and relative humidity will affect the capacitance micrometer error detectors and alter the balance conditions for the servo-control system. In practice the etalon will remain aligned providing all gap sensing capacitors are equally influenced.

The servo-control system cannot, however, compensate for changes in the optical spacing of the etalon due to the dependance of refractive index on these three parameters.

From the Fabry-Perot equation:

$$2 \mu t \cos\theta = \lambda n$$

and the Doppler relationship:

$$\delta\lambda/\lambda = \delta v/c$$

where 't' is the etalon spacing, 'λ' the wavelength, 'θ' the off-axis angle, 'n' the order of interference, 'μ' the refractive index of the medium between the plates and 'δv' the change in velocity corresponding to a change in the wavelength 'δλ' of the line, it follows that:

$$\delta(\mu t)/\mu t = \delta v/c$$

where 'c' is the velocity of light. A sensitivity of $1ms^{-1}$ is seen to require a stability of the etalon gap of 1 part in $3 \cdot 10^8$.

Using the expression relating refractive index to temperature and pressure given by Allen (1973), for an etalon with a nominal spacing of 100 µm the change in optical spacing is equal to $-\lambda/5000$ (λ = 500nm) for a 1 degree temperature increase at standard temperature and pressure.

A change in pressure of 1 mm Hg at standard temperature and pressure results in an optical spacing change of $+ \lambda/13000$, and an increase in relative humidity from 63% to 64% causes a change in optical spacing of $+\lambda/500,000$ (λ = 500nm).

These environmental effects can be eliminated by evacuating the etalon or by sealing it in a constant volume enclosure in a dry atmosphere. We have adopted the second option because it is the simpler of the two to implement.

Assuming a gas within a sealed chamber, the optical spacing will change by only $-\lambda/1,000,000$ (λ = 500nm) for a temperature increase of one deg.C at standard temperature and pressure. This is equal to a stability in the optical spacing of 1 part in 2.10^8 for a 100µm etalon gap and, from above, is equivalent to a velocity stability of ~ 1 ms^{-1}.

In practice apparent velocity changes caused by effects such as drift in the position of the filter passband and distortion of the etalon plates due to temperature gradients require the etalon to be controlled to ~ 0.01 deg.C.

4.2 Electronic Stability

Electronic noise generated in the CS100 servo-control system manifests itself as a random 'jitter' on the etalon plates. Measurements by the manufacturers of the system (Queensgate Instruments Ltd.) show that the noise has an amplitude of 1 nm peak-to-peak at a time constant of 0.2 ms. For an etalon of spacing 100 µm, this corresponds to a velocity uncertainty of ~ 40 ms^{-1} in a 1 s integration time, or 1 ms^{-1} in 1800 s (~ 30 min).

In practice it is photon shot noise which has determined our velocity sensitivity in observations so far (see section 7) and not instrumental noise.

5. CONTROL AND DATA ACQUISITION

An AIM65 microcomputer was used to control the CS100 servo electronics, to count pulses from the photomultiplier, to send the data to a digital cassette recorder and visual display unit via specially constructed interfaces.

5.1 Mode of operation

After setting on a star the stellar absorption line was located and identified by rapidly scanning the FP passband over a narrow spectral region centred on the line. The interference filter was tuned to be

symmetric with respect to the line, and a 'chop' sequence was
initiated in which the interferometer alternately sampled each of the
two points of inflexion of the line and the intensities I_a and I_b at
these points were measured. These values are recorded on digital
cassettes and the ratio $R = (I_a - I_b)/(I_a + I_b)$ was displayed on
chart as a direct measure of the radial velocity.

Rapid chopping between the points of inflexion helps reduce
fluctuations in I_a and I_b caused by 'seeing' and changes in overall
atmospheric transparency. In practice 'chop' frequencies of 5 Hz to
10 Hz are used depending on the brightness of the star.

5.2 Calibration

The ratio R is calibrated in terms of radial velocity by putting
known wavelength offsets into the CS100 controller, causing the
'chop' positions to be displaced from the points of inflexion of the
line hence simulating a change in radial velocity.

Computer simulations using the known line shape and instrument
passband function are used to cross-check the calibration.

6. OBSERVATIONS

The principle objective of the observations made so far has been to
commission the instrument. Eight stars have been observed, each for
a period of a few hours. Four of the eight have known radial
velocity periodicities and two (αCir and 44Tau) have photometric
periods. αCar is of spectral type F0 and has the virtue of being
very bright.

Selection criteria limited stars to cooler than spectral type
B to be sure of the presence of the 5317.4$\mathring{A}$ FeI line, and to stars
brighter than $m_v = 6$ to ensure an adequate supply of photons.

Table I lists the stars observed and summarises the data
obtained. These objects were observed on either the 1.9m
telescope at SAAO in February 1983 or on the 1.5m IRFC at the
Observatorio del Teide in October 1983.

Table 1. Observations

Star	m_V	Sp. Type	Known Periods /Amp.	Periods Found /Amp.	Noise (ms^{-1})
αCar	-0.9	F0II	none	none	30
αCir	3.19	F0p	6.8min(P)	6.8min(?)	160
ρPup	2.81	F5IIp	3.38hrs /∼10kms^{-1}	3.39hrs /9 kms^{-1}	60
44Tau	5.41	F2IV-V	3.48hrs(P) 2.51hrs(P)	2.88hrs /13 kms^{-1}	50
14Aur	5.06	A9V	2.1hrs	2.09, 2.5, 1.07hrs /2,9,0.48 kms^{-1}	64
δDel	4.43	F0IVp	3.24hrs /5.5 kms^{-1}	3.244hrs /4.9 kms^{-1}	58
βCas	2.27	F2II-IV	2.5hrs /2 kms^{-1}	2.5hrs /7.08 kms^{-1}	60
βAql	3.71	G8IV	none	none	42

(P) - photometric period only

6.1 Results and Discussion

(a) αCar. (Canopus) This star provided a good first object, being
of spectral type F0 and very bright. There are no known periodicities
in radial velocity. Total observing time was four hours and the
photon shot noise limited sensitivity was 30 ms^{-1}. At this level
there were no significant radial velocity periodicities on a
timescale down to one minute.

(b) αCir. A photometric period of 6.8 mins was discovered by Kurtz
and Cropper (1981) who classified αCir as a rapidly oscillating
Ap star.
    The radial velocity curve and its Fourier Transform are shown
in Figure 3. The observation lasted 2.7 hours and achieved a photon
shot noise limited sensitivity of 160 ms^{-1}. Although there is a peak
at 6.8 mins there are other equally significant looking peaks all
of which look like noise. It is sufficiently interesting, however,
to warrant further investigation.

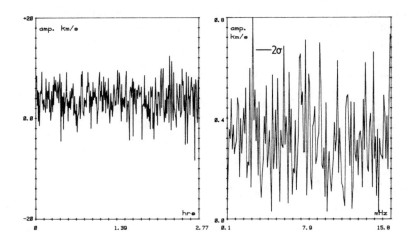

Figure 3. αCir. Radial velocity curve (left) and Fourier Transform
(right). Although there is a peak at 6,8 mins (the photometric
period) we cannot claim it to be significant. It is sufficiently
interesting, however, to warrant further investigation.

(c) ρPup. This star has a very regular 3,38 hour sinusoidal
oscillation. It has been observed by Reese (1903), Struve et al (1956),
Eggen (1956a), Danziger and Kuhi (1966) and Balona and Stobie (1983).
Our new data confirm these previous observations,

(d) 44Tau. There have been several photometric observations of
44Tau since Danziger and Dickens (1967) observed its variability.
Morguleff et al (1976) observed both light and radial velocity curves
but revealed no periodicities present in the pulsation of the star.
Percy and McAlary (1974) identified two periods, 3.48 and 2.51
hours,
    Our radial velocity curve is shown in Figure 4 together with
its Fourier Transform. The velocity spectrum is dominated by the
peak at 2,6 hours. This period agrees well with the shorter of the
periods found by Percy and McAlary. There is little evidence for a
3,48 hour oscillation.
    44Tau was the faintest star observed with the interferometer
and provided an excellent test of its performance. A precision of
50 ms^{-1} was achieved in a little over 5 hours,

(e) 14Aur. We observed 14Aur for 4.27 hours attaining an accuracy
of 64 ms^{-1}. We find a 2.09 hour period, (Figure 5), consistent with
previous observations by Danziger and Dickens (1967). There is also a
slope corresponding to a change of 6–8 kms^{-1} over the data. This is
consistent with the binary nature of this star.
    The Fourier Transform also shows peaks at 2.51 hours and at 1,0

hour at the 3σ level, This suggests that there are two further modes
present in the pulsation, a fact noted by Fitch and Wiśniewski (1979).

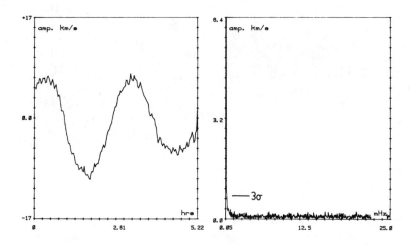

Figure 4.   44Tau.   Radial velocity curve (left) and Fourier Transform
(right), There is no evidence for periods shorter than the
fundamental at 2.88 hours.

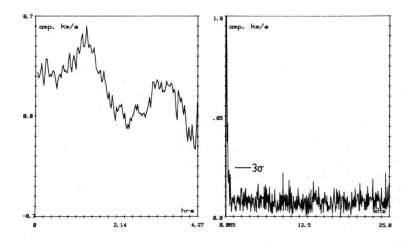

Figure 5.   14Aur, Radial velocity curve (left) and Fourier Transform
(right), The main peak in the Fourier Transform is at 2.09 hours and
there are other peaks at 2.5 and 1.07 hours,

(f) δDel. Eggen (1956b) reported the variability of δDel which
was subsequently observed by Wehlau and Leung (1964) and Kuhi and

Danziger (1967), Duncan and Preston (1979) concluded that δDel is a member of a binary system in which both companions are pulsating,
   Our new data shows an oscillation of period about 3,24 hours which agrees well with previously published values, The Fourier Transform of the velocity curve shows no other significant features.

(g) β Cas. The radial velocity curve of β Cas is of particular interest as it can be compared directly to the results of Yang et al (1982) who used the coude-Reticon method of Campbell and Walker (1979), Our new data have a precision of 60 ms^{-1} and a time resolution of about 40s, whereas Yang et al achieved an accuracy of 0.2 kms^{-1} with 900s integration time, Yang et al observed β Cas at the 3,6m CHFT giving them a 5.7 fold advantage in telescope area over the 1,5m IRFC used for our observations,
   Fourier analysis showed no other oscillations present in the data,

(h) β Aql, This star was observed for up to 3.28 hours on 4 separate nights, A precision of 42 ms^{-1} was achieved and no significant periods longer than 60 s were found in the data,

7. CONCLUDING REMARKS

We have demonstrated that FP-ISOS is sufficiently stable and luminous to measure periodic stellar radial velocity changes with amplitudes as small as a few metres per second. Although in its present state of development it is poised in accuracy between that of the coude-Reticon method and the ultra-high accuracy of the Michelson ISOS (Forrest 1982, Pietraszewski et al - these proceedings) or the resonant scattering method, it is intrinsically capable of achieving an accuracy better than 1 ms^{-1} on relatively faint (m$_V$∿5) stars.
   Our observations so far are all photon shot noise limited; we have not yet reached the limit imposed by electronic noise in the servo-control system. An immediate improvement in sensitivity could be achieved by collecting more photons! Using a 4m telescope would give a factor 4-7 increase in flux levels compared with 1.5-1.9m telescopes we have used so far. It is also realistic to observe for several nights rather than the 3-5 hours typical in the above data.
   Work is currently underway to improve the throughput of the instrument by up to 100-fold by replacing the interference filter pre-monochromator with an echelle spectrometer from which we will fibre feed the Fabry-Perot spectrometer with up to 100 spectral lines simultaneously, This technique will improve our present photon shot noise limited sensitivity by a factor ten. The combination of this with the use of larger telescopes and longer observing periods will enable us to achieve 1 ms^{-1} on m$_V$∿5 stars in ∿30 hours of observing time, Whilst this is some considerable way from the predicted sensitivity of, for example, the Connes 'Absolute Accelerometer' (Connes - these proceedings) it is a sensitivity we know we can achieve based on experience with a working instrument,

Finally, there are a number of developments in hand which will improve the intrinsic stability of the instrument.

A feasibility study has shown, for example, (see Butcher and Hicks - these proceedings) that the noise level in the CS100 control system can be reduced from its present equivalent level of 1 ms^{-1} in 1800 s to 1 ms^{-1} for integration times of 10 s or so. A super-CS100 should be available within a year.

We also propose to adopt a 'belt and braces' approach by adding an optical reference to the instrument. Periodic reference of the etalon spacing to a frequency stabilised laser will provide a cross-sheck on the capacitance servo-control system for those long term observing programs (e.g. searches for Jupiter-like planets) which will require a stability of metres per second over a decade or more.

ACKNOWLEDGEMENTS

K. A. R. B. Pietraszewski was in grateful receipt of a SERC studentship for the duration of this project.

REFERENCES

Allen C.W., Astrophysical Quantities, 1973, Athlone Press, University of London

Atherton, P.D., Reay, N.K., Ring, J. and Hicks, T.R., 1981, Optical Eng., 20, 806.

Balona, L.A. and Stobie, R.S., 1983, S.A.A.O. Circulars, 7, 19.

Baranne, A., Mayor, M. and Poucet, J.L. 1979, Vistas in Astronomy, 23, 279.

Brookes, J.R., Isaak, G.R. and van der Raay, H.B., 1976, Nature, 259, 92.

Campbell, B. and Walker, G.A.H., 1979, Publ. Astron. Soc. Pacific 91, 540.

Danziger, I.J. and Kuhi, L.V., 1966, Astrophys. J. 146, 743.

Danziger, I.J. and Dickens, R.J., 1967, Astrophys. J. 149, 55.

Duncan, D.K. and Preston, G.W., 1979, Bull. American Astron. Soc. 11, 729.

Eggen, O.J., 1956a, Publ. Astron. Soc. Pacific 68, 238.

Eggen, O.J., 1956b, Publ. Astron. Soc. Pacific 68, 541.

Fitch, W.S. and Wiśniewski, W.Z., 1979, Astrophys. J. 231, 808.

Forrest, A.K., 1982, Ph.D. Thesis, Univ. London.

Griffin, R.F., 1967, Astrophys. J. 148, 465.

Jacquinot, J.P., 1954, J. Opt. Soc. America, 44, 761.

Kuhi, L.V. and Danziger, I.J., 1967, Astrophys. J. 149, 47.

Kurtz, D.W., 1982, Mon. Not. R. astr. Soc. 200, 807.

Kurtz, D.W. and Cropper, M.S., 1981, Inf. Bull. Var. Star number 1987.

Morguleff, N., Rutily, B. and Terzan, A., 1976, Astron. Astrophys. Suppl. Ser. 23, 429.

Percy, J.R. and McAlary, C.W., 1974, The Observatory, 94, 225.

Reese, H.M., 1903, Lick Obs. Bull. 2, 29.
Struve, O., Sahade, J. and Zebergs, V., 1956, Astrophys. J. 124, 504.
Wehlau, W. and Leung, K.C., 1964, Astrophys. J. 139, 843.
Yang, S., Walker, G.A.H., Fahlman, G.C. and Campbell, B., 1982,
Pub. Astron. Soc. Pac. 94, 317.

INTERFEROMETRIC STELLAR OSCILLATION SPECTROMETRY

K. A. R. B. Pietraszewski, J. Ring and A. K. Forrest
Imperial College
Prince Consort Road
London
SW7 2BZ

ABSTRACT. The Michelson Interferometric Stellar Oscillation
Spectrometer (ISOS) has been developed and used over the past several
years. It has the capability of measuring the relative radial velocity
curve of a bright star to an accuracy of $\sim 1 ms^{-1}$ or less after a few
nights on a moderately sized telescope. This is barely sufficient to
detect oscillations in Solar type stars. A truly positive detection
of such oscillations may be achieved with the increased signal obtained
by using more than the one absorption line used at the moment.

We present here a description of the instrument, the method by
which it measures relative radial velocities, a discussion of some of
the data obtained with it and a brief outline of how the accuracy can
still be improved.

1. INTRODUCTION

The discovery and further studies of the Solar 5 minute oscillations,
such as those of Grec et al (1980), prompted the development of the
Michelson ISOS in order to search for similar oscillations in Solar
type stars.

At that time the usual method of obtaining radial velocity
curves of stars was the correlation spectrometer method, Coravel of
Baranne et al (1979), being a recent example. However, the mechanical
stability of such instruments is not good enough to allow their use in
the search for Solar type oscillations.

In recent years several accurate radial velocity spectrometers
have appeared, including the alkali metal vapour cell of Fossat et al
(1983) and the absorption cell method of Campbell and Walker (1979).

We have developed two interferometric systems; the Michelson
instrument described here and its sister instrument based on a
Fabry-Perot described in an accompanying paper elsewhere in these
proceedings.

D. O. Gough (ed.), Seismology of the Sun and the Distant Stars, 391–402.

2. THE MICHELSON INSTRUMENT

The interferometer used in the instrument is a solid, field-widened, constant path length, Michelson interferometer. An interferometer was chosen for the purpose rather than another system because of its advantages in luminosity (Jaquinot, 1954), flexibility in choice of resolution and hence application (choice of absorption line), and in the possibility of building a compact, stable and accurate instrument. The Michelson interferometer had a further advantage of being easily field-compensated. Starlight is fed to the instrument by a fibre optic cable from the Cassegrain focus of a large telescope.

The main features of the instrument are shown in Figure 1. The beam splitter is a coating enclosed between the hypotenuse faces of two large 45° prisms to form a cube. Cube corner retro-reflectors form the mirrors and displace the incoming and returning beams allowing both the input and both the output ports to be used while still using a single beam splitter/combiner. Four 45° prisms are used to steer the beams through the interferometer. The amount of glass in each arm has been arranged to give field-compensation, i.e. the path-difference is made insensitive to small changes in input angle.

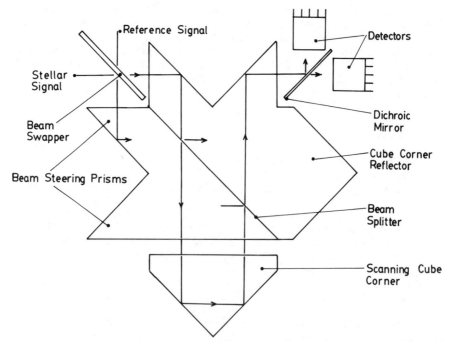

Figure 1. Main Features of the Michelson ISOS

Despite it being described as a constant path difference system the path difference of the interferometer can be either changed through about 6 fringes or kept constant with respect to a reference wavelength.

This is achieved by mounting one cube corner reflector on piezo-electric   transducers, which are employed as described later.

A rotating circular glass plate, one semi-circle of which is silvered, swaps the two input beams between the input ports at a rate of 25 Hz and eliminates inhomogeneity and differences between the two paths, differences in the sensitivities of the two sets of detectors and any relative drifts in these detectors. The electronic photon counting channels are also swapped at the same rate and the counts are directed to the correct counter. This has the effect of an average detector measuring the flux in each arm.

The stellar signal comprises a single absorption line isolated by an interference filter whose passband can be tuned by tilting it through a small angle in order to straddle the line. The signal and reference wavelengths are separated at the detectors by dichroic mirrors.

The general design of the block makes for a very stable instrument largely because of the solid construction of the interferometer which automatically compensates for such effects as thermal expansion.

The detectors are cooled RCA 31034/AO2 GaAs photomultipliers used in photon counting mode.

An onboard computer manages the functions of the instrument, the user interface, pre-reduction and data transfer and general housekeeping. A second computer calculates, on-line, the radial velocity curve and the Fourier transform of the most recent 2 hours of data.

3.1 Method of Operation

The ISOS instruments use the Doppler effect to measure the relative radial velocity curves of stars by following the changes in position of an absorption feature in the stellar spectrum. The essence of the technique is to monitor the position of the line by observing the intensity of the light at the points of inflection on either side of the line. Calculating the difference in intensity then gives a measure of the change in radial velocity.

In the case of the Fabry-Perot (FP) instrument this is done by chopping the FP passband between the points of inflection as shown in Figure 2. Rapid chopping of the passband is required to avoid fluctuations in seeing etc. influencing the results.

The action of the Michelson ISOS may be explained as follows:- If the path difference is held constant, then the transmission through the instrument as a function of wavelength is a set of pass-bands whose half-width is equal to one-half of their separation. If a few of these bands centred on a stellar absorption line are isolated by a narrow-band interference filter the transmitted spectrum is as shown in Figure 3.

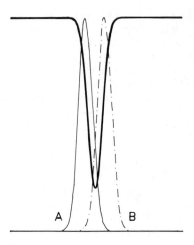

Figure 2.   Fabry-Perot pass-bands superimposed on an absorption line.

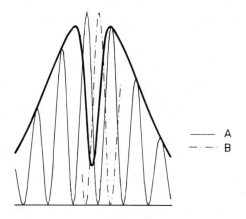

Figure 3.   Michelson pass-bands superimposed on an absorption line.

Since the energy entering the instrument is conserved, wavelengths
which are absent in one output beam (A) are transmitted by the
other (B).   The path difference of the interferometer may be varied
by the piezo-electric transducers until each output beam has a
transmission peak approximately centred on a point of inflection of
the absorption line.   The adjacent pass-bands in each beam transmit
light from the continuum which contributes photon-shot noise but
no signal.   This reduces the advantage of simultaneously observing
both points of inflection.

The reference wavelength referred to earlier (a monochromatic line from
a stabilized laser or stable spectrum lamp) is simultaneously
transmitted by other pass-bands close to its wavelength. The
path-difference of the interferometer is also chosen so that the two
output beams transmit equal intensities at the reference line. The
difference of these intensities is then used to control the voltage
applied to the transducers. The path-difference in the interferometer
is maintained constant as a number of wavelengths of the reference line,
thus removing thermal or mechanical drifts. The two criteria for
choosing the path difference can be satisfied within a range of tuning
of a few wavelengths.

3.2 The accuracy of measurements

The resolution of the interferometer is at present optimised for the
5317A line which has a half-width of 0.1 to 0.4A, depending mainly on
rotational broadening. The resulting path difference is about 5mm and
the order of interference is about 10^4. A change in radial velocity
of 1 in 10^4 ($\sim$30 kms^{-1}) will move the stellar line to the next order,
giving an identical condition to that shown in Fig. 3. The instrument
is therefore limited to measuring changes in velocity of a few
kilometres per second. In practice we are searching for changes of a
few metres per second. The change in the intensity of each output
beam as a function of Doppler shift depends on the profile of the
absorption line being observed, but is typically about 0.01% for a
shift of 15ms^{-1}. This change equals shot-noise if we count $\sim$10^8
photons.
     Using the above figures we estimate that we must observe a 0^m star
for 13 hours on a 4.2m telescope for a noise-equivalent velocity
sensitivity of 1ms^{-1}. In Sec. 4.1 we explain that oscillations are
best detected by computing the power spectrum of velocity versus
frequency of oscillation. The frequency resolution in this power
spectrum is inversely proportional to the total observing time and may
be as small as a few micro-Hertz. By smoothing the power spectrum
in frequency we may reduce the velocity noise by a substantial factor
and hence arrive at a sensitivity of $\sim$1ms^{-1} on fainter stars or
smaller telescopes.

4. OBSERVATIONS MADE WITH THE MICHELSON ISOS

The most recent observations with the Michelson instrument give a
good indication of the accuracy of the technique and are summarised
in Table 1.

Table 1. Most recent observations made with the Michelson
                                                                    ISOS

Star	Mv	Telescope	Date	Noise level
βAql	3.71	UKIRT	July 1983	2.4ms^{-1} (∿2 nights)
εCyg	2.46	INT	August 1984	68cms^{-1} (3 nights)
ζCyg	3.2	INT	August 1984	80cms^{-1} (2 nights)

4.1 Data reduction

The first step in this process is to correct the data for the effect
of the Earth's rotation etc. The relative radial velocity curve is
then plotted. In most cases this appears as a noisy and otherwise
featureless line. To extract any hidden periodicities in this curve
an amplitude or power spectrum is computed. It is preferable to
compute the amplitude spectrum rather than the power spectrum which is
not linear and not as reliably interpreted.

    A possible further stage of reduction is to follow the procedure
described by Fossat, (1984), which requires the computation of the
power spectrum of the power spectrum corrected for the effect of the
observational window function. The significance of this method is
discussed later.

    The on-line reduction mentioned in section 2 serves mainly to
show large, rapid variations in the signal and to show that the
instrument is functioning properly.

4.2 General Summary of the Results

(a) βAql. The smoothed power spectrum of the 19 hours of data taken
over 2 nights is shown in Figure 4. The resolution in the power
spectrum before smoothing was 15.26μHz, equivalent to an observation
65, 536 seconds long. There is a peak at frequency of 4.7mHz, or a
period of about 3.5 minutes, but we do not regard it as very
significant since it barely reaches the 3σ level when the modulus
of the power spectrum is plotted. The noise level in this observation
is unusually high, probably because an old optical fibre attenuated
the flux reaching the ISOS.

(b) εCygni. Figure 5 shows the smoothed power spectrum of the 20.5
hours of data obtained over 3 nights for εCyg. There are no
significant features to be seen in the power spectrum. The resolution
in the power spectrum before smoothing was 5.09 μHz equivalent to an
observation 196, 608 seconds long.

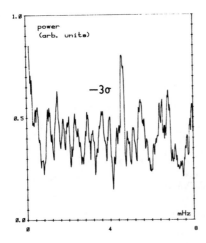

Figure 4. Power spectrum of combined data on βAq1 smoothed to 0.2 mHz.
σ = 2.4ms⁻1

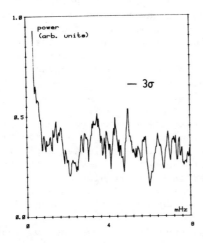

Figure 5. Power spectrum of combined data on εCygni smoothed to 0.2
mHz. σ = 68 cms⁻1

(c) ζCygni. This star was observed for a total of 14.2 hours over
a period of 2 nights and its smoothed power spectrum is shown in
Figure 6. Again no significant features can be seen. The resolution
in the power spectrum before smoothing was 7.06μHz, equivalent to an
observation of 131, 072 seconds duration.

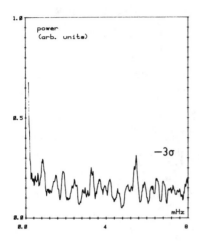

Figure 6, Power spectrum of combined data on Zeta Cygni smoothed to
0.2mHz, σ = 80 cms⁻1

We do not claim to have detected any significant peaks in the power
spectra derived from observations of these stars; in this respect we
are on a par with most other observers who have used a variety of
instruments. Our velocity calibration (derived from the calculations
mentioned in Sect. 3.2) showed that our technique is at least as
sensitive as any other and so it seemed worthwhile pursuing the more
elaborate data-processing methods which have been used by others who
have then claimed to detect stellar oscillations (Noyes, 1984, Fossat,
1984).

Fossat used the knowledge that the eigen-frequencies derived
from stellar models are almost equi-spaced – such a series of lines
has indeed been observed in the Solar observations. A power spectrum
of the power spectrum of velocity versus time, i.e. an analysis of
periodicities in the first power spectrum might thus show a peak at
the spacing frequency of the eigen-modes. Fossat carried out such
analysis and claimed to have detected a significant peak. Figure 7
shows his results for α Cen.

We carried out an identical procedure on our observational
data from Epsilon Cyg. and Figure 8 shows that we obtained a peak
which is of similar significance to Fossat's.

The comparison also seems to demonstrate the superiority of the
ISOS instruments since the velocity noise levels are similar despite
a twenty-fold lower photon flux.

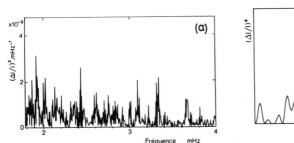
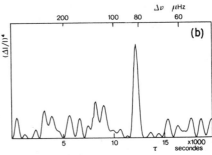

Figure 7. Fossat's results on α Cen (Fossat, 1984)
(a) Part of the power spectrum of the combined data and (b) the power
spectrum of the power spectrum in the range 2.3 to 3.85 mHz.

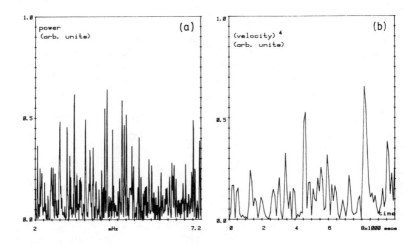

Figure 8. Authors results on ε Cyg.
(a) Part of the power spectrum of the combined data and (b) the power
spectrum of the power spectrum in the range 2 to 7.2 mHz.

We felt it important, however, to carry out tests of the significance
of such peaks in the second power spectrum. Since the ordinates in
these plots are in units of velocity to the fourth power, we
recalculated our data to show velocity versus time (Figure 9).

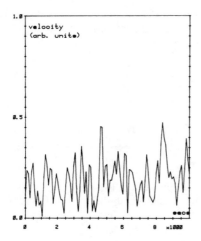

Figure 9. Velocity vs time plot of the power spectrum of the power spectrum in the range 2 to 7.2 mHz for εCyg.

The peak is seen to be less prominent, as might be expected.
    We next combined our three nights observations into two pairs of two nights and applied the same processes. Figure 10 shows that the peak is present in data from only one pair of nights.

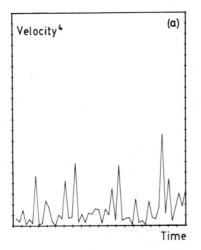
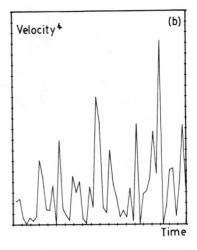

Figure 10. Three nights observations of εCyg, combined into two pairs of two nights (a)8th and 9th August 1984 and (b)9th and 10th August 1984.

Finally we produced simulated observations by randomly distributing photon-shot levels of noise around a mean level of photon counts which was identical to that of our observations and applying the same processes. Figure 11 shows that peaks occur in the second transform purely by chance - similar peaks occurred in almost half our trials.

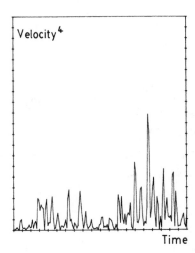

Figure 11. Power spectrum of power spectrum of simulated stellar observations. Typical result of about 50% of the trials.

We conclude, therefore, that such apparent detections of stellar oscillations may well be artefacts of the data-processing techniques and we are not convinced that any claims of such detection are to be relied upon. It is perhaps significant to note that Fossat's second series of observations of αCen (reported elsewhere in these proceedings) failed to reproduce his earlier results.

We must therefore improve our observational methods before we can hope reliably to detect stellar oscillations. There are several ways of doing so; we can use larger telescopes and (preferably) observe for more nights so as to also improve our temporal resolving power. The advent of much larger telescopes will perhaps permit us to make such observations although competition for observing time will be severe. It will be much more efficient to use more lines in the stellar spectrum simultaneously - we might then also be able to make differential measurements of oscillations at different stellar radii by selecting lines of different atomic origin.

We are presently modifying our ISOS instruments to use up to one hundred lines, with a consequent improvement of ten in our limiting velocity amplitude. Connes has described elsewhere in these proceedings a method for using all the information in the spectrum, which will be an even better approach if unforeseen instrumental difficulties do not appear; it may be that differential drifts between individual elements of C.C.D. detectors will limit the sensitivity of his technique.

We believe that we are on the brink of reliable detection of stellar oscillations and that instrumental developments will soon open up this new and highly important area of astronomical observations.

References

Baranne, A. et al, 1979, Vistas in Astronomy, 23, 279.
Campbell, B., and Walker, G.A.H., 1979, P. A. S. P. 91, 540.
Fossat, E. et al, 1983, The Messenger, 33, 29.
Fossat, E. et al, 1984, The Messenger, 36, 20.
Grec, A. et al, 1980, Nature, 288, 5791, 541.
Jacquinot, P., 1954, J. Opt. Soc. Am. 44, 761.
Noyes et al, 1984, Astrophys. J., 285, L23.

ABSOLUTE ASTRONOMICAL ACCELEROMETRY

Pierre CONNES
Service d'Aéronomie du CNRS
91371 - Verrières-le-Buisson France

ABSTRACT. The principle of our proposed accelerometers (first described at the ESA DISCO Utrecht Workshop three years ago) has now been discussed in two brief publications (IAU Symposium 112 on "Search for Extraterrestrial Life", 1984, and IAU Colloquium 88 on "Radial Velocities", 1984), plus a detailed one (Astrophys. and Sp. Sc., 110, 211, 1985) ; hence a mere abstract will suffice here. Essentially, one makes use of a variable path-difference FP interferometer, whose bandpasses are constrained by a first servo-loop to track the wavelength fluctuations of the studied lines (only one in the solar case, and all of them within a wide range in the stellar case). Through a second independent servo-loop, a tunable laser tracks the FP bandpasses. The final output is the beat frequency between this laser and a stabilized laser (of frequency N_0) ; one shows that if the measured radial velocity changes from V_1 to V_2 and the corresponding beat frequencies are BN_1 and BN_2, then $V_2 - V_1 = c(BN_2 - BN_1)/N_0$. The result is directly comparable with that given by Doppler radar ; one has basically transferred a problem from the incoherent to the coherent optics domain. All spectral or instrumental characteristics drop out ; the result is independent of wavelengths, calibrations, spectral or instrumental line profiles etc... The FP itself (together with the grating spectrometer required for error signal generation) and the detectors, are used solely as null-checking devices. Altogether, one may hope for an unusually low level of systematic errors both in the stellar oscillations frequency range, and in the ULF domain required for extrasolar planetary detection.

A second independent feature (in the stellar accelerometer case) is demonstrably optimal use of the radial velocity information available in the spectrum, for a given number of detected photons. The RMS velocity error from photon noise has been computed for a few spectral types ; it has been found lower by a factor of 30 than in the case of CORAVEL (so far our most efficient radial-velocity device) ;

403

D. O. Gough (ed.), Seismology of the Sun and the Distant Stars, 403–404.
© 1986 by D. Reidel Publishing Company.

this is equivalent to a 900 times reduction in observing time required for a given result.

These two fully compatible properties imply two practical consequences for the study of oscillations. First, (in the solar and stellar cases), the absence of all daily calibrations implies elimination of spurious effects which have led to the proposed long coherence 13 d solar oscillation (see other paper in the same Workshop), and may be in part responsible for the also-proposed 160 min oscillation. This advantage is shared by ACRIM (also an absolute device) used with great success in the search for solar oscillation through accurate photometry. The second point is that sensitivity of the stellar accelerometer is adequate for detecting solar type oscillations on a very large number of stars with small telescopes (i.e. in the 1m class). Considering the long observation times required (as seen from the solar case), this property is essential if stellar seismology is to become a widely useful tool.

EVIDENCE FOR P-MODES ON ALPHA CENTAURI AND PROCYON

E. Fossat; B. Gelly and G. Grec
Département d'Astrophysique
Université de Nice, Parc Valrose
F-06034 Nice Cedex

ABSTRACT. Helioseismology has proved to be a powerful tool to probe the internal structure of the Sun. With a new adapted optical resonance spectrophotometer, an extension has been attempted to two bright stars, namely Alpha Centauri A and Procyon. Results obtained from two observing runs on Alpha Centauri in may 1983 and may 1984 at LaSilla, and one observing run on Procyon in february 1984 at MaunaKea are presented. In both cases, solar-like pressure oscillations have been detected. The complete analysis is presented, which leads to determine essentially five parameters: the mean equidistance $\Delta\nu$, the frequency range, the amplitudes, the departure around equidistance (curvature of the Echelle diagram) and the fine frequency spacing between modes of degrees 0 and 2. For Procyon, all results are consistent with theoretical predictions, including excitation of oscillations, mass, radius and age of the star. For Alpha Centauri, all results are consistent together and are confirmed by the 1984 observation. They suggest that the star could be younger than estimated, consistent with a zero age main sequence model.

1. INTRODUCTION.

Last year (Fossat et al, 1984), we presented a first analysis of monochromatic photometric measurements of Alpha Centauri A, which revealed the presence of a p-mode-like set of discrete frequencies in the five-minute period range, similar to the result obtained on the Sun in spatially integrated light. We present here briefly the results of a more complete analysis of this observation, comforted by additional data obtained one year later. We also present similar results obtained independently on Procyon. A more detailed description of the analysis and of the results is being published elsewhere (Gelly et al, 1986).

D. O. Gough (ed.), Seismology of the Sun and the Distant Stars, 405–416.

2. DATA.

The measurements are made by means of a sodium cell monochromatic filter (Fossat et al, 1982), which sums the light from two 0.1 A bandwidths located in the red wings of the Na D1 and D2 resonance lines. A continuum reference channel is used for transparency and scintillation corrections. Alpha Centauri was observed at the Cassegrain focus of the 3.6 m telescope of ESO, LaSilla, Chile. 20 hours from 2 and half good nights in may 1983 and 7 hours only from one night in may 1984 have been analyzed. Procyon was observed at the Cassegrain focus of the 3.6 m CFH telescope on Mauna Kea, Hawaii. 27 hours from 4 consecutive nights in february 1984 have been analyzed. The monochromatic counting rate was of the order of 30 000 to 40 000 per second.

3. ANALYSIS.

With the experienced counting rate and the expected stellar p-mode amplitudes, the signal to noise ratio in the power spectrum must be of the order of 1. It is then necessary to use some a-priori known properties of this power spectrum to improve this situation and to make the detection possible. We use the fact that in the p-mode frequency range, the power spectrum is almost periodic in frequency. With the low resolution provided by one night of observation, one can detect alternately one $\ell=1$ mode, then one unresolved pair of $\ell=0$ and $\ell=2$ modes, then the next $\ell=1$ mode, ...
The best way to find this periodic pattern in the power spectrum appears to be the computation of the power spectrum of this interesting frequency range. It will have the dimension of the square of an autocorrelation function and must display also a set of equidistant peaks. They correspond to the inverse frequency spacing, with all harmonics. When several nights of data are used, the power spectrum contains sidelobes around each peak, and the power spectrum of power spectrum is multiplied by the window function autocorrelation squared. Many possibilities exist of filtering the information to retrieve the equidistant pattern with less noise. After testing several possibilities on the solar spectrum, we have adopted the following general procedure:
a. Low frequency filtering of the data, with a cut-off at about 0.3 to 0.5 mHz, to reduce the atmospheric and instrumental noise.
b. FFT power spectrum of this filtered data, taken as one single time series, filled with zeroes when observation is not available.
c. Search, in this power spectrum, for a frequency range whose Fourier analysis provides a Cha function, multiplied by the window function autocorrelation.
d. Calculation of the Fourier transform of the frequency range found in c. This is the squared autocorrelation function of the stellar signal filtered in this frequency range.

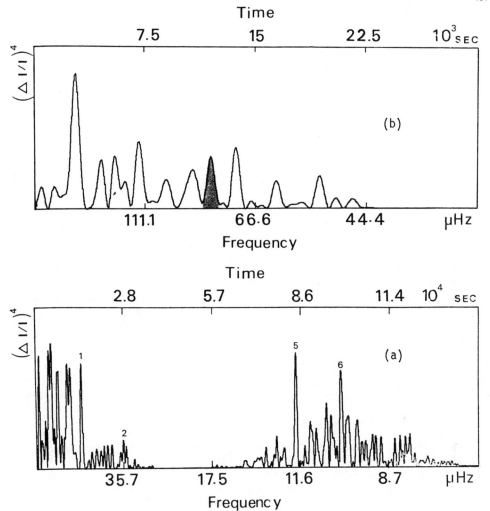

Figure 1.   Power spectrum of the 2.3 - 3.8 mHz frequency range of the
Alpha Centauri velocity power spectra obtained from 3 nights of measurements in 1983 (a) and one night in 1984 (b). Only two nights of the
window function autocorrelation are shown. The $(82\mu Hz)^{-1}$ peak is visible
with its harmonics n° 2, 5 and 6.

     e. Determination, along the autocorrelation time axis, of
narrow bands containing the Cha function defined in c.
     f. Fourier transform of the complete power spectrum, and
restriction of this Fourier transform to the narrow bands defined in
c by replacing the complex amplitudes by zeroes between these bands.
     g. Inverse Fourier transform of the function defined in f,
showing the discrete pattern searched for in the power spectrum.

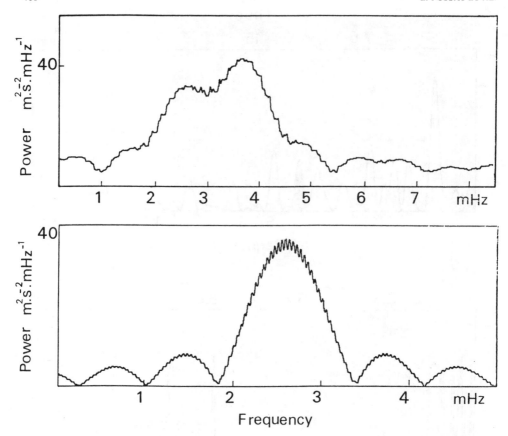

Figure 2. Alpha Centauri p-mode power spectrum envelope, respectively for the 1983 and 1984 measurements. The more significant 1983 result is very similar to the solar envelope.

h. From g, determination of the power envelope of the assumed p-mode set of discrete frequencies.
    i. Echelle diagram of the filtered power spectrum obtained in g, and measurement of the eventual curvature of the set of modes on this diagram.
    j. Straightnening of the Echelle diagram by a polynomial fitting of the curvature.
    k. Vertical sum on the Echelle diagram of the unfiltered power spectrum, to search for the eventual frequency separation of modes of degrees 0 and 2. This separation is not possible on the filtered spectrum, because the filtering method artificially makes it exactly periodic with the $\Delta\nu/2$ period.

    From all this analysis, the main five results which can be

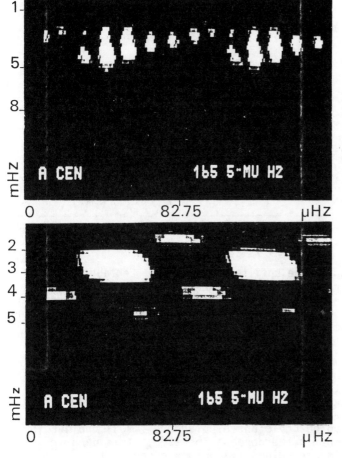

Figure 3. The two
Echelle diagrams of
Alpha Centauri p-mode
power spectra show
that despite its
lower significance,
the 1984 result con-
firm the exact posi-
tion of the peaks
found in the 1983
analysis.

compared with theoretical predictions are:
-The mean spacing $\Delta\nu$ between consecutive radial harmonics for a given degree (136 µHz in the case of the sun).
-The p-mode frequency range (2 to 5 mHz for the sun).
-The oscillation amplitudes (<20 cm/s for the sun).
-The curvature in the Echelle diagram, which is the departure from the exact equidistance   .
-The fine structure of the Echelle diagram, such as the frequency spacing between modes of degrees 0 and 2 (9 to 10 µHz for the sun).

The mean spacing is directly related to the sound speed travel time between the center and the surface of the star. At first order, it is a measure of the stellar radius.
The next two are related to the stellar mass and to the excitation mechanism.
The curvature depends, not simply, on the stellar structure.

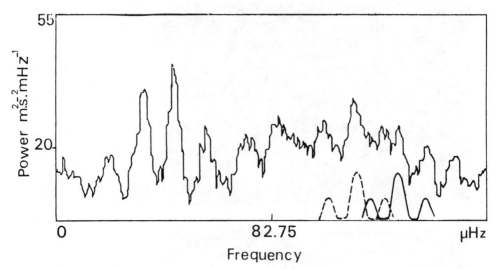

Figure 4. Vertical sum of the unfiltered 1983 Alpha Centauri Echelle diagram. The most contrasted part is assigned to modes of degree 1. A possible identification of the l=0 and l=2 fringe patterns is suggested, all the relative positions being consistent with the asymptotic equation The l=0 - l=2 separation would be 16 microherz.

It is predicted by the asymptotic equation (Tassoul, 1980) and its presence in the Echelle diagram gives us more confidence in the stellar p-mode detection.

The fine spacing between modes of degrees 0 and 2 (which must be consistent with the $\ell=0$ - $\ell=1$ spacing) is a measure of the parameter D∅ introduced by Christensen-Dalsgaard (1984). Related to the sound speed gradient in the core of the star, it is very sensitive to the stellar age along its main sequence life.

4. RESULTS FOR ALPHA CENTAURI.

Figure 1 shows the power spectrum of the 2.3 - 3.8 mHz range of the power spectrum, for the 1983 (a) and 1984 (b) data. Harmonics n° 1, 2, 5 and 6 of the $(82 \ \mu Hz)^{-1}$ peak are clearly visible in the Figure 1-a. The Figure 1-b, form only 7 hours of data, shows a non significant peak at the same spacing.

Figure 2 is the power envelope of the assumed existing discrete pattern with a 82 µHz mean spacing relatively for 1983 data (a) and 1984 data (b). It appears to be very similar to the solar one. The highest peaks of this envelope contain a power of about 3. 10^{-8} and this leads to an amplitude of 1 to 2 m/s in term of line of sight velocity. The total energy, in the five-minute range, corresponds to a r.m.s. amplitude close to 10 m/s. This value, about 10 times the solar one, may seem very large and will be discussed further.

In Figure 3, one can see that the two Echelle diagrams, modulo

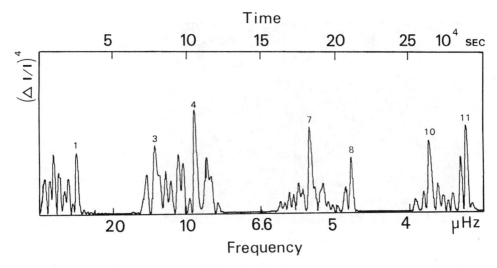

Figure 5. Power spectrum of the 1.2 - 1.65 mHz frequency range of the Procyon velocity power spectrum. The four night window function auto-correlation is clearly visible, together with a regular pattern of equidistant peaks, whose fundamental is at $(39 \mu Hz)^{-1}$.

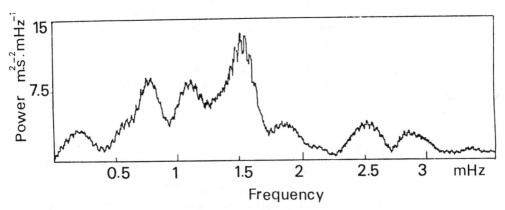

Figure 6. Procyon p-mode power spectrum envelope. The frequency range, about 0.7 to 1.7 mHz, corresponds to periods between 9 and 24 minutes.

165 μHz, are very exactly consistent, despite the very different frequency resolution. It means that the discrete pattern responsible for the $(82 \mu Hz)^{-1}$ peak exactly coïncide in the two power spectra (1983 and 1984). A slight curvature is also visible on the 1983 Echelle diagram. It is of the same order of magnitude than the curvature of the solar Echelle diagram, if restricted to the most powerful (2.5 -4.0 mHz) frequency range.

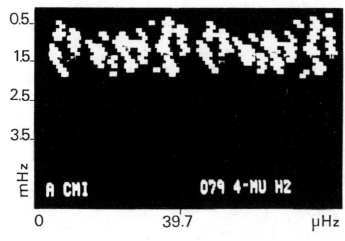

Figure 7. Echelle diagram of the Procyon p-mode spectrum. Although polluted by many fringes due to the discontinuous window function, it clearly shows, for one set of peaks, the curvature predicted by the asymptotic equation.

Figure 4 is a vertical sum of the unfiltered Echelle diagram. It is expected to show on one side the mode of degree 1 and its sidelobes, and on the other side a mixture of the two sidelobe systems of the modes of degrees 0 and 2. We would then assign the degree 1 to the left side and the degrees 0 and 2 to the right side, where the fringe contrast is much worse. Tentatively, an identification of the ℓ=0 (full line) and ℓ=2 (dashed line) fringe systems is suggested. The whole pattern is then consistent with the asymptotic equation. However, the ℓ=0 - ℓ=2 fringe pattern is quite confused, and this identification cannot be regarded as well established. The spacing (0-2) would be 16 µHz, giving Do $\simeq$ 2.6.

5. RESULTS ON PROCYON.

Figure 5 shows the power spectrum of the 1.2 - 1.7 mHz range of the power spectrum. A regular pattern limited to harmonics n° 1, 3, 4, 7, 8, 10 and 11 is clearly visible. It gives a mean spacing $\Delta\nu/2$ = 39.7 µHz.

Figure 6 is the power envelope of the filtered discrete pattern with a 39.7 µHz regular spacing in the power spectrum. The p-mode frequency range for Procyon appears to be 0.7 - 1.7 mHz.

The Echelle diagram, modulo 79.4 µHz, on Figure 7, is contaminated by many fringes. However, the curvature of one sequence of peaks is clearly visible and can be fit by the asymptotic equation. The best fit is for $A.\delta \simeq 1.8$.

The amplitude per single mode is found to be at most 70 cm/s (assuming that the monochromatic intensity oscillations on the stellar line wing are only due to Doppler shifts of this line), about 3 times the solar value.

On Figure 8, the vertical sum of the unfiltered Echelle

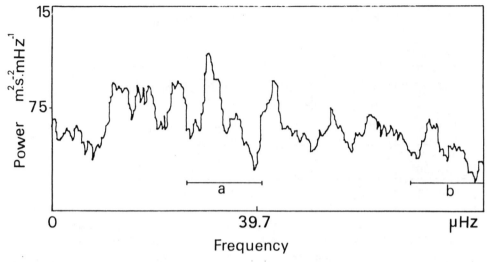

Figure 8.   Vertical sum of the unfiltered Procyon Echelle diagram, corrected for the curvature. The section (a) is the only one showing significant fringe contrast and is assigned to modes of degree 1 and their sidelobes. Although the figure is more confused on the other side, the similarity in the shape of the sections (b) and (a) and their separation by exactly 39.7 μHz suugest that the modes of degree 0 and 2 are not resolved.

diagram (after correction of the curvature by polynomial fitting) shows again one side with high contrast, and the other one without. Assigning again the high contrasted side to modes of degree $\ell = 1$, we could not find any fringe pattern on the other side that is consistent with the relative positions of the modes $\ell = 0$ and $\ell = 2$. The strong similarity between the sections (a) and (b) of this figure, which are exactly separated by 39.7 μHz, suggest that the peaks of modes $\ell = 0$ and $\ell = 2$ are almost coïncident (but not quite, because the lower fringe contrast). With the same caveat than in the case of alpha Centauri, we suggest that the spacing is less than 3 μHz, or $D\theta < 0.6$.

6. DISCUSSION

Two recent studies (Christensen-Dalsgaard & Frandsen, 1983 and Christensen-Dalsgaard, 1984) give us interesting points of comparison between our results and the theoretical predictions. The three main results of these studies are based upon the expected measurements of the two parameters $\Delta\nu$ and $D\theta$ in the asymptotic equation. First it is found that for any zero age main sequence model of star (ZAMS), $R\Delta\nu\simeq$constant, R being the radius of the star. Then, during its evolution in the main sequence, the star follows the homologuous scaling law $R^{3/2}\Delta\nu\simeq$constant. These laws can be calibrated against

	PROCYON			ALPHA CENTAURI		
$\Delta \nu$	79.4 µHz		very likely	165.5 µHz		very likely
period range	10 to 25 mn		"	3.5 to 8.5 mn		"
curvature	A. $\partial \approx 1.8$			slight curvature		"
maximum amplitude	70 cm . s^{-1}			150 cm.s^{-1}		"
spacing l=0-2	< 3 µHz		possible	16 µHz		possible
identification l and n	suspectable in l impossible in n			possible in l impossible in n		

Summary of numerical results.

the solar values.
     The third result is the diagnostic value of the parameter Dθ, which depends strongly on the gradient of the sound speed near the center. As the star burns its hydrogen, the central molecular weight increases, consequently the central sound speed and Dθ decrease with age.
     For Procyon, assuming a mass of 1.5 (in solar mass unit), the ZAMS model is computed to have $\Delta \nu \approx 95$ µHz (Christensen-Dalsgaard, 1984). With the calibration R$\Delta \nu \simeq 145$ µHz , the initial Procyon radius was 1.67 (still in solar unit). The hololognous scaling law gives then a present radius of 1.87. If some credit can be given to non-resolution of the l = 0 , l = 2 spacing, then Dθ<0.6 seems to indicate that Procyon is more evolved than the sun, consistent with the best model computed by Demarque who finds a star close to the end of its hydrogen burning main sequence life (Demarque et al., 1986).
     Regarding the frequency range and the amplitudes, two different possibilities of excitation have been suggested, namely the $\mathcal{K}$-mechanism (Unno et al., 1979), or the stochastic excitation by interaction with the convection (Goldreich and Keeley, 1977). The two suggestions can hardly be distinguished on the sun, because they both predict quite well the observations. In the case of Procyon, however, they are quite diverging and our result is strongly favouring the stochastic excitation (Christensen-Dalsgaard & Frandsen, 1983), agreeing both with the frequency range and the expected amplitudes.
     Now, if the general agreement between theory and observation,

is quite satisfactory on Procyon, ,the results obtained on alpha
Centauri are really puzzling. Being a member of the closest multiple
star system in our neighbourhood, this star has a well measured mass
of 1.09. Given the precise measurement of its temperature and of its
distance, it has an estimated radius of 1.23±0.04 (Blackwell and
Shallis, 1977). According to Christensen-Dalsgaard & Frandsen (1983)
and as already noticed by Gough (1985) $\Delta\nu$ should have been 142 µHz
when such a star was at zero age on the main sequence, and its radius
should have been 1.03 . With the measured value of 165.5 µHz, one can
deduce from homologuous scaling a present radius of 0.93. There is an
evident contradiction, as our result seems to imply that the star has
shrunk since its zero age. Even assuming that this star could be much
younger that the sun, down to zero age, we have to assume a mass of 1
to fit with the theory. There is indeed a general disagreement about
the age of alpha Centauri (Demarque, 1986 ; Flannery and Ayres, 1978
; Morel and Baglin, 1983) and the idea that that the star is at its
zero age evolution stage cannot be rejected. But can the knowledge of
the mass be wrong by 9% ? It must be noticed here, that if once again
some credit is given to the measurement of the spacing $\ell=0$, $\ell=2$,
then Do≃2.65 and the whole figure becomes exactly consistent with a
zero age solar mass star (Christensen-Dalsgaard, 1984, Fig. 10).
      The frequency range, centered around 3 mHz, is the only
quantity which agrees quite well with predictions. The amplitudes, 5
to 10 times too large are also puzzling. Now if we want to explain
all the results, the structure of alpha Centauri has really to be
quite different from expectations, and then a mismatch of
oscillations amplitude byh such a factor is not too surprising.

7. CONCLUSION

      Two stars have been observed. Two results are presented, with
about the same signal to noise significance. One is in quite good
agreement with expectations, the other one in very bad agreement. A
confirmation of these results by more and by other measurements is
highly required.
      Our instrument has not been photon noise limited yet, and it
has then the intrinsic capability to providing better results, or to
access to a few more stars. It is clear in our analysis that the
fringes due to the day-night cycle are severely limiting the amount
of reliable information obtained in such measurements. Combined
observations in complementary sites are desirable. We will not
discuss here any further how such future programs can be organized,
because it was the subject of another session of this workshop.

8. References

Blackwell D.E., Shallis M.J. : 1977, Mon. Not. R. Astr. Soc., **180**,
      177.
Christensen-Dalsgaard J., Frandsen S. : 1983, Solar Phys., **82**, 469.
Christensen-Dalsgaard J. : 1984, Space research prospects in stellar

activity and variability (eds A. Mangeney and F. Praderie,
    Paris Obs. Press), p. 11
Demarque P. et al. : 1986, these proceedings
Flannery B.P., Ayres T.R. : 1978, Astrophys. J., **221**, 175.
Fossat E., Decanini Y., Grec G. : 1982, Instrumentation For Astronomy
    With Large Optical Telescopes, (ed. C. Humphries, D. Reidel
    Publishing Company), p. 169.
Fossat E., Grec G., Gelly B. Decanini Y. : 1984, Comptes-Rendus Acad.
    Sci. Paris, Ser. 2, **229**, 17.
Gelly B., Grec G., Fossat E. : Submitted to Astronomy and
    Astrophysics.
Goldreich P., Keeley D.A. : 1977, Astrophys. J., **212**, 243.
Gough D. : 1985, Nature, **314**, 14.
Morel P., Baglin A. : 1983, Communication aux 5e journées de
    l'Observatoire de Strasbourg.
Tassoul M. : 1980, Astrophys. J. Suppl., **43**, 469.
Unno W., Osaki Y., Ando H., Shibahashi H. : 1979, Nonradial
    Oscillations of Stars (University of Tokio Press)

A REVIEW OF THE RAPIDLY OSCILLATING Ap STARS

D. W. Kurtz
Department of Astronomy
University of Cape Town

1.   Introduction

The Rapidly Oscillating Ap stars are cool magnetic Ap stars which oscillate with periods between 4 and 15 min with peak-to-peak light variations of $B \leq 0.016$ mag. There are, as of this writing, eleven of these stars known. In section 2 they are listed along with some of their observational characteristics and references to literature on them.

The short periods present in the light variations of these stars strongly suggest that their variations are due to high overtone p-mode oscillations. Kurtz (1982) has shown that the amplitude of the oscillations in some of them is modulated with the rotation period of the star with the times of maximum pulsation amplitude coinciding with the times of maximum extrema of the measured effective magnetic field strength. He has shown that the frequency pattern of these oscillations can be simply described by assuming that the oscillations are non-radial p-modes of low degree (ℓ small) with the axis of oscillation aligned with the magnetic axis of the star (the *oblique pulsator model*).

A problem with this model is that for oscillations about a magnetic axis which is inclined to the rotation axis of the star, advection is expected to cause the pulsation pattern to precess with respect to the magnetic axis. This is the same effect that lifts the degeneracy of the m-modes in a rotating star. This drift of the time of maximum pulsation amplitude with respect to the time of extremum of the magnetic field strength is not observed, however. In HR 1217 the times of pulsation amplitude maximum and effective magnetic field maximum have remained in phase with each other over 3 years of observations and in HR 3831 they have remained in phase over 4 years of observations. From calculations of $C_{n,\ell}$ (Shibahashi & Saio 1985a) we would expect to have already seen a drift in these phases in HR 1217 and HR 3831 if it were going to occur.

Dolez & Gough (1982) addressed this problem and suggested that the growth time for the oscillations in the Rapidly Oscillating Ap stars is short compared to the expected precession time of the pulsation axis for an oblique pulsator. Thus they suggest that only oscillations which are currently aligned with the magnetic field are excited to

417

D. O. Gough (ed.), Seismology of the Sun and the Distant Stars, 417–440.

observable amplitudes.   Shibahashi & Saio (1985a,b) agree
with Dolez & Gough on this point.   From calculations of the
effect of the magnetic field on the oscillations, Dziembowski & Goode (1985) suggest that the oscillations are dominated by the magnetic field rather than by rotation.   Hence
they expect that the pulsation axis will not drift with
respect to the magnetic axis.

Mathys (1984, 1985) has proposed an alternative to the
oblique pulsator model, the *spotted pulsator model*. He
suggests that the oscillations in the Rapidly Oscillating Ap
stars are aligned with the rotation axis of the star and
that the amplitude modulation of the oscillations is due to
surface inhomogeneities and a variable ratio of surface flux
to radius variations with cylindrical symmetry about the
magnetic axis of the star.   This explains the observed frequency splitting seen in these stars and allows arbitrary
amplitudes for the observed frequencies.   It also avoids the
problems discussed in the last paragraph.   It does, however,
have several free parameters which are unknown and currently
unmeasurable and hence it is neither applicable nor
testable.

The oscillation periods in the Rapidly Oscillating Ap
stars are so short that it seems likely that they are due to
very high overtone *p*-mode pulsation.   This raises the obvious questions why such high overtone pulsation modes should
be preferentially excited and what mechanism selects only
the few modes which are observed.   Shibahashi (1983) has
tried to answer these questions by suggesting that overstable magnetic convection is the driving mechanism in the
Rapidly Oscillating Ap stars.   Since this mechanism is most
effective at the magnetic poles of the star, it gives a natural explanation of why the oscillations should be aligned
with the magnetic axis.   Shibahashi also calculates that the
time scale for magnetic overstability in A stars is about 10
minutes; he therefore suggests that the observed modes are
simply those with frequencies closest to the time-scale of
the magnetic overstable convection which are resonantly
excited.   Cox (1984) has also discussed Shibahashi's idea.
A difficulty with the idea of excitation of· the Rapidly
Oscillating Ap stars by magnetic overstability is that it
introduces an additional pulsation mechanism where one is
already known; all of the known Rapidly Oscillating Ap stars
lie within the δ Scuti instability strip.   There are also a
few Ap stars which appear to be δ Scuti stars, a point discussed further in section 4.   Dziembowski (1984) favours excitation of the oscillations in the Rapidly Oscillating Ap
stars by the κ-mechanism; he suggests that the presence of
the magnetic field may enhance the instability of some of
the non-radial modes.

Definitive tests to distinguish between driving by the
κ-mechanism or by magnetic overstability are difficult to

devise. Finding a Rapidly Oscillating Ap star much hotter than the observed blue border of the δ Scuti instability strip would seem to be a good test, but that border is observationally not rigidly defined and theoretically may be dependent on the atmospheric structure and abundance distributions of the magnetic Ap stars which are not well known. In addition, it is clear that there are oscillating stars hotter than the blue border of the δ Scuti instability strip, such as the 53 Persei variables and the β Cephei variables, for which no pulsation mechanism is known. Finding a definitely non-magnetic rapidly oscillating star in the same part of the HR diagram as the Rapidly Oscillating Ap stars might disprove magnetic overstability— if it were certain that the star really was non-magnetic and if it were certain that yet another new class of variable stars had not been discovered.

The idea of the oblique pulsator model has given rise to several theoretical discussions. Cousens (1983) has independently developed the same idea which he calls the Obliquely Oscillating Magnetic Rotator and which he applies to a possible interpretation of the Blazhko effect in RR Lyrae stars. Aizenman et al. (1984) have calculated a formula for rotational mode splitting about an inclined axis which is discussed further by Pesnell (1985), but Shibahashi & Saio (1985b) dispute Aizenman et al.'s result. Dziembowski & Goode (1985) also state that Aizenman et al.'s result is not applicable to any real system. Basically, Shibahashi & Saio's argument is that a spherical harmonic aligned along an axis inclined to the rotation axis can be rewritten as a linear summation of spherical harmonics about the rotation axis by a simple rotation of coordinates. They conclude that the Ledoux (1951) rotational splitting formula must still apply, which gives rise to the previously mentioned effect of the expected, but not observed, precession of the pulsation pattern with respect to the magnetic axis. Dziembowski & Goode claim that the oscillation modes in the Rapidly Oscillating Ap stars are not normal modes and hence a coordinate transformation of the axis of symmetry of a set of spherical harmonics does not provide a formally correct description of those modes.

Models for the Rapidly Oscillating Ap stars have been calculated by Gabriel et al. (1985) and by Shibahashi & Saio (1985a). Shibahashi & Saio find that the observed frequencies of some of the Rapidly Oscillating Ap stars are above the critical frequencies of their models. When the frequency of oscillation is greater than the critical frequency then the wavelength of the oscillation is short compared to the height of the surface boundary and the wave penetrates that surface boundary and is not trapped. Shibahashi & Saio calculate that to increase the critical frequency to a value above the highest observed frequencies in the Rapidly

Oscillating Ap stars they must increase T_{eff}/T_{surf} to 1.5
times the value for standard A star models. This brings up
the interesting possibility that the very coolest of the
magnetic Ap stars, which appear to have spectral types
around F0, may have effective temperatures significantly
hotter than this spectral type implies.

2. Catalogue and basic data

Table 1. Strömgren indices for the Rapidly Oscillating Ap stars

HD	HR	V	b-y	m_1	c_1	β	δc_1	refs
6532		8.45	0.084	0.237	0.846		-0.142	1,2
24712	1217	5.99	0.183	0.212	0.634	2.744	-0.034	1,2
60435		8.889	0.132	0.234	0.843		+0.035	3
83368	3831	6.174	0.146	0.203	0.796	2.827	-0.035	3,1
101065		8.004	0.448	0.368	-0.014	2.641	-0.386	4
128898	5643	3.198	0.152	0.195	0.760	2.831	-0.077	3,1
134214		7.479	0.211	0.288	0.597	2.766	-0.115	6,7
137949		6.674	0.188	0.321	0.584	2.833	-0.256	3,1
201601	8097	4.68	0.147	0.238	0.760	2.819	-0.058	1,2
203932		8.820	0.169	0.196	0.736	2.814	-0.072	3,5
217522		7.520	0.289	0.215	0.487	2.701	-0.046	3,5

references: (1) Hauck & Mermilliod 1980; (2) Blanco et al.
1970; (3) Vogt & Faundez 1979; (4) Kurtz & Wegner 1979;
(5) Weiss (private communication); (6) Olsen & Perry 1984;
(7) Olsen 1983.

Table 1 gives a list of the 11 known Rapidly Oscillating Ap stars along with Strömgren and Hβ indices for them
where these have been measured. Most of the photometry has
been taken preferentially from the work of Vogt & Faundez
(1979) and secondarily from the compilation of Hauck & Mermilliod (1980). Werner Weiss has kindly provided unpublished Hβ measurements for the last two stars in the list.
Strömgren y magnitudes are listed in the column labelled V
where they are available, otherwise Johnson V magnitudes
from the compilation of Blanco et al. (1970) are given.

The c_1 indices calculated from Crawford's (1975, 1979)
F and A star calibrations are also given in Table 1. For
normal A and F stars a negative δc_1 index would be indicative of a subdwarf luminosity; for the Rapidly Oscillating
Ap stars these negative δc_1 indices are probably due to an
abnormal energy distribution in the continuum, in the spectral lines, or in both. It is worth remembering here Shibahashi & Saio's argument that the T-t structure of the
Rapidly Oscillating Ap stars must be abnormal in order for
them to oscillate with such high frequencies.

The effective temperature range found so far for the
Rapidly Oscillating Ap stars extends from about 7000 to

9000 K or from equivalent spectral types of about F0 to A5
judging by the b—y and Hβ indices. This presumes that these

Table 2. Magnetic field strengths and spectral types for the Rapidly Oscillating Ap stars.

HD	HR	H_e(G)	refs	Spectral Type	refs
6532				Ap SrCrEu	1
24712	1217	+300 to +1200	2	A5p	3
60435				Ap Sr(Eu)$_1$Fp SrEu	4,5
83368	3831	−700 to +700	6	Ap SrCrEu	7,5
101065		−2200	8	Controversial	9
128898	5643	−300(variable)	10,11,12	Ap SrEuCr,Ap SrEu	4,5
134214				F0, F0 SrEu	17
137949		+1400 to +1800	13	Fp SrCrEu	5
201601	8097	+500 to −800	14,15,16	F0p	3
203932				Ap SrEu$_2$	5,1
217522				Ap SiCr,Fp SrEu	1,5

references: (1) Houk 1982; (2) Preston 1972; (3) Hoffleit
1982; (4) Houk & Cowley 1975; (5) Bidelman & MacConnell
1973; (6) Thompson 1983; (7) Houk 1978; (8) Wolff & Hagen
1976; (9) Kurtz & Wegner 1979; (10) Wood & Campusano 1975;
(11) Borra & Landstreet 1975; (12) Borra & Landstreet 1980;
(13) Wolff 1975; (14) Babcock 1958; (15) Bonsack & Pilachowski 1974; (16) Scholz 1979; (17) Henry Draper Catalogue.

[1] Strong Sr
[2] May be Eu rather than Si

indices are indicative of effective temperature and, of
course, ignores the traditional argument about the effective
temperature of HD 101065. Certainly it is accurate to say
that Rapidly Oscillating Ap stars have been found so far
only amongst the coolest of the magnetic Ap stars. However,
no systematic surveys of large numbers of hotter stars or of
non—magnetic stars have yet been made. I have searched for
rapid oscillations in some hotter Ap stars and not found any
convincing evidence for light variations, but I am reluctant
to publish a list of stars so searched. There are several
reasons for this: 1) Because of the amplitude modulation in
the Rapidly Oscillating Ap stars, finding that a star shows
no variations during 1 or 2 hours of observation only means
that it was not varying during those 1 or 2 hours, not that
it is non—variable. 2) For some stars there appear to be
intriguing peaks in their amplitude spectra at a signal-to-
noise level of 2-3. The probability of this happening in a
given amplitude spectrum by chance can be significant, however, (see Scargle (1982) for a method to calculate such
probabilities) so that these stars need further observations

which I have not yet been able to make. 3) Finally, I fear
that such a list may discourage others from examining those
stars for variability and I do not believe that a few hours
of observations warrant that.

In Table 2 the Rapidly Oscillating Ap stars are listed
along with their magnetic field strengths and spectral
types. From Tables 1 and 2 it can be seen that all 11 of
the known Rapidly Oscillating Ap stars are very cool,
(almost certainly) magnetic Ap stars with SrCrEu spectral
peculiarities and with small or negative δc_1 indices. However, such characteristics may not even be necessary (and
they certainly are not sufficient) for rapid oscillations to
occur. For example, Weiss & Kurtz (in preparation) have
thoroughly examined β CBr, a prototypical cool magnetic Ap
star, for light variations and were unable to find any.

3. The frequencies of the Rapidly Oscillating Ap stars

Table 3 gives a list of the Rapidly Oscillating Ap
stars along with the frequencies and amplitudes so far derived for those stars. This table is only meant to be a
guide to the frequency patterns of these stars; in other
parts of this section further interpretation of these frequencies is given, but for complete details the original
papers should be consulted. Those references are also given
in Table 3.

It is not presently known what the lifetimes of the
oscillation modes in the Rapidly Oscillating Ap stars are.
It is certain that some of them are stable over an observing
season of about 100 days and it appears that the mode of
highest amplitude in HD 101065 has remained stable for 6
years. Given the present data, it is possible that all of
the frequencies in these stars are stable over time spans
much longer than that covered by the observations. Some of
the frequency spectra are extremely complex, however, and
have not yet been solved even given prodigious amounts of
data (by stellar astronomical standards). It is important
to consider the possibility that the lifetimes of at least
some of these modes may be short compared to the observational time spans. This is relevant to the suggestion of
Dolez & Gough which requires short growth times for the
oscillation modes to explain the coincidence of the pulsation axes and magnetic axes in the oblique pulsator model.
It is extremely difficult, however, to demonstrate frequency
instability at the signal-to-noise level that obtains for
the Rapidly Oscillating Ap stars. One always has the suspicion for stars which have unsolvable frequency spectra that
the lack of solution is due to the spectra being too complex for the available observations rather than due to short
lifetimes of the oscillation modes. This point is discussed
further in the sections on HD 60435 and HD 101065 below.

Table 3. Frequencies derived for the Rapidly Oscillating Ap stars.

HD	HR	P(min)	f(mHz)	A(mmag)	refs
6532		6.956	2.39612	1.01	1
		6.938	2.40210	0.73	
		6.922	2.40761	0.55	
24712	1217	6.126	2.7208	2.13	2,3,4
		6.283	2.6528	2.07	
		6.202	2.6875	1.10	
		6.048	2.7556	0.90	
		6.361	2.6200	0.60	
		5.966	2.7936	0.44	
60435		15.141	1.10077	2	5
		12.784	1.30371	6	
		12.327	1.35210		
		12.070	1.38088		
		11.841	1.40749		
		11.625	1.43364		
		6.06	2.75	1.5	
		3.994	4.17307	2	
83368	3831	11.705	1.423950	2.14	4
		11.638	1.432069	1.75	
		11.671	1.428011	0.38	
		5.836	2.856019	0.45	
		5.852	2.847906	0.20	
		5.819	2.864139	0.18	
101065		12.140	1.372865	5.40	6,7,8,9
		12.674	1.315079	0.67	9
		6.070	2.7459	0.26	9
128898	5643	6.825	2.442041	1.91	10
		6.832	2.4395	0.38	
134214		5.650	2.9496	3.23	14
137949		8.272	2.0148	1.39	4
201601	8097	12.448	1.339	0.86	11
203932		5.942	2.804789	0.66	12
217522		13.716	1.21510	2.	13

references: (1) Kurtz & Kreidl 1985; (2) Kurtz & Seeman 1983; (3) Kurtz, Schneider, & Weiss 1985; (4) Kurtz 1982; (5) Matthews, Kurtz, & Wehlau 1985; (6) Kurtz 1981; (7) Kurtz 1980; (8) Kurtz & Wegner 1979; (9) Kurtz unpublished; (10) Kurtz & Balona 1984; (11) Kurtz 1983a; (12) Kurtz 1984; (13) Kurtz 1983b; (14) Kreidl 1985b.

Some natural perversities complicate the task of deciphering the frequency patterns in the Rapidly Oscillating Ap stars. For high overtone p-modes Tassoul (1980) gives an asymptotic relation

$$\nu_{n,\ell} = \nu_0 (n + \ell/2 + \varepsilon) \tag{1}$$

where ε depends on stellar structure and slightly on the mode. Shibahashi & Saio (1985a) calculate for A star models that the separation of modes with quantum numbers (n,ℓ), $(n-1,\ell+2)$, $(n-2,\ell+4)$, etc. for ℓ=odd is about 6 μHz. Twice that number is very close to the daily alias frequency of 1 d^{-1}=11.6 μHz. This confusion has possibly arisen in HD 6532. A similar kind of problem occurs in another way in HR 1217 which has at least 5 frequencies separated by about 35 μHz. Shibahashi & Saio (1985a) calculate from Tassoul's asymptotic relation and A star models that 35 μHz is about the expected separation for alternating even and odd ℓ p-modes. It is also very close to 3 d^{-1}=34.7 μHz which gives rise to confusion between the frequencies and their overlapping aliases. Finally, the rotation periods of the cool magnetic Ap stars are often only a few days and, in some cases, may be very near to an integral number of days. This is a problem in HD 6532 which probably has a rotation period near 1.8 days, and it was thought to be a problem in α Cir which has been suggested to have a rotation period very close to 1 day (Lavagnino 1960).

The 1 d^{-1} alias problems that inevitably occur in observations made from a single site are exacerbated by the above mentioned effects which can give rise to natural frequency separations near to 1 d^{-1} (or a small multiple thereof) in the Rapidly Oscillating Ap stars. This often makes the solution of the frequency spectra extremely difficult and can only be ameliorated by obtaining contemporaneous observations from two or more observing sites. The vagaries of weather and the long time-spans of observations necessary to resolve the frequencies in these stars means that substantial commitments are demanded of several observers to obtain the requisite data. Fortunately many of the Rapidly Oscillating Ap stars are bright enough that this work can be done on small telescopes where large amounts of observing time are available. This sort of problem arises in several kinds of variable star research (including solar) and has given rise to a suggestion for the creation of a new IAU commission for coordinated multisite observations (IAU inf. bull. 53, 22).

The rest of this section is devoted to synopses of the current state of the individual frequency analyses of the Rapidly Oscillating Ap stars. Most of what follows comes from the papers referenced in Table 3, but some new and/or unpublished data and some work-in-progress are also discussed.

3.1 HD 6532

Kurtz & Kreidl (1985) derived three frequencies from 10 nights of observations of HD 6532, but the third frequency is separated from the first by a value so close to 1 d^{-1}

that it is not very secure. They also found a frequency present in their data at 1.17 mHz on two of their ten nights of observations, but the signal-to-noise ratio was too small to be certain of this identification. The frequency pattern in this star looks to be simple enough that it should be solvable, although contemporaneous observations from at least two observatories will be necessary to reduce the daily aliasing problem to a manageable level.

3.2 HD 24712 (HR 1217)

HD 24712 is one of the most observed of the Rapidly Oscillating Ap stars. Its apparent brightness, its relatively large light variations of up to 10 mmag in Johnson B, its well studied magnetic field variations, and its rich frequency spectrum all make this an important object to study.

Kurtz (1982) frequency analysed 63 hours of observations of this star obtained on 23 nights in the 1980-1981 observing season. He found two pairs of frequency triplets with spacings equal to the rotation frequency of the star. Severe daily aliasing problems made these identifications insecure, however. In order to reduce these aliasing problems Kurtz & Seeman (1983) reobserved HD 24712 for 119 hours on 19 nights late in 1981. For five of those nights centred on the time of pulsation amplitude maximum they had contemporaneous observations from both the South African Astronomical Observatory (SAAO) and the Cerro Tololo Interamerican Observatory (CTIO). In order to suppress the frequency sidelobes which arise due to the rotational amplitude modulation of the light variations in this star, they chose to frequency analyse only three nights of contemporaneous SAAO and CTIO data obtained centred on the time of pulsation amplitude maximum. This allowed them to derive the six frequencies listed in Table 3. The amplitudes given for those frequencies in Table 3 are the amplitudes at the time of pulsation amplitude maximum. At other times those amplitudes are smaller.

Because of the nearly equal spacing of the six frequencies, Kurtz & Seeman suggested that they are due to pulsation in modes of consecutive overtone, n, which implies that the values of n are near 80. They could offer no explanation of the relative amplitudes of these frequencies. Shibahashi (1984) and Shibahashi & Saio (1985a) calculated that the frequency separation of consecutive overtones in a model appropriate to HD 24712 is about $\nu_0=65$ µHz. Since this is about twice the frequency separation of 33 µHz actually observed in HD 24712, they suggested that these frequencies are due to modes of alternating even and odd ℓ. A look at equation 1 explains this suggestion. This leads to a best estimate for the overtones of the frequencies in

HD 24712 of $39 \leq n \leq 42$ (Shibahashi, private communication). Gabriel *et al.* (1985) have made similar calculations which agree with Shibahashi & Saio's results. Shibahashi & Saio also found that theoretically

$$\nu(n,\ell=1)-\nu(n,\ell=0) \;<\; \nu(n+1,\ell=0)-\nu(n,\ell=1) \qquad\qquad (2)$$

whereas

$$\nu(n,\ell=1)-\nu(n-1,\ell=2) \;>\; \nu(n,\ell=2)-\nu(n,\ell=1). \qquad\qquad (3)$$

This led them to the suggestion that if f_1 and f_2 are due to $\ell=1$ modes in HD 24712, then f_3, f_4, and f_5 are more likely to be due to radial ($\ell=0$) than quadrupole ($\ell=2$) modes. This suggestion is easy to test since radial oscillations undergo no amplitude modulation as a function of aspect since they are spherically symmetric. Kurtz, Schneider, & Weiss (1985) made this test by observing HD 24712 at the time of pulsation amplitude minimum. They obtained 20 hours of observations from CTIO and SAAO contemporaneously on 4 nights in 1983 centered on the time of pulsation amplitude minimum. If any of the modes in HD 24712 are due to radial pulsation then the amplitude of those modes should be the same at all times. Kurtz, Schneider, & Weiss found that all of the frequencies in HD 24712 are amplitude modulated so that none of them can be due to radial pulsation.

This leaves an important outstanding problem. Shibahashi & Saio's calculations of the frequency spacings are straightforward and agree with the relative spacings of the frequencies of modes of even and odd ℓ in the sun. Gabriel *et al.* obtain similar results from their calculations. Yet it is apparent from the observations that there are no frequencies in the light variations of HD 24712 which behave as if they were due to radial pulsation modes. One may make many speculations as to the reason for this, but at present an answer must await more detailed observations and new theoretical developments.

I currently have collected 210 hours of observations of HD 24712 obtained by my collaborators and me on 51 nights from 1980 to 1983. Although these observations are well spread over the rotational phases of HD 24712, the aliasing problems due to the very low fraction of the total time span that the observations cover precludes the possibility of obtaining a definitive frequency solution from them. I have tried to frequency analyse the entire data set with the following results. The two frequencies of highest amplitude, f_1 and $f_{2,-1}$ can be derived with essentially no ambiguity; their 1 yr $^{-1}$ =32 nHz aliases are the only possible problem. The frequency f_3 is less certain, but possibly correctly identified. With caution those three frequencies are given in Table 4 along with their separations and the earlier estimates derived from the values given in Table 3 from Kurtz & Seeman. The inequality $f_1-f_3<f_3-f_2$ still appears to obtain.

Table 4.   The frequencies f_1, f_2, and f_3 derived from 51 nights of observations of HD 24712.

f(mHz)	A(mmag)
±0.000004	±0.03
2.720907	1.16
2.652927	1.05
2.687404	0.63

$f_1-f_3=33.503$ µHz (Kurtz & Seeman: 33.3 µHz)
$f_3-f_2=34.477$ µHz (Kurtz & Seeman: 34.7 µHz)

After the frequencies f_1, f_2, and f_3 have been removed from the data the highest peaks in the amplitude spectrum are centred on the frequencies f_1 and f_2. It is not possible, however, to tell if these peaks are associated with frequency triplets, quintuplets, or something else because of the severe aliasing problems.   My earlier claim of frequency triplets about f_1 and f_2 (Kurtz 1982) is not necessarily correct.   It is still clear, however, that the pulsation amplitude modulation remains in phase with the magnetic field variations to an accuracy of better than 0.2% (Kurtz & Seeman 1983).   This means that the frequencies f_1 and f_2 are amplitude modulated with the rotation period of the star; it is only the form of the amplitude modulation which is unknown.

A complete frequency solution for HD 24712 is extremely desirable.   In order to obtain such a solution, the following observations are necessary.   Because the separations of the frequencies in HD 24712 are so close to 3 d^{-1}, contemporaneous observations from at least two observatories, and preferably three, well separated in longitude are needed. Obvious choices are observatories in Chile, South Africa, and Australia or New Zealand, but northern hemisphere observatories are not excluded since the declination of HD 24712, $\delta=-12^0$, is not too far south.   Because the rotation period of HD 24712 is 12.4564 day and because it is necessary to cover the rotation cycle by at least 1.5 cycles and preferably 2 cycles in order to resolve the rotational frequency sidelobes fully, it will be necessary to obtain these multisite observations as close to continuously as possible over a period of about 25 days. With 50% photometric weather, that may provide sufficient observations to solve the frequency spectrum.

3.3 HD 60435

　　HD 60435 has the most complex frequency spectrum of any of the Rapidly Oscillating Ap stars. At times it pulsates with periods near 15, 12, 6, and/or 4 min including both the longest and shortest periods yet discovered in these stars. The principal oscillation modes are the ones with periods near 12 min. At times the peak-to-peak light variation in HD 60435 reaches about 16 mmag making it also the largest amplitude Rapidly Oscillating Ap star.

　　Matthews, Kurtz, & Wehlau (1985) have frequency analysed 147 hours of observations obtained on 17 nights at

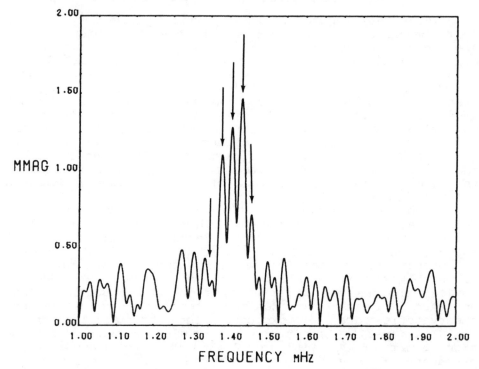

Figure 1

SAAO and the Carnegie Southern Observatory (CARSO) in 1984, much of it contemporaneously from the two sites, in addition to 56 hours of observations obtained on 16 nights at SAAO in 1983 by Kurtz (1984). Some of the frequencies they found seemed to be amplitude modulated on a time scale of about 7-8 days during one observing run, but then were either not

seen again or rarely seen again during other observing runs.
Mean light observations of HD 60435 support the contention
that the rotation period of this star is about 7.7
days.

On the night of JD2445728 Matthews, Kurtz, & Wehlau
obtained a light curve 13.53 hours long which gave the
interesting amplitude spectrum shown in Fig. 1. There are
at least three, and probably four, frequencies present in
this amplitude spectrum with a uniform spacing of about
26 μHz. Table 5 gives those frequencies and amplitudes in
addition to the single frequency found on the night of
JD2445383 when HD 60435 was at its highest amplitude. It can

Table 5. Frequencies derived on JD2445383 and JD2445728 for
HD 60435.

JD	f (mHz)	A (mmag)
2445383	$f_1=1.353\pm0.020$	6.13
2445728	$f_2=1.382\pm0.005$	1.04
	$f_3=1.408\pm0.005$	1.33
	$f_4=1.434\pm0.005$	1.52
	$f_5=1.459\pm0.005$	0.68

$$f_5-f_4=25 \ \mu Hz$$
$$f_4-f_3=26 \ \mu Hz$$
$$f_3-f_2=26 \ \mu Hz$$
$$f_2-f_1=29 \ \mu Hz$$

be seen that the frequency on JD2445383 was 29 μHz less than
f_2 on JD2445728, yet it is conspicuously absent in Fig. 1.
Thus there is the suspicion that some of the frequencies in
HD 60435 may have lifetimes which are short on the time-
scale of the observations. Matthews, Kurtz, & Wehlau give
other arguments in support of this idea. Whether the modes
associated with these frequencies are transient or whether
the frequency spectrum of HD 60435 is so complex that we
have not been able to decipher it yet is uncertain.

This series of frequencies roughly equally spaced at an
interval of 26 μHz is consistent with modes of alternating
even and odd ℓ as was observed in HD 24712. That gives
$\nu_0=52$ μHz which is consistent with a slightly evolved A
star (Shibahashi & Saio 1984a, Gabriel et al. 1985). It is
interesting to note in this context the δc_1 index for
HD 60435 given in Table 1 is the largest of any of the
Rapidly Oscillating Ap stars. HD 60435 may be the most
evolved of these stars.

Simple calculations using the asymptotic formula in
equation 1 and the frequencies in Table 3 indicate that the
12 min periods in HD 60435 are probably due to modes with
$n\cong20$. Similar calculations indicate that the 15 min periods

are probably due to $n \cong 17$. This brings up the very interesting question of where the intermediate modes with $n=18$ and
19 are. The 6 min and 4 min periods in HD 60435 may be
associated with harmonics of the 12 min periods, but why
should the 15 min periods be present and not ones intermediate between the 15 and 12 min periods?

The complexity of the variations in HD 60435 make its
frequency spectrum very difficult to decipher, but that very
complexity promises a great deal of information once it is
sorted out. This star deserves substantial observing
efforts.

3.4 HD 83368 (HR 3831)

Of all the Rapidly Oscillating Ap stars, HD 83368 presents the best case for the oblique pulsator model. Kurtz
(1982) found the six frequencies listed in Table 3 from an
analysis of 135 hr of observations obtained on 37 nights in
1980-1981. He suggested that the low frequency triplet is
due to $\ell=1$ oblique pulsation and that the high frequency
triplet is due to $\ell=2$ oblique pulsation. That led to the
prediction of a rotation inclination, $i=86^{o}$, a magnetic
obliquity, $\beta=36^{o}$, a polarity reversing magnetic field, and a
rotational velocity of v sin $i=32$ km s^{-1}. Thompson (1983)
has since confirmed that HD 83368 has a polarity reversing
magnetic field and Carney & Peterson (1985) have measured
the rotational velocity to be v sin $i=33\pm3$ km s^{-1}.

The frequency pattern in HD 83368 has been solved down
to a noise level of 0.15 mmag for the low frequency triplet
and 0.10 mmag for the high frequency triplet. Although this
is the best case for the oblique pulsator model, the inequality of the amplitudes of f_1 and f_2 in the low frequency
triplet is a problem for that model. It was just this problem which led Mathys (1984, 1985) to propose the spotted
pulsator model in which the amplitudes and phases are free
parameters.

The relative amplitudes of the frequencies in HD 83368
are arbitrary if it is not an oblique pulsator. Kurtz
(1982) suggested that the reason for the inequality of the
amplitude of f_1 and f_2 in this star is because of the
presence of either one or both of the $m \neq 0$ modes for the low
frequency $\ell=1$ oscillation. These m-modes would have frequencies which are very close to, but not coincident with,
f_1 and f_2 so that over a sufficient length of time they
could be resolved from f_1 and f_2. I have tried to resolve
these frequencies partially by examining the phase of the
low frequency oscillation as a function of rotational phase.

In the oblique pulsator model the low frequency triplet
in HD 83368 results from amplitude modulation of a single
$\ell=1$, $m=0$ mode with the rotation period of $P_{rot}=2.85204$ day
(Renson et al. 1984). (These authors prefer half that

period based on the mean light variations alone, but it is clear from both the rapid oscillations and the magnetic field variations that the longer period is the correct one.) Thus HD 83368 actually oscillates with only one frequency, $f_3 = 1.428011$ mHz. The other low frequencies, f_1 and f_2, arise because of the amplitude modulation of f_3 with rotation; the high frequencies are possibly due to the first harmonic of f_3 or a mode resonantly excited by the first harmonic of f_3.

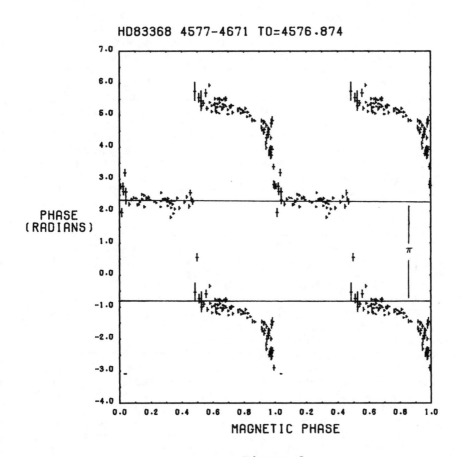

Figure 2

I have fitted f_3 to one hour segments of the observations of HD 83368 by least squares. Fig. 2 shows the phase of f_3 as a function of magnetic (rotation) phase. The error bars are the formal 1σ phase errors from the least squares fit. The magnetic phase has been calculated using the ephemeris of Thompson (1983) and the above given P_{rot}. If f_3 were a purely obliquely pulsating $\ell=1$ mode, then the phases in Fig. 2 would phase-flip by π radians at quadrature, but otherwise would be constant. That the phases in Fig. 2 do undergo reversal at quadrature (particularly at magnetic phase 0.5) strongly supports the contention that f_3 is due to an $\ell=1$, $m=0$ mode. However, the deviations from the expectations of a pure $\ell=1$, $m=0$ mode need explanation.

The phases in Fig. 2 can be modelled exactly by using the amplitudes and phases of f_1, f_2, and f_3 given in Table 3. The explanation for Fig. 2 in the spotted pulsator model would be that f_1, f_2, and f_3 just happen to have the amplitudes and phases that they do, since those are free parameters in that model. In the oblique pulsator model, on the other hand, it is necessary to propose that $m\neq0$ modes are present with frequencies very near to f_1 and f_2. It is the presence of these $m\neq0$ modes which causes the phase in Fig. 2 to drift between magnetic phase 0.5 and 1.0. I find I can get a reasonable fit to the phases in Fig. 2 by assuming the presence of an $\ell=1$, $m=0$ mode at f_3 and an $\ell=1$, $m=1$ mode at f_1 with an amplitude of 0.1 that of f_3. This second frequency is, of course, not resolved from f_1. Thus it provides an explanation within the oblique pulsator model why f_1 and f_2 do not have equal amplitudes and why the oscillation phase is not constant between magnetic phase 0.5 and 1.0 in Fig. 2.

Within the oblique pulsator model $f_3=f_1-\Omega$ where Ω is the rotation frequency. The frequency of the proposed $\ell=1$, $m=1$ mode would be $f=f_1-(1-C_{n,\ell})\Omega$ where $C_{n,\ell}$ is very small (Shibahashi & Saio (1985a) estimate $C_{n,\ell}\leq0.001$ for the Rapidly Oscillating Ap stars). Thus $f-f_1=C_{n,\ell}\Omega$. We would, therefore, expect the amplitudes of f_1 and f_2 to change with time and we would expect the magnetic phase of the deviation from a pure $\ell=1$, $m=0$ mode in Fig. 2 also to drift with time if this model is correct.

In order to test this hypothesis, I have obtained 43 hr of new observations of HD 83368 on 11 nights in 1985. These observations are separated from the observations of Kurtz (1982) by 4 yr which should be sufficient time to see the expected effect, even if $C_{n,\ell}$ is as small as Shibahashi & Saio suggest. Fig. 3 shows the phases of the new observations plotted against magnetic phase using the same magnetic ephemeris as in Fig. 2. The zero point of the oscillation phase is arbitrary in Figs. 2 and 3 because f_3 is not known well enough to jump the time-gap between the 1980-1981 and the 1985 observations. It is clear that Fig. 3 is subtly, but significantly, different from Fig 2. Such a change could be due to some arbitrary change in the pulsation modes of HD 83368, but I think that it can be explained more satisfactorily in terms of oblique pulsation in the manner outlined above. A thorough analysis of the new observations and formal modelling of the phases in Figs. 2 and 3 is in preparation.

Finally, it should be noted that Dziembowski & Goode (1985) expect that the amplitudes of f_1 and f_3 should be unequal due to the the effect of the magnetic field. In fact, their model can be inverted and the inequality of f_1 and f_3 can be used to calculate the strength of the internal magnetic field in HR 3831. Within their model, however, there appears to be no explanation for the slight difference between Figs. 2 and 3.

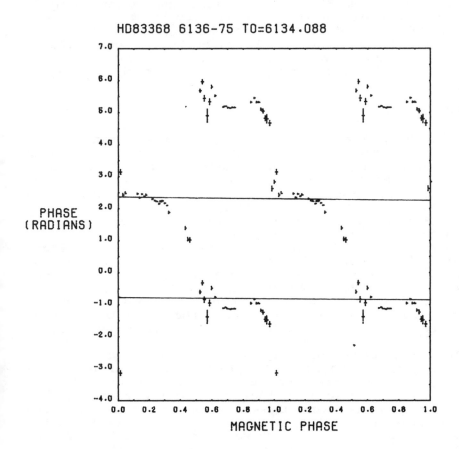

HD83368 6136-75 TO=6134.088

PHASE (RADIANS)

MAGNETIC PHASE

Figure 3

Kurtz (1982) rejected the interpretation of the high frequency triplet in HD 83368 as simply the first harmonic of f_3 because of the expectation that the harmonic frequency should be amplitude modulated in the same way as the fundamental frequency; *i.e.* because the high frequency triplet should be split like an $\ell=1$ mode which is contrary to what is observed. Osaki (private communication) has pointed out that if the high frequency triplet is a harmonic of the low frequency triplet, then the second order terms in the pulsation equations lead to the expectation that the harmonic should behave like a quadrupole mode, as is observed.

Further observations are needed to examine this question in more detail. First harmonic frequencies have now been detected in HD 60435, HD 83368, and HD 101065. On one night of the new HD 83368 observations the second harmonic is clearly visible in the amplitude spectrum. Matthews, Kurtz, & Wehlau (1985) have detected what appears to be the second harmonic (P=4 min) of the highest amplitude oscillation in HD 60435 (P=12 min), but the fundamental was not present during the week that the P=4 min period was seen in that star.

3.5 HD 101065

HD 101065, Przybylski's star, was the first of the Rapidly Oscillating Ap Stars to be discovered. This star has a long-standing controversy surrounding it. Estimates of its temperature range from about 6300 K to 7500 K and there is disagreement whether it is an Ap star. This review is not the place for a rediscussion of these problems; see Kurtz & Wegner (1979) and Wegner *et al.* (1983) for their opinion on this problem and Przybylski (1982) for his.

There is, however, no doubt that this star is light variable. Its principal light variation has a period of 12.140 min and a peak-to-peak amplitude in B of just over 0.01 mag. In 1984 I obtained an additional 79 hr of B observations of HD 101065 spread over 23 nights. An analysis of those observations along with the 71 hr of observations spread over 20 nights in 1978, 1979, 1980 (Kurtz 1981) gives the following results.

The frequency of highest amplitude, $f_1=1.372865$ ±0.000001 mHz, is the first frequency listed for HD 101065 in Table 3. This frequency can be independently derived from the yearly subsets of the data (within their resolution). It does, however, appear to be amplitude modulated on an unknown time-scale. A least squares fit of f_1 to the yearly subsets of the data shows a small, but significant variation in amplitude from year to year. Whether this is due to a real variation in the amplitude of the mode which gives rise to f_1 or whether it is due to beating between f_1 and some nearby frequency cannot be determined from the present data. When f_1 is removed from the entire data set, there is amplitude well above the noise level left in the amplitude spectrum around f_1. The actual frequencies of those secondary peaks in the amplitude spectrum are different for the different data sets, though, so I have not been able to obtain a definitive frequency solution.

The second frequency listed for HD 101065 in Table 3, $f_2 = 1.315079 \pm 0.000001$ mHz is well separated from f_1. This frequency can be independently derived from the yearly subsets of the data. Its alias pattern is substantially asymmetrical for all of the data subsets in the sense that its -1 d$^-$ is always higher than its $+1$ d$^-$ alias. The frequency which perturbs the f_2 alias pattern has an amplitude too small to be seen above the noise level when the data are prewhitened by f_2. The difference $f_1 - f_2 = 57.786 \pm 0.002$ µHz is about what we expect from Shibahashi & Saio's (1985a) and Gabriel et al.'s (1985) A star models for pulsation modes of consecutive overtones, n, of the same degree, ℓ. If that interpretation of f_1 and f_2 is correct, then HD 101065 is near the main sequence.

The third frequency given for HD 101065 in Table 3, $f_3 = 2.7459 \pm 0.0001$ mHz, is probably the first harmonic of f_1 since $f_3 = 2f_1$ with the errors. The f_3 given here was determined from 10 nights of data spanning 21 days in 1984. It stands well above the noise level and can be seen on many nights so there is little doubt that it is real. It can also be seen in the 1979 HD 101065 data; the only reason I did not report its existence earlier is that I never thought to look for it carefully at the time I was doing the earlier frequency analyses. Thus three of the Rapidly Oscillating Ap stars, HD 60435, HD 83368, and HD 101065 show the presence of the first harmonic of their frequency of highest amplitude.

In earlier work on this star (Kurtz 1980, 1982) I suggested, based on the phase shift between the colour and light curves, $\Delta\phi(V,B-V)$, that f_1 may be due to an $\ell = 2$ oscillation mode. I now think that this suggestion was premature. Further work on these phase shifts in HD 83368 (Kurtz 1982), α Cir (Kurtz & Balona 1984), and HD 6532 (Kurtz & Kreidl 1985) shows that secure mode identifications from $\Delta\phi(V,B-V)$ is not yet possible in the Rapidly Oscillating Ap stars. The degree of the oscillation mode for f_1 in HD 101065 is therefore unknown.

3.6 HD 128898 (HR 5463; α Cir)

At $V = 3.198$ α Cir is the brightest of the Rapidly Oscillating Ap stars. Kurtz & Balona (1984) frequency analysed 138 hr of observations obtained on 38 nights in 1981 and 1983 including two nights of contemporaneous observations from SAAO and CTIO. They found only the two frequencies listed in Table 3 above a noise level of 0.2 mmag. The separation of those two frequencies is $f_1 - f_2 = 2.5$ µHz which indicates that they may be due to pulsation in modes of

overtone and order (n, ℓ) and $(n-1, \ell+2)$ with $n \cong 25$. This
small separation is more consistent with $\ell=$even than $\ell=$odd
modes according to the models of Shibahashi & Saio (1985a).
Kurtz & Balona also measured the phase shift $\Delta\phi(V,B-V)$ for
α Cir, but they found that no consistent mode identifications are possible from this phase shift for all of the
Rapidly Oscillating Ap stars for which it has been
determined.

3.7   HD 134214

HD 134214 was discover to be a Rapidly Oscillating Ap
stars by Kreidl (1985b). It oscillates principally with one
of the shortest periods yet found in these stars, 5.650
minutes, and with a peak-to-peak amplitude in B of about 7
mmag. Kreidl discovered the variability in this star only
shortly before this paper was written, so little more can be
said at the present time.

3.8   HD 137949 (33 Lib)

Kurtz (1982) obtained 44 hr of observations of
HD 137949 on 17 nights in 1981. A frequency analysis of the
observations showed only one frequency present in the light
variations at an amplitude greater than the 0.2 mmag noise
level in the amplitude spectrum. He showed that this single
frequency is consistent with the oblique pulsator model and
the few magnetic observations which are available for this
star.

3.9   HD 201601 (HR 8097; γ Equ)

Kurtz (1983a) obtained 36 hr of observation of γ Equ on
11 nights in 1981. The amplitude of the single frequency
detected in the light variation of this star is only 0.86
mmag, but it clear that this single frequency is amplitude
modulated. However, any other frequencies present in the
light variations are lost in the 0.2 mmag noise. The
magnetic variations in this star are well known for their
apparently very long period. Bonsack & Pilachowski (1974)
have suggested that the rotation period of γ Equ is around
72 yr if the magnetic variations are due to oblique rotation. Kurtz (1983a) pointed out that if γ Equ is also an
oblique pulsator, then its rotation period could be, in
principle, derived from observations of the pulsational
light variations. This could be done by detecting the rotational sidelobes of the main frequency. The problem in
practice is that the light variations in γ Equ are so small
that detection of further frequencies with even smaller
amplitudes will be very difficult.

3.10 HD 203932

HD 203932 is the smallest amplitude Rapidly Oscillating Ap star yet found. From 29 hr of observations obtained on 12 nights in 1983 Kurtz (1984) found the single frequency given in Table 3, f_1=2.804789 mHz, present in the data above the noise level of 0.2 mmag.

3.11 HD 217522

Kurtz (1983b) observed this star for 97 hr on 18 nights in 1982. He found a single frequency present in the light variations, f_1=1.21510 mHz, which he was able to show is clearly amplitude modulated on an unknown time-scale. He pointed out that further observations of this star, particularly contemporaneous observations from at least two observatories, may lead to a frequency solution, but that, given the very low amplitude of the variations in HD 217522, effort would be more profitably spent on some of the larger amplitude Rapidly Oscillating Ap stars.

4. Low overtone pulsating Ap stars

Weiss (1983) discussed five stars which are classified as Ap stars and which are also δ Scuti variables. He concluded that at least some of these stars very probably are Ap stars, which means that, under some conditions, the Ap phenomenon can occur in a δ Scuti pulsator. Some of these stars are problematical, however. HD 108945 (HR 4766; 21 Com) has been observed many times over a time span of more than 20 yr. Some observers claim to have detected a period near 30 min while others find no light variability at all. Whether some of the observations are in error or whether this star has a highly variable amplitude is not known; Weiss et al. (1980) discuss the observations. Another star which has this same problem is HD 65339 (HR 3109; 53 Cam). Burnashev et al. (1983) reported periods of 20, 27, and 79 min in their data whereas Kreidl (1985a) more recently could find no light variability at all.

It is difficult to conclude anything at present from these stars, but their existence certainly has a bearing on the problem of the excitation mechanism in the Rapidly Oscillating Ap stars. These low-overtone oscillating Ap stars appear to be δ Scuti stars which we think are excited by the κ-mechanism. If we believe that the oscillations in the Rapidly Oscillating Ap stars are excited by overstable magnetic convection, then we must accept that there are two pulsation driving mechanisms operating under (presumably only slightly) different conditions in Ap stars which are otherwise very similar to each other.

5. Mean light observations and the rotational periods of the Rapidly Oscillating Ap stars

For many of the Rapidly Oscillating Ap stars the best method for determining the period of rotation is by observing the mean light variations. For some of these stars this has been done, but for many there is little known about the rotation period. For interpretation of the magnetic observations in the oblique rotator model and the rapid oscillations in the oblique pulsator model or the spotted pulsator model, it is imperative that the rotation period be known. I therefore encourage observers of mean light variations in Ap stars to put the Rapidly Oscillating Ap stars on their observing lists. In particular HD 60435 and α Cir need accurately determined rotation periods. The brightness of α Cir probably accounts for the lack of previous mean light observations. Information about mean light observations already made can be found in the individual papers listed in Table 3.

Acknowledgements

I would like to thank Dr. Hiromoto Shibahashi for his comments on this paper, Drs. Wojtek Dziembowski and Phil Goode for enlightening discussions, Dr. Tobias Kreidl for information about the oscillations in HD 134214 prior to publication, and Dr. Werner Weiss for measurements of the Hβ indices in HD 203932 and HD 217522.

References

Aizenman, M. L., Hansen, C. J., Cox, J. P., & Pesnell, W. D., 1984. *Astrophys. J.*, **286**, L43.
Babcock, H. W., 1958. *Astrophys. J. Suppl.*, **3**, 141.
Bidelman, W. P., & MacConnell, D. J., 1973, *Astr. J.*, **78**, 687.
Blanco, V. M., Demers, S., Douglass, G. G., & Fitzgerald, M. P., 1970. *Publs U. S. Naval Obs.*, **11**, 1.
Bonsack, W. K., & Pilachowski, C. A., 1974. *Astrophys. J.*, **190**, 327.
Borra, E. F., & Landstreet, J. D., 1975. *Publ. astr. Soc. Pacific*, **87**, 961.
Borra, E. F., & Landstreet, J. D., 1980. *Astrophys. J. Suppl.*, **42**, 421.
Burnashev, V. I., Polosukhina, N. S., & Malanushenko, V. P., 1983. *Sov. Astr. Letters*, **9**, 151.
Carney, B. W., & Peterson, R. C., 1985. *Mon. Not. R. astr. Soc.*, **212**, 33P.
Cousens, A., 1983. *Mon. Not. R. astr. Soc.*, **203**, 1171.

Cox, J. P., 1984. *Astrophys. J.*, **280**, 220.
Crawford, D. L., 1975. *Astr. J.*, **80**, 955.
Crawford, D. L., 1979. *Astr. J.*, **84**, 1858.
Dolez, N., & Gough, D. O., 1982. in *Pulsations in Classical and Cataclysmic Variables*, eds. J. P. Cox and C. J. Hansen, JILA, Boulder, p. 248.
Dziembowski, W., 1984. in *Theoretical Problems in Stellar Stability and Oscillations*, Proc. 25th Liege coll., eds. M. Gabriel and A. Noels, in press.
Dziembowski, W., & Goode, P. R., 1985. preprint.
Gabriel, M., Noels, A., Scuflaire, R., & Mathys, G., 1984. *Astr. Astrophys.*, **143**, 206.
Hauck, B., & Mermilliod, M., 1980. *Astr. Astrophys. Suppl.*, **40**, 1.
Hoffleit, D., 1982. *Bright Star Catalogue*, 4th edn, Yale University Observatory.
Houk, N., 1978. *Michigan Spectral Catalogue*, vol. 2, Department of Astronomy, University of Michigan, Ann Arbor.
Houk, N., 1982. *Michigan Spectral Catalogue*, vol. 3, Department of Astronomy, University of Michigan, Ann Arbor.
Houk, N., & Cowley, A. P., 1975. *Michigan Spectral Catalogue*, vol. 1, Department of Astronomy, University of Michigan, Ann Arbor.
Kreidl, T. J., 1985a. *Mon. Not. R. astr. Soc.*, in press.
Kreidl, T. J., 1985b. preprint.
Kurtz, D. W., 1980. *Mon. Not. R. astr. Soc.*, **191**, 115.
Kurtz, D. W., 1981. *Mon. Not. R. astr. Soc.*, **196**, 61.
Kurtz, D. W., 1982. *Mon. Not. R. astr. Soc.*, **200**, 807.
Kurtz, D. W., 1983a. *Mon. Not. R. astr. Soc.*, **202**, 1.
Kurtz, D. W., 1983b. *Mon. Not. R. astr. Soc.*, **205**, 3.
Kurtz, D. W., 1984. *Mon. Not. R. astr. Soc.*, **209**, 841.
Kurtz, D. W., & Balona, L. A., 1984. *Mon. Not. R. astr. Soc.*, **210**, 779.
Kurtz, D. W., & Kreidl, T. J., 1985. *Mon. Not. R. astr. Soc.*, in press. 216, 987.
Kurtz, D. W., Schneider, H., & Weiss, W. W., 1985. *Mon. Not. R. astr. Soc.*, in press.
Kurtz, D. W., & Seeman, J., 1983. *Mon. Not. R. astr. Soc.*, **205**, 11.
Kurtz, D. W., & Wegner, G., 1979. *Astrophys. J.*, **232**, 510.
Lavagnino, C. J., 1960. *Bol. Assoc. Argent. Astron.*, **2**, 40.
Ledoux, P., 1951. *Astrophys. J.*, **114**, 373.
Matthews, J., Kurtz, D. W., & Wehlau, W., 1985. *Astrophys. J.*, in press.
Mathys, G., 1984. in *Theoretical Problems in Stellar Stability and Oscillations*, 25th Liege International Astrophys. Symposium, in press.
Mathys, G., 1985. *Astr. Astrophys.*, in press.
Olsen, 1983. *Astr. Astrophys. Suppl.*, **54**, 55.
Olsen E. H., & Perry, C. L., 1984. *Astr. Astrophys. Suppl.*, **56**, 229.

Pesnell, W. D., 1985. *Astrophys. J.*, in press.
Preston, G. W., 1972. *Astrophys. J.*, 175, 465.
Przybylski, A., 1982. *Astrophys. J.*, 257, L83.
Renson, P., Manfroid, J., Heck, A., & Mathys, G., 1984. *Astr. Astrophys.*, 131, 63.
Scargle, J., 1982. *Astrophys. J.*, 263, 835.
Shibahashi, H., 1984. Mem. Soc. Astron. Ital., 55, 181.
Shibahashi, H., 1983. *Astrophys. J.*, 275, L5.
Shibahashi, H., & Saio, H., 1985a. *Publ. astr. Soc. Japan*, in press.
Shibahashi, H., & Saio, H., 1985b. *Publ. astr. Soc. Japan*, in press.
Scholz, G., 1979, *Astr. Nachr.*, 300, 213.
Tassoul, M., 1980. *Astrophys. J. Suppl.*, 43, 469.
Thompson, I. B., 1983. *Mon. Not. R. astr. Soc.*, 205, 43p.
Vogt, N., & Faundez, M., 1979. *Astr. Astrophys. Suppl.*, 36, 477
Wegner, G., Cummins, D. J., Byrne, P. B., & Stickland, D. J., 1982. *Astrophys. J.*, 272, 646.
Weiss, W. W., 1983. *Astr. Astrophys.*, 128, 152.
Weiss, W. W., Breger, M., & Rakosch, K. D., 1980. *Astr. Astrophys.*, 90, 18.
Wolff, S. C., 1975. *Astrophys. J.*, 202, 127.
Wolff, S. C., & Hagen, W., 1976. *Publ. astr. Soc. Pacific*, 88, 119.
Wood, H. J., & Campusano, L. B., 1975. *Astr. Astrophys.*, 45, 303.

Asterseismology for Certain Ap Stars

W. Dziembowski
N. Copernicus Astronomical Center
Warsaw, Poland
        and
Philip R. Goode
Department of Physics
New Jersey Institute of Technology
Newark, New Jersey, USA

ABSTRACT. The discovery by Kurtz of phase coherent, rapid
oscillations in certain Ap stars holds great promise for
the new field of asterseismology. We discuss here a
generalized oblique pulsator model for these stars which
allows us to accurately compare the effects of rotation and
magnetism. A central role for mode coupling and amplitude
limitation in the observed period doubling is suggested.

1. Introduction

    We discuss here a seismology for certain Ap stars which
are observed to oscillate rapidly with extremely stable
periods between 4 and 15 minutes. Each of these stars is a
cool magnetic Ap star. Kurtz (1982) attributed the
oscillations to high order, low degree acoustic modes
because of their short periods and their spacing in
frequency. The amplitude of many of these oscillations are
modulated with the rotation period of the star with the
maxima of the magnetic field and oscillation amplitude
coinciding. Further, some of these frequencies are
separated from their nearest neighbors by precisely the
rotation frequency of the star. Kurtz (1982) has explained
these phenomena by assuming that the large amplitude
oscillations are aligned with the magnetic field, which is
inclined to the axis of rotation, and each of these
oscillations, as reported by an inertial frame observer,
will generate (2ℓ+1)-frequencies spaced by the rotation
rate of the star. This is the so-called oblique pulsator
model. Kurtz's model ignores advection. Primarily on this
ground a series of more complicated models were proposed,

D. O. Gough (ed.), Seismology of the Sun and the Distant Stars, 441–451.

each with its own fundamental shortcoming associated with
its failure to account for the phase coherence of the
oscillations. For a discussion of these models, see
Dziembowski and Goode (1985) and the review by
Kurtz (1986). Dziembowski and Goode (1985) pointed out
that the oblique pulsator model is generalized by treating
the effects of advection and magnetism together. In a
simple and natural way, the frequency spacing and phase
coherence of the data are explained. Also, this
generalization allows a fairly detailed seismology of the
stars from other apparent anomalies in the data with regard
to the original oblique pulsator model. We emphasize that
the data are so precise as to provide us a unique situation
in asterseismology. We can hope, for instance, to
determine detailed properties of stellar magnetic fields
which are critical to understanding stellar evolution, in
general.
        We develop here the theory behind the generalized
oblique pulsator model including specific results for one
star. Some general expectations of the seismology,
including diagnostic roles for driving, damping, mode
coupling and amplitude limitations, are suggested.

2.   The Generalized Oblique Pulsator Model

        We show here a generalization of Kurtz's model to
account for advection. Formally, we treat the problem of a
nonaxisymmetric steady perturbation in the intrinsic frame
which we assume co-rotates with the field. Because the
perturbations do not have the same axis of symmetry, the
normal modes are coupled which implies the use of
degenerate perturbation theory. Our treatment is similar
to that of Gough and Taylor (1984) and Dziembowski and
Goode (1984) and consistent with that of Dicke (1982).
        We write $\vec{\xi}_{n\ell m}(r,\Theta,\phi,t)$ abbreviated as $\vec{\xi}_m$, the
unperturbed oscillation in the intrinsic frame with respect
to the axis of rotation of the field and star. The
perturbation is steady in this frame. Defining O to be the
nonaxisymmetric perturbing operator, we may write O_{jm}, one
of its matrix elements, as

$$O_{jm} = \int \vec{\xi}_j \cdot O \vec{\xi}_m d^3x.$$ (1)

        The perturbed eigenfunction in this frame is a linear
combination of the $\vec{\xi}_m$ where

$$\vec{\xi}^{(s)} = \sum_{m=-\ell}^{\ell} a_m^{(s)} \vec{\xi}_m$$ (2)

and

$$(\delta_{j,m}\omega_1^{(s)} - O_{jm})a_m^{(s)} = 0.$$ (3)

For each $(n\ell)$-multiplet, there are $(2\ell+1)$ eigenvalues, $\omega_1^{(s)}$ and eigenvectors, $\vec{\xi}^{(s)}$. The $(2\ell+1)$ eigenvalues and eigenvectors are labeled by s. The matrix element O_{jm} has components due to rotation and magnetism,

$$O_{jm} = O_{jm}^{rot} + O_{jm}^{mag}$$ (4)

The rotation matrix is diagonal in the chosen frame and is given by

$$O_{jm}^{rot} = \omega_{1,m}^{rot} \delta_{j,m}$$ (5)

If the rotation rate, Ω is constant, as we assume, then the Ledoux (1951) formula applies and the advective splitting, $\omega_{1,m}^{rot}$ is given by $cm\Omega$ where c is a constant for each multiplet. O^{mag} is not symmetric in the chosen frame. Assuming the field has axial symmetry, its effect is most easily evaluated in its diagonal frame. That is, we evaluate O^{mag} after transforming to the frame in which the z-axis is the field's axis of symmetry and then transforming the result back to the original frame.

The transformation to the field's axisymmetric system is obtained by applying Weiner functions to the spherical harmonics of the eigenfunction, following Edmonds (1960), we write

$$Y_\ell^m(\theta,\phi) = \sum_{m'} d_{m'm}^{(\ell)}(\beta) Y_\ell^{m'}(\theta',\phi')$$ (6)

where

$$d_{m'm}^{(\ell)}(\beta) = d_{mm'}^{(\ell)}(-\beta)$$ (7)

and where β is the angle between the rotation and magnetic field axes. Using Equations [1] and [6], we write

$$\vec{\xi}_m(\theta) = \sum_{m'} d_{m'm}^{(\ell)}(\beta) \vec{\xi}_{m'}(\theta')$$ (8)

and thus

$$O_{jm}^{mag} = \sum_{m'} d_{m'j}^{(\ell)} d_{m'm}^{(\ell)} \omega_{1,m'}^{mag}.$$ (9)

If, for instance, the field were a simple dipole, we would have

$$\omega_{1,m'}^{mag} = \frac{\ell(\ell+1) - 3(m')^2}{4\ell(\ell+1) - 3} \omega_1^{mag},$$ (10)

where the coefficient of ω_1^{rot} follows from a $|Y_\ell^{m'}|^2$ - weighted integration over the P_2-distortion caused by the field and ω_1^{mag} from the radial properties of the field as sampled by the oscillation. The splitting is due to the $(m')^2$-term. Even if the magnetic field is relatively weak so that the shift it causes in the oscillation frequency is quite small, its effect may well dominate that of rotation because

$$\frac{\omega_1^{rot}}{\omega_0} \sim \frac{mc\Omega}{\omega_0} \sim 10^{-2} \frac{\Omega}{\omega_0} \sim 10^{-5} - 10^{-6}, \tag{11}$$

where ω_0 is the unperturbed oscillation frequency.

The relative change in luminosity due to oscillations, as given by Dziembowski (1977), is

$$\Delta L^{(s)} \sim \sum_m Re \left[Y_\ell^m(\theta,i)e^{j\omega t} \right] a_m^{(s)} \tag{12}$$

where θ and i are the inclination angles of the rotation axis with respect to the observer. The transformation to the observer's frame includes

$$i = \alpha - \Omega t. \tag{13}$$

Thus, the splitting is described by

$$\Delta L^{(s)} \sim \sum_m a_m^{(s)} P_\ell^m(\cos\theta) \sqrt{\frac{(\ell-|m|)!}{(\ell+|m|)!}} \cos\left[(\omega_0 - \omega_1^{(s)} - m\Omega)t + \alpha^{(s)} \right], \tag{14}$$

where the α's are arbitrary phase constants. Therefore, if only one oscillation is excited in an $(n\ell)$-multiplet the observer will report $(2\ell+1)$-frequencies from it, each of which is separated by Ω from its nearest neighbors. We emphasize, thus, that advection does not preclude the phase coherence of the oscillations or upset the agreement between observed and calculated frequency spacings.

The reasons we expect that only a single mode is excited will be discussed in a subsequent section. In the remainder of this section we apply the generalized oblique pulsator model to HD83368.

To present our results in a simple form, we re-write Equation [3] as

$$\left[(\sigma - m\sigma_r)\delta_{j,m} + b_{jm} \right] a_m = 0, \tag{15}$$

where σ now includes constant offsets and

$$b_{jm} = \sum_{m'} d_{m'j} d_{m'm} (m')^2, \tag{16}$$

because we assume a dipole field. We also have

$$\sigma = \omega_1 / _\lambda$$

$$\sigma_r = \omega_1^{rot} / _\lambda$$

and

$$\lambda = \frac{+3}{4\ell(\ell+1)-3} \omega_1^{mag} \qquad (17)$$

Thus, σ_r is the ratio of the rotation effect to the magnetic effect on frequencies. In Kurtz's model, Ironically, the strength of the field can be determined only in the circumstance it is weak enough that its effect on frequencies is comparable to that rotation.

In our calculations for HD83368, we used i and β equal to 33.5 and 87.8 which are consistent with Kurtz's choice of 36 and 85.9 , respectively. The difference is due to our fitting ℓ=1 and 2 together while Kurtz fit only ℓ=1. Our results, in Kurtz's model, are indistinguishable from his.

In this star, all three members of a particular ℓ=1, (nℓ)-multiplet, spaced by the rotation frequency of the star, are identified and three of the five members of an ℓ=2 multiplet, spaced by twice the rotation frequency of the star are identified. Table 1 is taken from Table 3 of Kurtz (1986).

Table 1. Frequencies and Amplitudes for ℓ=1 and
          ℓ=2 Modes in HD83368

P(min)	f(hr)	A(m.mag)		
11.705	1.423950	2.14	0.04	
11.638	1.432069	1.75	0.04	
11.671	1.428011	0.38	0.04	(A_0)
5.836	2.856019	0.45	0.04	(A_0)
5.852	2.847906	0.20	0.04	
5.819	2.864139	0.18	0.04	

The first three frequencies in the Table belong to an ℓ=1 mode and the second three to an ℓ=2 mode. The rotation period of the star is 2.85204 days. For ℓ=1, the predicted ratio of the excitation amplitude of the side peaks to the central peak is

$$\left|\frac{A_{\pm1}}{A_0}\right| = \left|\frac{a_{\pm1}}{a_0}\right| \frac{\tan i}{\sqrt{2}} \xrightarrow{\sigma_r=0} \frac{\tan\beta \ \tan i}{2} \qquad (18)$$

and for l=2

$$\left|\frac{A_{\pm1}}{A_0}\right| = \left|\frac{a_{\pm2}}{a_0}\right| \sqrt{\frac{3}{2}} \frac{\sin^2 i}{(3\cos^2 i-1)} \xrightarrow{\sigma_r=0} \frac{3}{2} \frac{\sin^2\beta}{(3\cos^2\beta-1)} \frac{\sin^2 i}{(3\cos^2 i-1)} \qquad (19)$$

and $\left|\frac{A_{\pm1}}{A_0}\right|$ for ℓ=2 i.e. not the ratio given by $\left|\frac{A_{\pm2}}{A_0}\right|$ Equation [18], must be small with respect to .

We see, from Table 1, that the observed difference between $\left|\frac{A_{+1}}{A_0}\right|$ and $\left|\frac{A_{-1}}{A_0}\right|$ for ℓ=1 is statistically significant

while they would be predicted to be identical in the old
oblique pulsator model.

To use the generalized oblique pulsator model, we vary
and solve Equation [15], until the ratio of the A's, for
the side peaks to the central peak are the same as those
observed.

Our result for σ_r is

$$\sigma_r = 0.10 \pm 0.02 \qquad \text{for } \ell=1$$

and   (20)

$$\sigma_r = 0.03 \pm 0.06 \qquad \text{for } \ell=2.$$

The difference between the two values of σ_r is
a reflection of the period difference between the $\ell=1$ and
$\ell=2$ oscillations. Namely, c is inversely proportional to
the square of the frequency of the oscillation. More
precisely, from Shibahashi and Saio (1985), $c \sim 8 \times 10^{-3}$
for $\ell=1$ and $c \sim 2.5 \times 10^{-3}$ for $l=2$. Since,

$$\sigma_r = \frac{\omega_1^{rot}}{\lambda} \sim \frac{c\Omega}{\omega_1^{mag}}$$ (21)

and since c for $\ell=1$ is about three times that for $\ell=2$ --
the difference is accounted for. For $\ell=1$,

$$\sigma_r = 0.1 = \frac{(8 \times 10^{-3})\,(4\mu Hz)}{\omega_1^{mag}}$$ (22)

implying that $\omega_1^{mag} = 0.32\,\mu Hz$.

Dziembowski and Goode (1984), have shown for high
order p-modes that

$$< \frac{\omega_1^{mag}}{\omega_0} > \sim < \frac{P^{mag}}{P_0} >$$ (23)

where the right hand side of Equation [23] is the ratio of
the magnetic to the hydrostatic pressure. Using Equation
[23], we have

$$< \frac{P_{mag}}{P_0} > \sim < \frac{\omega_1^{mag}}{\omega_0} > \sim \frac{0.32}{1.4 \times 10^3} \sim 3 \times 10^{-4}$$ (24)

which provides a measure of the magnetic field.

If the field were intense rather than weak, another
way to determine the magnetic field might be available.
This would involve a coupling of modes, for instance of $\ell=0$
and 2 modes. Then, an inertial frame observer could report
ten frequencies, instead of the six that could be seen in a
weak field. The pattern of the ten peaks would be used to
determine the field. The role of mode coupling as a
diagnostic will be discussed in a subsequent section.

3.   Frequency Spacing of the Oscillations as a Diagnostic

HD24712 is the only one of these stars for which
oscillations having consecutive n-values have been

observed. Following Shibahashi (1984), the nearly
equidistant spacing between peaks in the frequency spectrum
is interpreted as two interlaced sets of oscillations which
are of consecutive orders and differ by one in degree.

Observationally, the spacing between consecutive
n-values is about 67μHz   This difference may be
accurately estimated using the asymptotic approximation of
Tassoul (1980), in which

$$\nu_{n+1,\ell} - \dot{\nu}_{n,\ell} \simeq \left[2 \int_0^{r_\odot} \frac{dr}{c} \right]^{-1}. \tag{25}$$

The 67μHz value is also consistent with that derived from
stellar models assuming the expected mass range and
provided that the star is somewhat evolved, as discussed by
Gabriel, et al (1984) and Shibahashi and Saio (1985). As
pointed out by Shibahashi and Saio (1985), there is a
consistency problem for these models because the observed
frequencies are higher than the critical frequencies in
these models. The solution they suggest is a modification
of $T=T(\tau)$ relative to standard models because of the
effect of the magnetic field and chemical peculiarities of
the star. The suggestion consists of lowering T (o) which
results in a higher critical frequency. The veracity of
this suggestion may be determined from a search for
spectroscopic evidence reflective of lower temperatures.
Another possible resolution of this problem is the
existence of a strong, turbulent magnetic field in the
stellar interior which would result in a lower speed of
sound relative to that assumed above.

The spacing in the frequency spectrum of HD24712 is
not precisely equidistant. In particular, Kurtz (1982)
points out that

$$\nu_{\ell_1,n_1+1} - \nu_{\ell_2,n_2} < \nu_{\ell_2,n_2} - \nu_{\ell_1,n}. \tag{26}$$

where the subscripts 1 and 2 refer to the higher and lower
amplitude sequences, respectively. Using Tassoul's (1980)
formula, one can derive

$$S_\ell = \frac{\nu_{\ell+1,n} - \nu_{\ell,n}}{\nu_{\ell,n} - \nu_{\ell,n+1}} - \frac{1}{2} = (\ell+1)W \tag{27}$$

where

$$W \simeq \frac{1}{2\pi^2\nu_0} \int \frac{dc}{dr} \frac{dr}{r} \tag{28}$$

and W is negative. These results taken together imply that

$$\ell_1 = \ell_2 + 1$$

and

$$n_1 = n_2 - 1 \tag{29}$$

This led Shibahashi and Saio (1985) to suggest that $\ell = 0$. Subsequently, Kurtz, Schneider and Weiss (1985) showed that the modes labeled by 1 could not be radial modes because their amplitudes are clearly modulated by rotation.

Another possible interpretation is that $\ell = 1$ and $\ell = 2$. However, this suggestion leads to a contradiction between observation and theory. Namely, from Kurtz's data and Equation [27], we calculate $s = -7.16 \times 10^{-3}$ while the results of Shibahashi and Saio (1985) allow one to deduce $s = -10^{-1}$.

It is worth emphasizing the extreme accuracy with which s is determined for these Ap stars. In particular, s is much more accurately determined that for the Sun. Equivalently, Kurtz (1986) gives some Ap star frequencies to $10^{-6} \mu H_z$, while for the sun one can only hope to realize accuracies to $0.1 \mu H_z$. Theoretical results are, perhaps, even less accurate. It is important to note, however, that the theoretical values are primarily sensitive to the deep interior structure of the star-independent of the validity of the asymptotics - and these accurate observations provide new, independent and important constraints on stellar models.

Christensen-Dalsgaard (1986) outlines some of the theoretical problems in more general terms, especially those which remain to be studied. First, what is the degree of validity of the asymptotics? After all, one can easily reach accuracies of say 10^{-3} in the determination of W, but to reach this level of accuracy in the estimates of s one needs to calculate frequencies with a relative precision of 10^{-6} . Second, what are the possible effects of magnetism on the value of s? We note here that a magnetic perturbation of the frequencies in the order of $1 \mu H_z$ has to be considered to be large for the problems discussed here. We elaborate on this point in the next section.

4.   The Driving Mechanism and Amplitude Limitations as
      Diagnostics

We discuss here how studies of nonadiabatic and nonlinear effects may be used to diagnose the internal structure of Ap stars - especially in their outer layers. The fundamental questions here are why do the stars rapidly pulsate, and, in particular in high-order acoustic modes possessing essentially the same symmetry as the magnetic field?

The observed longevity of the modes suggest that they are overstable rather than being stochastically excited by some sort of turbulence. Dolez and Gough (1982) calculated that high-order p-modes are unstable with a sharp maximum in the driving rate at $\nu \simeq 1.5_m H_z$ in a normal star with $M=1.5M_0$ and $T_{eff} = 8000\,^0K$. Convection was included in their model ($\alpha = 2$) and its effect on the oscillations was treated using the method of Baker and Gough (1979). They predicted no instability for these modes in models in which they assumed the helium was depleted in the outer layers. Dziembowski and Krolikowska (unpublished) predicted no instabilty using models which ignored convection altogether ($\alpha = 0$), but they did find an instability pattern similar to that of Dolez and Gough when they included convective flux in the equilibrium model but ignored its Lagrangian perturbation.

From these numerical experiments, it follows that convection and chemical stratification play decisive roles in predicting the instability of the high-order p-modes. Thus, a successful description of instability will provide an important test for the theory of these two, otherwise difficult to study, phenomena. In particular, it may help to reveal their connection to the stellar magnetism.

If one assumes that the field has a dipole-like structure and that it effects only nonadiabatic quantities, then one can obtain [Dziembowski (1984)], the following formula for the dependence of the driving rate on ℓ and m, namely,

$$\gamma_{\ell m} = \gamma_{00} + \gamma' \cdot \frac{\ell(\ell+1)-3m^2}{4\ell(\ell+1)-3} \qquad (30)$$

The γ_{00} term represents the driving rate for the radial mode and $\gamma'(>0)$ represents the unknown, destabilizing effect of the magnetic field. We assume here that the axis of the magnetic field defines the z-axis in the reference system and that the modes are described by a single Y_ℓ^m - harmonic. A generalization of this formula was given by Dziembowski and Goode (1985) for the case of an oblique rotator. From Equation [30], one can see the degree preference for the excitation of $m = 0$ and low ℓ (>0) modes.

A study of linear instability alone does not provide conditions to tell us whether a mode may develop to an observable level. Such conditions follow from considering nonlinear effects which lead to mode coupling and amplitude limitation.

Since the light amplitudes in Ap stars are so much lower than in Cepheids it seems quite improbable that the dominant nonlinear effect is the saturation of driving mechanism. Observations in fact give us the strong hint that there is a 2:1 resonance between unstable modes and higher order, stable ones which play the crucial role. Indeed, in at least two of these stars, the period ratios of excited modes are, to observational accuracy, 2:1. Kurtz (1982) points out that the periodograms cannot be interpreted in these cases in terms of nonlinear distortion of the light curve because the amplitudes are too low. However, his suggestion that the modes are independently excited is also very unlikely to be true, because it requires that the period ratio be very exactly 2 and poses the question why none of the numerous modes at intermediate frequencies are excited.

The theory of nonlinear mode coupling offers not only a natural explanation of the observations but also provides formulae which allow a quantitative verification of parameters derived from linear nonadiabatic theory.

Let us consider a linearly unstable mode characterized by angular frequency ω_1 and linear driving rate γ_1 and a linearly stable mode with

$$\omega_2 = 2\omega_1 + \Delta\omega \text{ and } \gamma_2 < 0, \tag{31}$$

where the frequency mismatch $\Delta\omega$ is small in the sense that

$$|\Delta\omega| << \omega_1, \text{ but } |\Delta\omega| \sim |\gamma_2|, |\gamma_1|. \tag{32}$$

Dziembowski (1982) showed that the nonlinear interaction between such modes leads to a stable, constant amplitude solution if $\gamma_2 + 2\gamma_1 < 0$, and then the amplitudes are given by

$$\tag{33}$$

where
$$A_1 = \sqrt{\frac{-\gamma_1\gamma_2}{c_1 c_2}} (1+q^2) \text{ and } A_2 = \frac{\gamma_1}{c_1}\sqrt{1+q^2}$$

c_1 and c_2 are coupling coefficients which can be calculated for any pair of modes using their linear adiabatic eigenfunctions. We will not reproduce here lengthy formulae for the c's, but just remark that they are nonzero if

$$\ell_2 = 2\ell_1, 2\ell_1 - 2, \ldots \text{and } m_2 + 2m_1 = 0. \tag{34}$$

Thus, for instance, an unstable $\ell = 1$ mode may be coupled to $\ell = 0$ or 2 modes. The coupling causes small frequency shifts such that the modes appear to be in exact resonance. The phases of the modes satisfy equation

$$\tan(\phi_2 - 2\phi_1) = q. \tag{35}$$

The amplitudes and phases in Equations [33] and [35] refer to the surface distortion,

$$\left(\frac{\delta r}{r}\right)_{surface} = A Y_\ell^m \exp(j\omega t). \tag{36}$$

They can be connected to the disk averaged observed radial velocity amplitude if angles ρ and i are known. Thus if such amplitudes were known we could calculate γ_1, γ_2 and $\Delta\omega$ using these equations. With only photometric data available one has to carry out linear nonadiabatic calculations to obtain the complex coefficient, Ψ , in the relation

$$\frac{\delta T_{eff}}{T} = \Psi \left(\frac{\Delta r}{r}\right)_{surface})$$ (37)

together with γ . Then Equations [33] and [35] may be used as a test.

We wish to thank Don Kurtz for comments on this paper and suggesting we study HD83368 and providing us with an early copy of his review.

REFERENCES

Baker, N.H. and Gough D.O., 1979, Ap.J. 234, 232.
Christensen-Dalsgaard, J., 1986, Review, these proceedings.
Dicke, R.H., 1982, Nature 300, 693.
Dolez, N. and Gough, D.O., 1982, "Pulsations in Classical and Cataclysmic Variables", eds. J.P. Cox and C.J. Hansen, Boulder, JILA, 248.
Dziembowski, W., 1977, Act. Astron. 27, 203.
Dziembowski, W., 1982, Act. Astron. 32, 147.
Dziembowski, W. and Goode, P.R., Ap.J. Lett., in press.
Dziembowski, W. and Krolikowska, M., unpublished.
Edmunds, A.R., 1960, Angular Momentum in Quantum Mechanics, Pricneton University Press, Princeton, NJ.
Gabriel, M., Noels, A., Scuflaire, R. and Mathys, G., 1984, Astron. Astrophys. 206, 143.
Gough, D.O. and Taylor, P.P., Mem. Soc. Astron. Ital. 55, 215.
Kurtz, D.W., 1982, MNRAS 200, 807.
Kurtz, D.W., 1986, Review, these proceedings.
Kurtz, D.W., Schneider, H. and Weiss, W.W.,1985, MNRAS, 213, 773.
Shibahashi, H., 1984, Mem. Soc. Astron. Ital. 55, 181.
Shibahashi, H. and Saio, H., 1985a, Publ. Astr. Soc. Japan, in press.
Shibahashi, H. and Saio, H, 1985b, Publ. Astr. Soc. Japan, in press.
Tassoul, M., 1980, Ap.J. Supp. 43, 469.

NONRADIAL MODES IN LINE-PROFILE VARIABLE STARS

Yoji Osaki
Department of Astronomy, Faculty of Science
University of Tokyo, Bunkyo-ku, Tokyo, Japan

ABSTRACT. Nonradial oscillation modes in line-profile variable stars
are discussed. The line-profile modeling based on nonradial oscillations is first reviewed and then general observational properties of
line-profile variable stars are presented. A question is raised about
the mode identification of nonradial oscillations based on the conventional line-profile modeling. It is demonstrated that line-profile
variations of "traveling bumps" observed in the ζ Oph type stars, which
were usually explained by nonradial spheroidal modes of intermediate
sectoral harmonics ℓ=6∿10 with dominantly radial velocity fields, can
equally well be reproduced by sectoral Rossby modes, whose velocity
fields consist solely of horizontal components. It is suggested that
wave phenomena observed in line-profile variable stars could be superficial phenomena confined more or less in the atmospheric layers of
these stars.

1. INTRODUCTION

A new class of variable stars called line-profile variable B stars have
recently been discovered owing to the development of high precision spectroscopy with the use of solid-state detectors. The stars exhibit characteristic variations in absorption-line profiles but with little or no
detectable variation in radial velocity (i.e. line centroid) or in
brightness. Variability in line-profiles in some B stars had been noticed before, but its full significance as a distinct class of variable
stars was first recognized by Smith (1977). Two sub-classes of line-
profile variable B stars are presently known: the one is sharp-lined
variables called 53 Per stars (Smith 1977) and the other is rapidly
rotating B or Be stars called ζ Oph stars (Vogt and Penrod 1983). The
line-profile variations in these stars are now understood to be caused
by nonradial pulsations (hereafter abreviated as NRP) in rotating stars.
    In section 2 of this paper, we shall briefly describe the standard
procedure of line-profile modeling by NRP, and we review the basic observational properties of line-profile variable stars in section 3. We then
discuss the so-called K-problem and suggest Rossby modes as a possible

453

D. O. Gough (ed.), Seismology of the Sun and the Distant Stars, 453–463.
© *1986 by D. Reidel Publishing Company.*

solution in section 4. Section 5 presents a concluding remark.

2. LINE PROFILE MODELING BY NONRADIAL MODES

Since the main characteristic of line profile variable stars is, of
course, profile variations, we shall briefly describe the standard procedure of line profile modeling by NRP below. The possibility of NRP as
a cause of profile variation was first suggested by Ledoux (1951) in
connection with observed variations of line widths in Beta Cephei stars.
Osaki (1971) presented extensive calculations of line profiles produced
by NRP of traveling-wave type in a rotating star. Smith (1977, 1978),
Vogt and Penrod (1983), and Baade (1984) have shown that observed profile
variations in the line-profile variable B stars can be matched by model
profiles produced by NRP velocity fields.

The standard procedure of line profile modeling by NRP is as follows. We first choose a certain stellar absorption line and calculate
its intrinsic profile for a given effective temperature T_{eff} and surface
gravity g appropriate to the star of interest by using the standard stellar atmosphere computer code. We then choose a particular nonradial
modes. The velocity field for a single NRP mode with spherical harmonic
indices ℓ and m can be written in the spherical polar coordinates (r, θ,
ϕ) as

$$V_{osc} = A\ (1,\ k\ \frac{\partial}{\partial\theta}\ ,\ k\ \frac{1}{\sin\theta}\ \frac{\partial}{\partial\phi})\ P_\ell^m\ (\cos\theta)\ e^{i(\sigma t + m\phi)} \tag{1}$$

where A is the velocity amplitude in the radial (vertical) direction and
k measures the ratio of the horizontal to the vertical velocity amplitudes. The constant k is not a free parameter but its value is related
to the frequency of oscillation (Osaki 1971) such that

$$k\ =\ \frac{\frac{GM}{R^3}}{\sigma^2}\ =\ \frac{1}{\omega^2}\ =\ (\frac{Q}{0.116})^2 \tag{2}$$

where $\omega = \sigma/(GM/R^3)^{1/2}$ is the dimensionless frequency of oscillation, G the
gravitational constant, M and R are the stellar mass and radius and
$Q[=\text{Period}(\bar\rho/\bar\rho_\odot)^{1/2}]$ is the pulsation constant. In general, k<<1 for p-
modes but k≳1 for low-frequency g-modes. We also specify the equatorial
rotational velocity V_e and the inclination i of the axis of rotation
relative to the line of sight. We usually assume that the pulsation axis
is aligned with the rotation axis. We then divide the stellar visible
disk into many surface elements and calculate the velocity of each element due to combined effect of rotation and oscillation. The line profiles are constructed by integrating the intrinsic profile over finely
divided surface elements of the visible disk with due account of a
Doppler shift and of a weight appropriate to the limb-darkening law. We
usually neglect variations in surface area and in brightness due to oscillation. Thus, profile variations are assumed to be generated solely by

Doppler effects of oscillations.

We summarize parameters necessary to construct a single line-profile due to a single NRP mode: (1) equatorial rotational velocity V_e, (2) inclination i, (3) NRP mode specified by ℓ and m, (4) its amplitude A, (5) the ratio of the horizontal to vertical velocity amplitude, k, although it is not a free parameter, and (6) the phase of the pulsation mode, ϕ_p. (7) We sometimes include effect of broadening due to the macroturbulence, M. If a single coherent oscillation mode is involved, all parameters must remain constant all the time except the pulsation phase ϕ_p, which should increase linearly with time.

The conventional procedure of line-profile fitting is then performed by calculating many line-profile with different combination of parameters and searching for best-fit model profiles to observed profiles by visual inspection. If successful, this process assigns a pulsation phase ϕ_p to each profile and several profiles with different epochs can in principle determine the period of the pulsation mode.

Observed profiles usually exhibit migratory signature across the profile, and they are reproduced by NRP modes with $m \neq 0$, which represent waves traveling around the equator with the phase velocity

$$(\frac{d\phi}{dt})_{ph} = - \frac{\sigma}{m} \tag{3}$$

Thus, modes with m<0 represent waves propagating in the same direction to rotation, and they are called "prograde" modes, while those "retrograde" modes with m>0 are waves propagating in the opposite direction to rotation.

It may be noted here that the frequency of oscillation of non-axisymmetric NRP mode with $m \neq 0$ and its sense of "prograde" or "retrograde" depend on the coordinate system observed, that is, either the inertial system or the corotating system of the star. The frequency σ_c of oscillation in the corotating frame of reference of the star is related to that of the inertial frame, σ_i, by

$$\sigma_i = \sigma_c - m\Omega \tag{4}$$

where Ω denotes the angular velocity of rotation of the star and the term $m\Omega$ represents the well-known Doppler effect of waves in the moving system.

Some typical model profiles produced by a single NRP mode may be found for low ℓ modes in Unno et al. (1979) and Smith (1980) and for high ℓ sectoral modes in Vogt and Penrod (1983) (see also figure 1a). Actual fitting to observed profiles may be found in many papers by Smith and by Baade and in Vogt and Penrod (1983).

A question now arises whether or not this profile modeling gives a unique solution, particularly concerning the NRP mode, because there are many parameters involved in modeling line profiles. In fact, Osaki (1971) already noticed that a NRP mode with $\ell=2$ and m=-1 seen near the inclination $i \sim 45°$ yielded profiles very similar to those produced by a NRP mode with $\ell=2$ and m=-2 seen nearly from equator-on (i.e., $i \sim 90°$). In order to avoid this kind of ambiguity, it is a usual practice in most investigations of line-profile modeling to adopt an assumption that NRP modes involved are sectoral modes with $m=\pm\ell$.

Nonradial eigenfunctions given in equation (1) and described by the spherical harmonics are correct only in a non-rotating star. When a star is rotating, its nonradial oscillations are affected by rotation, particularly when angular frequencies of oscillations are comparable to that of rotation. Besides that, there appear new modes called Rossby modes (or "r-modes"). They are basically of horizontal eddy motion and they are described by

$$V_{Rossby} = B(0, \frac{1}{\sin\theta} \frac{\partial}{\partial\phi}, - \frac{\partial}{\partial\theta})P_\ell^{\ m}(\cos\theta)e^{i(\sigma t+m\phi)} \tag{5}$$

where B stands for velocity amplitude of Rossby waves, and the angular frequency of oscillation σ is given by (Papaloizou and Pringle 1978)

$$\sigma = m\Omega \ (\ -1 + \frac{2}{\ell(\ell+1)} \ ) \ . \tag{6}$$

Rossby waves are prograde in the inertial frame (except $\ell=1$) but retrograde in the corotating frame of the star. Smith (1982) has calculated line-profiles produced by Rossby modes and he has concluded that Rossby modes are unlikely to be responsible for oscillations in line-profile variable stars. A different conclusion is obtained in this paper but its discussion will be deferred to a later section.

3. OBSERVATIONAL PROPERTIES OF LINE PROFILE VARIABLE STARS

Line profile variable stars are those early-type stars which exhibit variations in profiles of photospheric absorption lines with a typical time scale of hours and days, but little or no accompanying variation in brightness or in radial velocity. Some of Beta Cephei stars, which are pulsating variables of early spectral type, also show variations in line-profile but their main variations are those of brightness and of radial velocity. For the moment we assume that line-profile variable stars and Beta Cephei stars are distinct classes of variable stars. Line profile variable stars range in spectral types from O8 to B5 and in luminosity classes from V to II so that they surround Beta Cephei variables in the HR diagram. It is suggested (Smith 1980) that the classical Beta Cephei stars may represent the most easily observable "tip of the iceburg" of a much wider class of line-profile variable stars. However, the exact relation between these two classes of variables is still unknown. It is suspected that most of stars in this spectral range may be variables in line-profile if careful spectroscopic observations are performed.

Line-profile variable stars are found both among slow rotators (V_e sin i$\lesssim$50 km/s) and rapid rotators (V_e sin i$>$50 km/s). Sharp-lined variables of apparently slow rotators are called 53 Per stars, and their main characteristic of profile variation is periodic variations in line-width and line-skewness. They are usually explained in terms of NRP modes of traveling-wave type with a low ℓ spherical harmonics (i.e., $\ell=2\sim3$). The

line-profile variations in rapidly rotating B stars and Be are characterized by "traveling bumps", a continuous migratory signature of quasi
absorption/emission bumps, usually from blue to red, across the rotationally broadened profiles (e.g., ζ Oph by Walker et al. 1979; μ Cen
by Baade 1984). These features are explained by nonradial modes of
traveling-wave type with intermediate ℓ ($\ell=6\sim10$) and the sectoral mode
($m=\pm\ell$) (see Vogt and Penrod 1983).

It was originally thought that nonradial modes with high ℓ ($\ell\gtrsim6$)
could not be observed in distant stars because of the cancellation effects of high ℓ modes over the visible disk. It was also thought that
line-profile variations would not be observed in rapid rotators because
of their diffuse appearance of absorption lines. However, it has turned
out that rapid rotation rather helps to enhance the visibility of intermediate and high ℓ nonradial modes with $\ell\gtrsim6$ because it tends to resolve
the stellar disk into different parts within the absorption line profiles
through rotational Doppler shift (a phenomenon called "Doppler imaging").
Then, observational manifestation of low ℓ NRP modes in 53 Per stars and
of high ℓ modes in ζ Oph stars may simply reflect different visibility
of different modes depending on stellar rotational velocities, and there
exists a good possibility that these two sub-classes of stars could be
basically similar in oscillations.

Periods of NRP modes estimated from profile variations in these
stars range from a few hours to a few days with a typical period of about
half a day. On the other hand, the typical period of the radial fundamental mode for a B star is estimated to be several hours (cf. the observed pulsation periods of the Beta Cephei stars, which are thought to
be radial fundamental or first-harmonic pulsators, are 4∼6 hours). Then,
this indicates that most of NRP modes in line-profile variable stars may
correspond to nonradial g-modes, except of those very short period oscillations which could be nonradial p-modes.

NRP waves observed in line-profile variable stars are in most cases
prograde modes in the inertial frame, as their characteristic features
move across profiles from blue to red. However, they could be either
prograde or retrograde on the surface of the star because they depend on
relative speed between waves and rotational angular velocity Ω (see equation [4]). Observations indicate that NRP waves tend to be prograde in
slow rotators but retrograde in rapid rotators when seen from the corotating frame of the star. This suggests that the term $m\Omega$ in equation
(4) and hence rotation play a dominant role in rapid rotators and that
Rossby waves could be responsible for "traveling bumps" seen in some of
ζ Oph stars. On the other hand, retrograde propagating features even in
the inertial frame have been observed in two stars, 22 Ori (Smith 1977)
and μ Cen (Baade 1984).

When more than two modes exist simultaneously in a single star,
their pattern speeds given by equation (3) tend to be the same. In other
words, we can define a superperiod by

$$\text{"superperiod"} = m \times \text{Period} \simeq \text{constant} ,$$

where the same superperiods result for various modes with different m in
a star. This aspect has been observed in μ Cen (Baade), ε Per (Smith

1985a), α Vir (Smith 1985b) and other stars.

4. THE K-PROBLEM AND A POSSIBILITY OF ROSSBY MODES

The present interpretation of line-profile variable stars in terms of
NRP modes is as follows. Stellar NRP eigenmodes, particularly those of
sectoral traveling-wave type with m=-ℓ or +ℓ are somehow excited in
these stars and they manifest themselves observationally as periodic
variations in line-profile. Judging from rather long periods of observed oscillations, most of NRP modes are thought to be nonradial g-modes.
Although this model is successful in reproducing profile variations,
there are several problems and difficulties with it and they are summarized as nonradial oddities by Smith (1985c). Here we shall examine the
so-called K-problem which remains one of the biggest difficulties in the
NRP interpretation.

In the standard line-profile modeling, there exists a parameter
defined by k in equation (1), that measures the ratio of the horizontal
velocity amplitude to the vertical velocity amplitude at the stellar
surface. It has repeatedly been stated (see Smith and McCall 1978;
Smith 1978, 1980, 1982; Vogt and Penrod 1983) that the k value has to be
chosen very small (i.e., k≲0.15) in order to reproduce observed line
profile variations. On the other hand, it is known that the k-value is
not a free parameter but it is theoretically related to the oscillation
frequency by equation (2). Since observed frequencies of oscillations
are rather low particularly in the corotating frame of the star, the k-
value must be large (i.e., k≳1) for most cases. It apparently contra-
dicts with the requirement imposed from the line-profile modeling, and
this is the so-called K-problem.

We first consider the theoretical relation of equation (2). This
relation is derived from the surface boundary condition for nonradial
oscillations. It reflects wave behaviour in the stellar atmosphere.
Although one may relax some of restrictive assumptions, it is yet an unavoidable conclusion that the horizontal velocity fields dominate over
the vertical velocity fields in the case of low frequency oscillations.

Let us then examine the restriction imposed on the k-value by line-
profile modeling. Smith (1980) notes that NRP velocity fields with
dominantly horizontal components always produce line profiles with wings
that are too strong. This phenomenon occurs because transverse oscillatory velocity fields produce no Doppler effects near the disk center
but their effect coincides fully with that of rotation at the stellar
limbs, producing large variations in wings. Such large variations in
wings are not observed in line-profile variable stars. This argument
looks at the first sight quite reasonable and it seems to provide a good
reason to exclude the case of NRP modes with dominantly horizontal velocity fields. However, there exists an oversight in this argument. It
turns out that this applies only to sectoral spheroidal modes described
by equation (1). To show this, we present a counter example below.

To do so, we consider a sectoral Rossby mode with m=-ℓ which is described by

$$v_r = 0 , \tag{7a}$$

$$v_\theta = A_H \sin^{\ell-1} \theta \, \sin(\sigma t - \ell\phi) , \tag{7b}$$

and

$$v_\phi = -A_H \sin^{\ell-1} \theta \, \cos\theta \, \cos(\sigma t - \ell\phi) , \tag{7c}$$

where A_H stands for the horizontal velocity amplitude of the Rossby mode. This velocity fields have no vertical component and they are confined more or less in the equatorial belt for intermediate and high ℓ with $\ell \gtrsim 6$. We may find a good visualization of velocity fields for such a sectoral Rossby mode with $\ell=-m=8$ in Figure 16 of Vogt and Penrod (1983) where we reinterpret their velocity map on the sphere as stream lines of the corresponding sectoral Rossby mode. We first note that the θ-component of the sectoral Rossby mode becomes maximum at the equator but the ϕ-component becomes zero there. That is, the rotational velocity vector is perpendicular to the oscillation velocity vector at the equator. This mode can therefore avoid large variations in line wings and it provides a possibility to solve one of difficulties mentioned above. Another difficulty associated with modes of dominantly transverse velocity fields concerns the problem of no Doppler shifts near the disk center and of traveling signature across the profile. This is certainly a problem in the case of the equator-on geometry with i=90°. However, if we tilt the rotation axis slightly, say the inclination i to be some 60°, we can see a significant line of sight component for the oscillatory transverse fields of the equatorial belt. This suggests that "traveling bumps" observed in line-profile variable stars may equally well be reproduced by the sectoral Rossby mode.

To confirm this, we have calculated line profiles produced by the sectoral Rossby modes and compared them with those of the sectoral spheroidal mode of the same m=-ℓ with purely vertical velocity component, which is described by

$$v_r = A_V \sin^\ell \theta \, \cos(\sigma t - \ell\phi) , \tag{8a}$$

$$v_\theta = v_\phi = 0 . \tag{8b}$$

In other words, we have equated k to be zero in equation (1). Figure 1a and 1b illustrate an example of line-profile variations produced by the spheroidal mode and the Rossby mode, respectively. The modes considered here are prograde sectoral modes with -m=ℓ=8, the inclination i is 60°, the amplitudes of oscillations measured in units of the equatorial rotational velocity, V_e, are A_V/V_e=0.1 and A_H/V_e=0.173, respectively. The amplitudes A_V and A_H are so chosen that they give equal amplitudes in the line of sight component at the equator, that is, A_H=A_V tan i. We adopt the standard phase convention of Osaki (1971) with ϕ_p=$\sigma t/2\pi$ for the spheroidal mode but we use the phase definition given by equation (7) for the Rossby mode. The symmetric line-profiles then occur at ϕ_p= 0.25 and 0.75 for the spheroidal mode but at ϕ_p=0.0 and 0.50 for the Rossby mode. In figure 1 we have aligned line-profiles in such a way that each line of the spheroidal mode corresponds to that of the Rossby

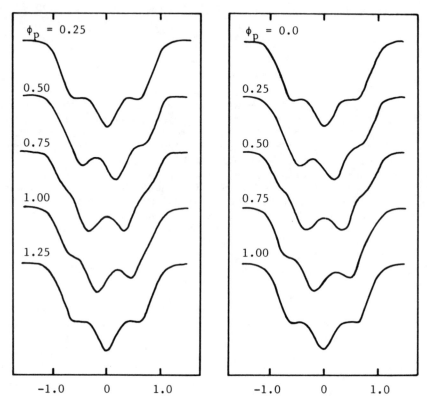

Figure 1.   Model line profiles produced by NRP modes with $-m=\ell=8$ of (a) the spheroidal mode and of (b) the Rossby mode. The inclination i is 60°, and amplitudes of oscillations are (a) $A_V/V_e=0.1$ and (b) $A_H/V_e=0.173$. The abscissa is $V/V_e \sin i$ and the ordinate uses an arbitrary scale.

mode.
       From figure 1, we see that line profiles produced by the spheroidal mode and by the Rossby mode are surprisingly similar and that both of them can reproduce very well "traveling bumps" observed in line-profile variable stars. We have examined several other cases with different combination of parameters in spherical harmonic number $m=-\ell$, inclination i, and oscillation amplitude A/V_e. We have obtained essentially similar results as shown in figure 1, provided that $30° < i < 80°$ and $A/V_e \lesssim 0.2$. We therefore conclude that the line-profile modeling does not exclude NRP modes with dominantly transverse velocity fields and that no K-problem exists any more. We suggest that some of NRP modes in line-profile variable stars could be sectoral Rossby modes. However, we do not claim any unique mode identification because we have not examined extensively various possible cases with different parameters. Particularly we have not examined any tesseral modes with $|m| < \ell$.
       Once we admit Rossby modes as possible modes for line-profile vari-

ability, then a completely different interpretation of observed modes becomes possible. To demonstrate this, we consider the case of Spica discussed by Smith (1985b). He has found two commensurate, equatorial modes responsible for traveling bumps in Spica; (1) an $\ell=|m|=8$ mode with an apparent period (in the inertial frame) of 6.52±0.08 hrs and (2) an $\ell=|m|=16$ mode with an apparent period 3.20±0.016 hrs. Their superperiods are then 52.16 hrs and 51.2 hrs, respectively. He has presented an argument that Spica rotates "bisynchronously" with respect to the orbital angular velocity, that is, $\Omega_{rot}\sim2\Omega_{orb}$ where Ω_{rot} and Ω_{orb} are rotational angular velocity and orbital angular velocity, respectively. Since the orbital period of Spica's system is 4.0 days, the rotation period of Spica would be about 2 days or 48 hrs. Comparison of two superperiods 52.16 hrs and 51.2 hrs to the rotation period of about 48 hrs shows that apparent periods of two modes are essentially due to the Doppler term $m\Omega$ of equation (4) and that oscillation frequencies of the two modes are very low in the corotating frame of the star.

By adopting the conventional interpretation of NRP modes, Smith (1985b) has identified these two modes as nonradial g-modes. This interpretation, however, encounters difficulty. Firstly, if these modes are conventional g-modes, they must be very high overtone g-modes with the radial order n as high as 100, which is rather unlikely possibility. Secondly, if these modes are sectoral spheroidal g-modes, then the K-problem discussed above remains still a difficulty.

On the other hand, if we identify these two modes to be the sectoral Rossby modes, then it is found that observations are perfectly in accord with this picture. To show this, we first use the observed period of the $\ell=|m|=16$ mode to estimate the rotation period of Spica from equation (6), which yields $P_{rot}=50.8$ hrs~2.1 days. This period is consistent both with the bisynchronous nature of Spica and with observed $V_e \sin i$. Then, by assuming $P_{rot}=50.8$ hrs as the surface rotation period, we predict an apparent period of an $\ell=|m|=8$ sectoral mode by equation (6), which results P=6.53 hrs. This period is perfectly in agreement with observed period of 6.52±0.08 hrs.

5. A CONCLUDING REMARK

We have shown in the previous section that some of NRP modes observed in the line-profile variable stars could be Rossby modes. Since the Rossby modes are thought to be more or less surface phenomenon, this suggests that wave phenomena observed in line-profile variable stars could, in general, be superficial phenomena confined more or less in the atmospheric layers of these stars. It is supposed in this picture that waves are somehow generated somewhere in the sub-photospheric layers and they are propagated upwards to emerge in the stellar atmosphere. They are amplified there because of decreasing density of the atmosphere with height and they are observed as oscillatory velocity fields with dominantly transverse components. Thus they are not stellar global eigenmodes but rather transient propagating waves in the atmosphere.

This picture differs fundamentally from the conventional one in which observed NRP modes are thought to be stellar global eigenmodes

whose main part of energy is located in the deep interior and that waves
observed are a tip of the global oscillations appearing in the atmosphere
as evanescent waves.

One of problems discussed by Smith (1985c) concerns the transient
appearance of observed NRP modes. He has noted that mode growth/decay
and mode switching often occur over a time-scale as short as 2 days. If
these modes were really global stellar eigenmodes, oscillation energy
involved in the transient process would be enormous. This will certainly
pose a problem in the conventional NRP interpretation. However, there
is no problem in our picture because the observed wave phenomenon is only
a superficial phenomenon and the oscillation energy is minute in this
picture. Furthermore, wave transience is rather normal than exceptional.
It will therefore worth examining all available observations of line-
profile variable stars from this standpoint.

REFERENCES

Baade, D. 1984, 'Discovery and preliminary identification of two retro-
    grade nonradial pulsation modes in the Be star μ Centauri', Astron.
    Astrophys., 135, 101.
Ledoux, P. 1951, 'The Nonradial Oscillations of Gaseous Stars and the
    Problem of Beta Canis Majoris', Astrophys. J., 114, 373.
Osaki, Y. 1971, 'Non-Radial Oscillations and the Beta Canis Majoris
    Phenomenon', Publ. Astron. Soc. Japan, 23, 485.
Papaloizou, J., and Pringle, J.E. 1978, 'Non-radial oscillations of
    rotating stars and their relevance to the short-period oscillations
    of cataclysmic variables', Mon. Not. R. astr. Soc., 182, 423.
Smith, M.A. 1977, 'Nonradial Pulsations in Early to Mid-B Stars',
    Astrophys. J., 215, 574.
Smith, M.A. 1978, 'Nonradial Pulsation in Iota Herculis and 10 Lacertae
    (1975-1977)', Astrophys. J., 224, 927.
Smith, M.A. 1980, 'The Line Profile Variable B Stars', Goddard Conference
    on Current Problems in Stellar Pulsation Instabilities, eds. J.R.
    Lesh, W.M. Sparks, and D. Fischel, p.391.
Smith, M.A. 1982, 'Nonradial Pulsations in Early-Type B Stars: g-Modes
    or r-Modes?', Astrophys. J., 254, 708.
Smith, M.A. 1985a, 'Pulsational Mode Typing in Line-Profile Variables.
    VI. Nonradial Modes in the Remarkable B Star Epsilon Persei',
    Astrophys. J., 288, 266.
Smith, M.A. 1985b, 'The Nonradial Oscillations of Spica. I. Commensurate,
    Equatorial Modes and Spin-Orbit Angular Momentum Distribution',
    Astrophys. J., in press.
Smith, M.A. 1985c, 'Nonradial Pulsations in Massive Stars: Observations
    and Oddities', Publ. Astron. Soc. Pacific, in press.
Smith, M.A., and McCall, M.L. 1978, 'Undulations of a B-Type Star: 53
    Persei', Astrophys. J., 223, 221.
Unno, W., Osaki, Y., Ando, H., and Shibahashi, H. 1979, 'Nonradial Oscil-
    lations of Stars', University of Tokyo Press, Tokyo, p.30.
Vogt, S.S., and Penrod, G.D. 1983, 'Detection of High-Order Nonradial
    Oscillations on the Rapid Rotator Zeta Ophiuchi and Their Link with

Be Type Outbursts', Astrophys. J., 275, 661.
Walker, G.A.H., Yang, S., and Fahlman, G.E. 1979, 'Spectral Variations in ζ Ophiuchi Related to Rotation?', Astrophys. J., 233, 199.

Observational Properties of Nonradial Oscillations in Early-Type Stars
and their Possible Effect on Mass Loss: the Example of ζ Puppis (O4 If)

Dietrich Baade
The Space Telescope European Coordinating Facility
European Southern Observatory
Karl-Schwarzschild-Str. 2
D-8046 Garching b. München
W. Germany

ABSTRACT. Thanks to the fast rotation which via the Doppler effect
permits one to indirectly resolve the stellar disk in one dimension and
amplitudes often reaching 10 km/s or more, nonradial oscillations with
degrees up to m = 20 are easily detected in high S/N (better than 200)
observations of early type stars. Recent work by Smith (1981, 1985),
Vogt and Penrod (1983), Penrod and Smith (1985), Penrod (1985), Baade
(1982, 1984, and unpublished data), and Baade and Ferlet (1984) show the
ubiquity of nonradial oscillations in the upper left part of the HRD.
With the increasing amount of data some general properties of these
pulsations are emerging, and there are even first indications of
systematic differences between different groups of stars. The latter is
encouraging in the context of this workshop because it could mean that
from the analysis of the oscillations additional insight can be gained
into the stars' internal structure, i.e. seismology of early-type stars
may in fact become a reality. From the above references the following
can be extracted as the most important results obtained thus far:
o Virtually all "normal" stars with spectral type B6 or earlier are
  nonradial oscillators, at least occasionally (for supergiants the
  spectroscopic evidence is still less complete).
o Observations have not so far succeeded in restricting the range of
  possible excitation mechanisms suggested by various theoretical
  models. But the rotation rate may play an important role.
o Simultaneously excited modes are are often coupled: P_m x m = const
  (m = mode degree; the constant is different for different stars).
o The modes with the longest wavelengths have typical periods between
  0.5 and 2 days. If they are g-modes, very high radial overtones are
  implied. On the other hand, the nonradial modes observed in some of
  the radially pulsating ß Cephei stars all fall into the range 0.13 to
  0.3 day and thus are almost identical to those of the radial modes.
o For modes with periods much longer than the radial fundamental mode,
  in some slow to moderately fast rotators a horizontal-to-vertical
  amplitude ratio, K, of only 0.15 has been deduced. This is in conflict
  with the characteristics of g and r modes as the only ones with so
465

D. O. Gough (ed.), Seismology of the Sun and the Distant Stars, 465–466.

long periods. At least in most Be stars, K appears to be much larger.

o The mode spectra of Be and Bn stars seem to differ in that symmetry variations of the overall profiles have been found only in Be stars - but also in the narrow lined 53 Per stars (however, the modes causing these variations are probably different in 53 Per and Be stars).

o Mass loss episodes from Be stars are correlated with temporarily decreased (often vanishing) pulsation amplitudes.

o The V/R ratio of the double emission in the HeI λ6678 profile of many Be stars varies with the period of the slowest pulsation mode.

o In the case of multi-mode oscillations amplitude ratios can be variable. In some stars also a mode switching has been diagnosed.

o In many stars pulsation cycles are not always all alike. Various deviations from the mean occur. At least in some cases these are very unlikely to be due to multi-mode beating.

Some of the above points can be illustrated with the example of ζ Puppis. Since there is no star in the sky which is both brighter and hotter ζ Pup is often taken as the "mass loss standard star", and any peculiaritiy of its wind, i.e. the effect of any parameter other than radiation pressure, would therefore be important to know.

Series of spectra with high spectral (R = 80,000) and temporal (14 min) resolution and good S/N (400) obtained with the European Southern Observatory's Coude Echelle Spectrograph fed by the 1.4-m Coude Auxiliary Telescope show variable profiles of the CIV λλ 5802 and 5812 lines that are typical of the nonradial oscillations of other early-type stars. Most conspicuous is the splitting of the profiles into two components. Every 0.178 day a new component develops in the blue wing, and it takes about twice that long for one feature to traverse the entire profile from blue to red. However, spectra taken 0.178 days apart are not identical but differ in their overall asymmetry thus disclosing the presence of a second mode whose period turns out to be 0.3561 day. Small wiggles in the velocity curve of the two features mentioned may be indicative of a third, still shorter, period of 0.09 day. If tentatively identified with m=2, m=4, and m=8 modes, the case of ζ Pup thus provides another example of commensurate periods ($2 \times P_2 = 4 \times P_4 = 8 \times P_8$).

Double emission components straddle each of the two CIV lines. The strength of at least the blue component of the λ 5802 line changes with phase of the 0.356-day period. From observations of the blue displaced absorption component in Hα Moffat and Michaud (1981) deduced a 5.075-day period and proposed an oblique magnetic rotator model as the explanation. The Nyquist period of that data is about 2 days. If one nevertheless extends the period search to shorter periods, the strongest feature in a PDM (Stellingwerf 1978) spectrum appears at a period of 0.356 day, i.e. the longest of the pulsation periods. These results suggest that the oscillations have some effect at the base of the wind. Numerous space-UV observations of resonance lines have documented a relatively small but significant variability on a wide range of time scales also of the outer wind. Time resolution and coverage, however, are insufficient to conclude whether the phase coherence of the suggested modulation of the wind at its base is still maintained at larger distances from the star.

(A detailed account of the ζ Puppis data will be published elsewhere.)

SOUTHERN ZZ CETI STARS

Darragh O'Donoghue
Department of Astronomy
University of Cape Town

ABSTRACT. We present preliminary results of recent observations of the low amplitude southern ZZ Ceti, MY Aps (= L 19-2) which confirm the inequality in the separation of the frequency triplet comprising the 192s oscillation seen in this star. This inequality is in reasonable agreement with the calculation by Chlebowski of 2nd order effects in the removal of pulsation mode degeneracy by rotation. We also present an initial analysis of the frequency structure of a large amplitude southern ZZ Ceti, VY Hor (= BPM 31594).

1. INTRODUCTION

The ZZ Ceti stars are a physically homogeneous class of variable luminosity DA white dwarfs (see reviews by Winget and Fontaine, 1981; Robinson, 1980). They possess normal DA spectra dominated by broad absorption lines and their effective temperatures cluster near 10000K with a dispersion of only a few hundred Kelvin. Their luminosity variations are almost certainly due to non-radial g-mode pulsations.

We present preliminary results of (i) recent observations of a well-studied low amplitude southern ZZ Ceti, MY Aps (= L 19-2) (O'Donoghue and Warner, 1982) and (ii) an extensive study of the large amplitude variable VY Hor (= BPM 31594). Full details will be published elsewhere.

2. NEW OBSERVATIONS OF MY APS (= L 19-2)

O'Donoghue and Warner (1982) found that MY Aps possesses five distinct oscillation modes, all of which are comprised of closely-spaced multiple components, presumably due to rotational removal of the m-degeneracy (Unno *et al.*, 1980). The most prominent mode near 192s was of particular interest since it was resolved into 3 almost equally-spaced components. Furthermore, O'Donoghue and Warner found evi-

467

D. O. Gough (ed.), Seismology of the Sun and the Distant Stars, 467–472.
© *1986 by D. Reidel Publishing Company.*

dence for a slight inequality in the splitting in accordance
with second order theory (Chlebowski, 1978).

In order to confirm this result, we obtained new
observations and, combining these with those published previously, fit the following model to the data: consider three
sinusoids of amplitudes A_o, A_1 and A_2 and frequencies f_o, f_o
+ F + ε and f_o - F where F is the first order splitting and
ε the second order perturbation. Adding and expressing the
result as a single sinusoid of frequency f_o and time-vari-
able amplitude A_t and phase ϕ_t, it can easily be shown that:

$$A_t^2 = (A_o^2+A_1^2+A_2^2) + 2A_o(A_1\cos2\pi(F+\varepsilon)t+A_2\cos2\pi Ft) +$$

$$+ (2A_1A_2\cos4\pi(F+\varepsilon)t) \qquad (1)$$

Since the measured A_1 and A_2 in MY Aps are small compared to
A_o, the last term may be neglected. Equation 1 predicts
that the size of the usual amplitude modulation resulting
from 3 equally spaced beating components is itself modulated
between $2A_o(A_1+A_2)$ and $2A_o(A_1-A_2)$ with frequency ε.

We measured the 192s amplitude from short data sections, typically 1-2 hr in length. These amplitudes were
squared and folded on the 0.89 day beat period (=1/F) and a
sinusoid plus constant was fit by least squares. The constant, k_1, can be identified with $A_o^2+A_1^2+A_2^2$ in equation 1
while the sinusoid amplitude, k_2, measures the size of the
amplitude modulation. The results are presented in table I.

Table I: Amplitude Modulation In MY Aps

Data	Time	k_1	k_2
McGraw	1976.58	42±2	26±3
Hesser	1976.6	35±5	25±9
Week 3	1979.5	41±1	16±2
Week 5	1979.9	42±3	15±3
Week 6	1980.1	40±2	15±2
Week 7	1980.2	29±2	18±3
Week 8	1980.4	38±2	9±2
Wks 10-12	1980.7	39±1	13±2
Latest	1983.5	40±2	5±2

The constancy of k_1 (except for week 7) lends reliability to the model. The values of k_2 indicate that the size of the amplitude modulation has decreased by about a factor of 4.

We confirmed the result by directly measuring the 3 component frequencies using non-linear least squares. The value of ε obtained is

$$\varepsilon = (2.0\pm0.4) \times 10^{-6} \text{ mHz}$$
$$= 1.0 / (16.7 \text{ years})$$

From Chlebowski (1978), the theoretical value is given by

$$\varepsilon = \frac{-4\pi \ (\ C^{II}_{klm} - C^{II}_{klo} \ ) \ F^2}{m^2 \ (\ 1 - C^{I}_{kl} \ )^2}$$

Chlebowski has calculated values of the C's for white dwarf models of uniform composition. For k=1, l=1, m=1 and using the observed 12.9×10^{-3} mHz for F, the above yields a value of $\varepsilon = -93\times10^{-6}$ mHz. The best agreement we could find from Chlebowski's models is provided by k=1, l=2 and m=1,2 for which $\varepsilon = 6\times10^{-6}$ mHz. It should be noted that the sign and magnitude of ε are very sensitive to the values of the C's and that these are not calculated from the more realistic stratified models of Winget et al. (1981), Dolez and Vauclair (1981) and Dziembowski and Koester (1981).

We have thus confirmed the earlier detection of second order effects in the rotational splitting of the 192s mode and find reasonable agreement with the theoretical values.

3. VY HOR (= BPM 31594)

The low amplitude stars are usually extremely stable, linear pulsators. In contrast, the large amplitude stars have a multiplicity of frequency components in apparently non-stationary power spectra. Confusingly, McGraw et al. (1981) found two counter-examples to this amplitude-complexity relationship: a large amplitude, apparently stable pulsator and a low amplitude variable with an unusually complex power spectrum.

3.1 The Principal Pulsation Frequency

Reports of mode changes in three ZZ Ceti stars have been made by McGraw (1976), Robinson et al. (1978) and Fontaine et al. (1980). In the first paper reporting a mode change, VY Hor was discovered by McGraw (1976) to have a dominant pulsation of period 308s on its discovery night.

However, the two following nights, together with an isolated observation some months later, showed a dominant pulsation at 617s. In the 200 hours of photometry we have obtained, we failed to detect a repetition of the behaviour during the discovery run. A re-examination of Dr. McGraw's raw data shows that he erroneously reduced the first night's data. The reported mode change is therefore fictitious.

Although the analysis is not yet complete, VY Hor appears to be a stable pulsator because:

(1) The 617s mode has remained constant in frequency within the errors of measurement during 15 weeks observation spanning 2.5 years.

(2) Combining all data from a 125 day observing season during September 1984 – January 1985, we obtained a power spectrum in which the width of the 617s peak is entirely consistent with a strictly periodic signal.

It should be mentioned that when trying to combine all 3 seasons together, there was a substantial reduction in the amplitude of the 617s dominant frequency component. We are currently investigating whether this is due to either: (1) growth/decay of the mode or (2) resolution of a very closely-spaced pair centred on the dominant frequency or (3) a timing error in the data.

3.2 Other Frequencies In VY Hor

A power spectrum of a 7-hour run on VY Hor reveals the following set of frequencies:

mHz	Ratio
1.620	1.00f
2.492	1.53f
3.240	2.00f
4.108	2.53f
4.864	3.00f
5.733	3.53f
6.484	4.00f
7.354	4.53f
±5	±.01f

A very similar pattern is seen in the northern hemisphere 'twin' of VY Hor, BG CVn (= GD 154). We can learn more about this frequency structure by using the particularly rich data set from the last season to obtain better resolution. Considering first the fundamental mode at 617s (1.62 mHz), the power spectrum obtained after prewhitening by the main peak reveals at least 2 additional low amplitude components which altogether form an equally-spaced triplet. By carrying out a similar process of successive prewhitening, the individual components near the frequencies in the above

list may be derived and are summarised in table II (? indicates doubtful significance, alternatives indicate that
aliasing problems were encountered):

Table II: Frequency Components In VY Hor

No.	Mode	Amp	f (mHz)	Δf (mHz)
		±10	±2	±3
1.	1.00f	200	1.61968	0.00120
2.		575	1.61848	
3.		120	1.61729	0.00119
4.	1.53f	100	2.48329	0.00695
5.		100	2.49024	
6.	2.00f	120	3.23695	0.00120
7.		145	3.23815	
			(or 3.23801)	
			(or 3.23829)	
8.	2.53f	50	4.10871	0.00574
9.		35	4.10297?	
10.	3.00f	50	4.85675	0.00738
11.		40	4.86413	
12.	3.53f	25	5.72131	0.00575
			(or 5.73290)	
13.		20	5.72706?	

While these results must await confirmation from data obtained from other seasons, we can make the following
comments:
    (1) The detection of a triplet structure for the domi-
nant mode results in a rotation period of order 10 days indicating that VY Hor is another slowly rotating white dwarf.
    (2) Frequency number 6 in the above table is twice the
frequency of the dominant mode at 617s to within the errors
of measurement. This presumably indicates that it is not an
independently excited g-mode but appears in the power spectrum as a 'pulse shape' component. The existence of a
nearby frequency (number 7) separated from this precise harmonic by a 'significant' frequency separation (cf. the
1.00f group) points to the fact that excitation of the
former is clearly associated with the presence of the
dominant pulsation mode.
    (3) $f_7 = f_1 + f_2$; $f_8 = f_2 + f_5$; $f_{10} = f_2 + f_7$ (3rd alias choice).
These relationships indicate clearly that non-linear effects
are important in the pulsations of VY Hor.

(4) No member of the 3.00f group is related to the 1.00f group in a simple way. Neither are there any obvious other simple relationships between the frequency components. This suggests that there are many independently excited g-modes which are grouped into narrow frequency intervals connected with the dominant pulsation. This pattern may be related to stratification of chemical composition in the white dwarf envelope which gives rise to 'trapped' modes (Winget et al., 1981). This results in a preference to excite those g-modes whose radial wavelength is related to the thickness of the various layers in the envelope.

ACKNOWLEDGEMENT

This work is a joint project with Brian Warner. We are grateful to Drs. J. McGraw, B. Lasker and J. Hesser for generously making their data available to us.

REFERENCES

Chlebowski, T.: 1978, *Acta Astr.*, 28, 441.
Dolez, N. and Vauclair, G.: 1981, *Astr. Astrophys.*, 102, 375.
Dziembowski, W. and Koester, D.: 1981, *Astr. Astrophys.*, 97, 16.
Fontaine, G., McGraw, J.T., Coleman, L., Lacombe, P., Patterson, J. and Vauclair, G.: 1980, *Astrophys. J.*, 239, 898.
McGraw, J.T.: 1976, *Astrophys. J. (Letters)*, 210, L35.
McGraw, J.T., Fontaine, G., Dearborn, D.S.P., Gustafson, J., Lacombe, P. and Starrfield, S.G.: 1981, *Astrophys. J.*, 250, 349.
O'Donoghue, D.E. and Warner, B.: 1982, *Mon. Not. R. astr. Soc.*, 200, 563.
Robinson, E.L.: 1980, in *Proceedings of IAU Colloquium No. 53: White Dwarfs and Variable Degenerate Stars'* eds. H.M. Van Horn and V. Weidemann (Rochester, N.Y.: University of Rochester Press), p. 343.
Robinson, E.L., Stover, R.J., Nather, R.E. and McGraw, J.T.: 1978, *Astrophys. J.*, 220, 614.
Unno, W., Osaki, Y., Ando, H. and Shibahashi, H.: 1979, *Nonradial Oscillations of Stars* (Tokyo: Tokyo University Press).
Winget, D.E., Van Horn, H.M. and Hansen, C.J.: 1981, *Astrophys. J. (Letters)*, 245, 33.
Winget, D.E. and Fontaine, G.: 1982, in *Pulsations in Classical and Cataclysmic Variables*, ed. J.P. Cox and C.J. Hansen (Boulder: University of Colorado), p. 142.

HOW WELL DO WE KNOW STELLAR PARAMETERS?

Søren Frandsen
Institute of Astronomy
University of Aarhus
DK-8000 Aarhus
Denmark.

ABSTRACT. A small table is presented containing estimates of the standard errors in the best determinations of fundamental parameters for stars. Only for a small sample of late type main sequence stars do we have values for the mass, the radius and the luminosity or the combination of these parameters the surface gravity g and the effective temperature T_e that are known with a precision getting near 1 percent. It should be investigated in detail, how much asteroseismology (the extension of helioseismology to stars) can improve the situation, once the sensitivity of the instruments have improved so much, that regular observations of stellar oscillations are possible.

1. INTRODUCTION

It would be a surprise for most astronomers if the Sun was the only oscillating late type star. Christensen-Dalsgaard has discussed (1984 and these proceedings) what one might learn from asteroseismology. One of the most interesting aspects is that it opens up a road, that gives access directly to the interior. Also the fundamental stellar parameters can be determined without the complications of the stellar atmosphere. Asteroseismology provides an independent technique and serves as a very important check on the results reached by other methods. It might turn out to be more precise or to be useful for a number of stars, where the traditional methods fails. In this short contribution a summary of the precision quoted in earlier work is given.

2. THE TWO-DIMENSIONAL DIAGRAMS

Assuming the same chemical composition and the same dynamical state and the same age, stars can be sorted after two independent parameters like luminosity L and effective temperature T_e. Observationally the immediate result is a colour-luminosity diagram or some other photometric indices. Christensen-Dalsgaard has shown that from oscillation frequency data one possibility is to construct a similar diagram, by plotting $\Delta\nu_0$ and D_0, which are both quantities in an asymptotic fitting formula to

473

D. O. Gough (ed.), Seismology of the Sun and the Distant Stars, 473–476.
© *1986 by D. Reidel Publishing Company.*

the pulsational frequencies. In such a diagram it is possible to
distinguish between stars of different types. If a two-dimensional
diagram can be calibrated in terms of luminosity L, mass M and radius R,
locating a star in the diagram immediately gives the fundamental parameters. The calibration depends on other parameters like chemical
abundances and dynamical state and maybe more parameters, and here
the main uncertainty is often found, where systematic errors occur and
the risky assumptions are made. the accuracy of the resulting luminosities, masses and radii depends on the calibration and the observational
error. Often the observations (photometry f.ex.) are very precise, and
consequently the main problem lies in calibrating the data. Asteroseismology also needs to calibrate $\Delta \nu_0$ and D_0 in terms of the fundamental
parameters, but this involves the interior structure and not the atmospheric structure.

The calibration of one or another system is not discussed here,
but as a reference for future work on using asteroseismology a small
table has been compiled. Table I gives an estimate of the accuracy with
which different techniques obtain one or more of the fundamental parameters. The table concerns only late type main sequence stars in the
spectral range FO to MO, which are the primary candidates for observation of solar type p-mode oscillations. After the table some comments
follow addressing some of the entries in the table. Often the parameter
determined by a method is not one of L, M or R, but the combination T_e,
($L = 4 \pi R^2 \sigma T_e^4$) or $g = GM/R^2$. The standard error quoted is for the best
cases and is the error claimed by the observers. Be aware, that this
accuracy is obtained only in a few cases and often for nearby stars.

Table I   Minimum standard errors in the fundamental
parameters for stars using different observing techniques.

Observational Techniques	L	M	g	T_e	R	Comment
1. Spectroscopy and photometry of single stars	0.25		0.25 $\Delta \log g$ = 0.1	0.03 ΔT_e=150K		
2. Trigonometric parallaxes for nearby stars	0.02					d<10 parsec only 126 stars with s.e<0.12
3. Visual binaries		0.05 -0.10				αCen αC Min
4. Eclipsing binaries		0.02			0.02	
5. Intensity interferometri					0.03	αC Min
6. Rotating spotted stars					0.15	
Abundance determinations:		typical $\Delta \log X = 0.1$				

Ad 1. L is found by a calibration of the photometric system using nearby stars, where trigonometric parallaxes have been measured. T_e and g are the result of a detailed model atmosphere analysis. The errors are typical discrepancies between different modellers. For T_e determinations see Bohm-Vitense (1981). A thorough discussion of the calibration of photometric indices is presented by Heyn Olsen (1984) for the Strömgren uvby system.

Ad 2. For some statistics see Gliese (1978). Parallax measurements are constantly improving. Errors as small as a few milliarcs are now quoted for ground based work. (Gatewood and Stein, 1983, Monet and Dahn, 1983) and similar or better quality is expected from the HIPPARCHOS satellite.

Ad 3. Only few systems are near enough. The values are best cases from the review by Popper (1980).

Ad 4. Popper (1980) gives examples of systems with mass and radius determinations in the percent range (δ Equ, F7, M/M = 0.01, R/R = 0.02). A long series of observations by the Copenhagen group obtain similar good results (see f.ex. Andersen et al. 1984) for late type stars and early type stars.

Ad 5. Only very few interferometric observations of late type near main sequence stars exist, see Code et al. (1976) and Hanbury Brown (1982).

Ad 6. The radius found depends on the measured Vsini velocity and the Ca II H and K emission modulation. The period is well determined, so the main part of the error comes from the error in Vsini. Vsini can be measured to a precision of $\sim$ 1 km/s (Soderblom, 1982) and sini is probably close to 1 in most cases, where modulation is clearly observed.

A small supplementary table has been added after a suggestion by Bob Noyes. Table II contains numbers for the errors in rotational velocities observed with two techniques, high resolution spectroscopy (Soderblom, 1982) or from Ca II H and K line emission modulation (Vaughan et al.,1981, Noyes et al.,1984). The rotational velocity has been shown to be correlated with stellar activity and with age of the star. If the rotation is fast enough the line splitting of oscillation frequencies should be observable, if long time series of stars can be obtained.

Table II. Minimum errors in rotational velocities

Technique	V	Comment
Spectroscopic	0.10	Depends critically on Vsini
Ca II emission	0.01	or better, depending on the regularity of the emission.

3. CONCLUSION

When everything is added together it becomes clear, that reliable parameters only exist for a very small sample of late type main sequence
stars. Any new technique (astereoseismology?) capable of improving the
situation is therefore extremely important for the whole understanding
of stellar evolution. The advantage would be even greater, if one could
reach beyond the immediate solar neighbourhood. No population II star
has been observed with a precision better than the photometric.

REFERENCES

Andersen, J., Clausen, J.V., and Nordstrøm, B., 1984, *Astron.Astroph.*
    137, 281
Böhm-Vitense, E., 1981, *Ann.Rev.Astron.Astroph.* 19, 395
Christensen-Dalsgaard, J., 1984, *Proc. of Workshop on Space Research
    in Stellar Activity and Variability*, Ed. Mangeney, A. and Prade-
    rie, F., 11
Code, A.D., Davis, J., Bless, R.C., and Hanbury Brown, R., 1976, *Ap.J.*
    203, 417
Gatewood, G., and Stein, J., 1983, *IAU Coll. 76 The Nearby Stars and
    the Stellar Luminosity Function*, 75
Gliese, W., 1978, *IAU Symp. 80 The HR diagram*, 79
Hanbury Brown, R., 1983, *Understanding the Universe*, ed. West, R.M.,
    D. Reidel, 73
Heyn Olsen, E., 1984, *Astron.Astrophys.Suppl.* 57, 443
Monet, D.G., and Dahn, C.C., 1983, *IAU Coll. 76 The Nearby Stars and
    the Stellar Luminosity Function*, 91
Noyes, R.W., Harmannn, L.W., Baliunas, S.L., Duncan, D.K., and Vaughan,
    A.H., 1984, *Ap.J.* 279, 763
Soderblom, D.R., 1982, *Ap.J.* 263, 239
Vaughan, A.H., Baliunas, S.L., Middelkoop, F., Hartmann, L.W., Mihalas,
    D., Noyes, R.W., and Preston, G.W., 1981, *Ap.J.* 250 276.

INDEX